A Practical Approach to Corporate Networks Engineering

RIVER PUBLISHERS SERIES IN COMMUNICATIONS

Consulting Series Editors

This series focuses on communications science and technology. This includes the theory and use of systems involving all terminals, computers, and information processors; wired and wireless networks; and network layouts, procontentsols, architectures, and implementations.

Furthermore, developments toward new market demands in systems, products, and technologies such as personal communications services, multimedia systems, enterprise networks, and optical communications systems.

- Wireless Communications
- Networks
- Security
- Antennas & Propagation
- Microwaves
- Software Defined Radio

For a list of other books in this series, visit
http://riverpublishers.com/river_publisher/series.php?msg=Communications

A Practical Approach to Corporate Networks Engineering

António Manuel Duarte Nogueira
Department of Electronics, Telecommunications and Informatics
Institute of Telecommunications
University of Aveiro, Portugal

and

Paulo Jorge Salvador Serra Ferreira
Department of Electronics, Telecommunications and Informatics
Institute of Telecommunications
University of Aveiro, Portugal

River Publishers
Aalborg

ISBN 978-87-92982-09-4 (hardback)

Published, sold and distributed by:
River Publishers
P.O. Box 1657
Algade 42
9000 Aalborg
Denmark

Tel.: +4536995319
www.riverpublishers.com

To my wife, Manuela Gomes, and our children Miguel and Matilde, for all their unconditional love and inspiration.
To my parents, António and Felismina, who have sacrificed so much for my happiness.
António Nogueira

To my wife, Fernanda Gonçalves, for her unlimited love and "making me want to be a better man", and my mother Conceição, for her loving support.
Paulo Salvador

Contents

Preface

Welcome to "A Practical Approach to Corporate Networks Engineering". This book discusses relevant concepts that should be taken into account when designing and implementing local corporate IP networks. A large set of topics will be covered: the design impacts of the underlying network layout, addressing issues (including all the details of the IPv4 and IPv6 addressing paradigms), data link layer design requirements (with special emphasis on Virtual Local Area Networks (LANs) and Spanning-Tree family protocols), network layer design issues (including the most used unicast and multicast routing protocols), quality of service assurance (discussing different queuing mechanisms and the Integrated Services and Differentiated Services quality of service models), network security (with special emphasis on access control and secure network communications), configuration and optimization of some of the most useful corporate network services.

Each topic will be illustrated with practical examples implemented using the open source Graphical Network Simulator (GNS), version 3, publicly available from *http://www.gns3.net*. GNS allows the emulation of complex networks using the Cisco Internetwork Operating Systems (IOS). Basically, GNS3 is a graphical front end to a product called Dynagen, while Dynamips is the core program that allows IOS emulation. Dynagen runs on top of Dynamips to create a more user friendly, text-based environment. GNS3 works over the Windows or Linux operating systems, and supports several router platforms and PIX firewalls. Since GNS3 runs the Cisco IOS, the experience that we can get is almost the same of using real equipment running the IOS. Of course, due to licensing restrictions, users need to purchase their own Cisco IOSs to use with GNS3.

Besides these focused examples, a transversal example of the design, engineering and implementation of a local area corporate network will be used throughout the different book chapters. This example, also implemented in GNS3, will allow readers to have a global view of the different topics that have to be addressed when designing and configuring such an infrastructure, besides allowing them to use the same network layout when exploiting the different network mechanisms and protocols.

Organization

The book is organized in 6 chapters, following a bottom-up approach when looking at hierarchical network reference models.

Chapter 1 covers several network design topics, discussing the most relevant network design principles, the advantages and disadvantages of the hierarchical and flat network design approaches, the relevance of network redundancy at different levels, the Internet Protocol (IP) in its 4 and 6 versions (including addressing issues and functional aspects of their operation), IPv4 to IPv6 transitions mechanisms, the Domain Name System network service, the TCP and UDP transport layer protocols, private addressing and translation mechanisms.

Chapter 2 is dedicated to data link layer design, covering Virtual LANs and different link layer protocols that are used to avoid logical loops: the basic Spanning Tree Protocol (STP) and the Rapid and Multiple STP variants. At the end of the chapter, the basic principles of Wireless LANs (specially those conforming to the IEEE 802.11 standard) are presented and discussed: the architecture, the notion of channel and association, the Media Access Control Layer specifications and the frame format are covered in detail.

Chapter 3 discusses network link layer design topics, covering static and dynamic unicast routing, as well as multicast routing. The Routing Information Protocol (RIP), as a representative example of a distance vector protocol, and the Open Shortest Path First Protocol (OSPF), as an example of a link state protocol, will be covered in detail, since they represent important and widely deployed unicast routing solutions for corporate networks. Regarding multicast, the Internet Group Management Protocol (IGMP) and its IPv6 counterpart, the Multicast Listener Discovery (MLD) protocol, will be discussed in detail, as well as the most important multicast routing protocols, with special emphasis on the Protocol Independent Multicast (PIM) routing mechanism.

Chapter 4 is dedicated to Quality of Service issues, discussing different queuing mechanisms and two approaches that have been used to provide QoS in IP networks: the Integrated Services and the Differentiated Services models. Illustrative examples of these methodologies will be presented and discussed.

```
// Some unnumbered configurations ...
```

Box 1 Unnumbered configuration box.

```
// Some numbered configurations ...
```

Box 2 Numbered configuration box.

Chapter 5 discusses network security issues, being mainly focused in two particular aspects of this broad problematic: access control and secure communications. Regarding access control, Access Control Lists and IPtables will be discussed in detail, while IPSec will be explored as an appropriate technology to build a secure communications environment.

Finally, Chapter 6 will discuss some of most important services for corporate networks, giving a brief explanation of their functioning principles and presenting examples of their configuration and deployment.

Layout Items

Whenever some idea needs to be strengthened throughout the book, information boxes similar to the one shown below will be used:

**Information example**

This idea is very important ...

If there is a need to draw attention to some particular issue, a warning message like this one will be used:

**Warning example**

Pay attention to this!

A box similar to Box 1 will be used whenever some configuration code has to be shown separately and there is no need to mention a specific line of the configuration. Thus, these configuration boxes are unnumbered.

If it necessary to identify specific command lines of the configuration, a configuration box similar to Box 2 will be used because command lines are numbered.

In order to present unnumbered command inputs, outputs or configuration commands in line with the text, a box similar to this one will be used:

```
// Some in line unnumbered configurations ...
```

Finally, a box similar to this one will be used to present numbered command inputs, outputs or configuration commands in line with the text:

```
// Some in line numbered configurations ...
```

All terminal and router commands start with character #. Different indentation levels correspond to the different levels of the command insertion procedures.

Web Site

We intend to maintain a public web site (*NetConfs.com*) containing all figures included in the book, the exercises that are proposed, the configuration details of the different experiments, the results of the different exercises that are included and additional technical material.

Acknowledgments

We would like to acknowledge the Department of Electronics, Telecommunications and Informatics, from University of Aveiro, for its institutional support and Instituto de Telecomunicações - Aveiro Pole for the excellent logistics conditions it provides to its collaborators.

1

Architectures and Basic Operational Protocols

1.1 Introduction

This chapter presents relevant concepts about corporate network design that will be intensively used and discussed throughout the book. The major topological considerations that have to be taken into account in corporate network design are identified and discussed, together with the basic functionalities of the IP protocol (in both of its versions), the main characteristics of the transport layer protocols, IPv4 and IPv6 addressing issues, the functioning details of the Domain Name System service, network address translation, among other important topics. So, this chapter essentially covers architectural concepts and basic networking protocols.

1.2 The OSI and TCP/IP Reference Models

The software modules of each communication host can be seen as vertically stacked in layers [94, 64]. Each layer is responsible for solving a part of the problem. Conceptually, sending a message from an application at the sending host to an application at the receiving host involves the transfer of the message down through the successive software layers of the sending host, sending the message over the network and the transfer of the message up through the successive layers of the receiving host. Figure 1.1 illustrates this communication paradigm.

The Open Systems Interconnection (OSI) reference model defines a seven-layer model of data communications, as shown in the left part of Figure 1.2. Each layer provides a set of functions to the layer above and relies on the functions provided by the layer below. Although messages can only pass vertically through the stack from layer to layer, from a logical point of view each layer communicates directly with its peer layer on other nodes.

The different layers and their basic functionalities are the following [57]:

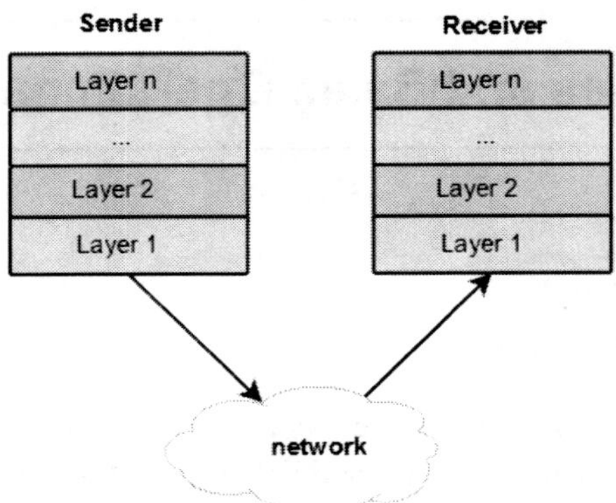

Figure 1.1 Layered organization.

	OSI model		TCP/IP model
7	Application	4	Application
6	Presentation		
5	Session		
4	Transport	3	Transport
3	Network	2	Network
2	Data Link	1	Network Access
1	Physical		

Figure 1.2 The OSI and TCP/IP reference models.

- Application - This layer gives the user access to all the lower OSI functions; its purpose is to support semantic exchanges between applications.
- Presentation - This layer is concerned with the representation of user or system data, including the necessary conversions and code translation.
- Session - The session layer provides mechanisms for organizing and structuring interaction between applications and/or devices.
- Transport - This layer provides transparent and reliable end-to-end data transfer, relying on lower layer functions for handling the specificities of the transfer medium. TCP (Transmission Control Protocol) and UDP (User Datagram Protocol) are two transport layer protocols.

- Network - Provides the means to establish connections between networks, including also procedures for the operational control of the communications and for information routing through multiple networks. IP (Internet Protocol) is the most popular network layer protocol.
- Data Link - This layer provides the functions and protocols that are necessary to transfer data between network entities and to detect (and possibly correct) errors that may occur in the physical layer.
- Physical - This layer is responsible for physically transmitting data over the communication link, providing the mechanical, electrical, functional and procedural standards to access the physical medium.

The development of the Transport Control Protocol / Internet Protocol (TCP/IP) protocol suite was based on a public Request for Comments (RFC) policy, and this fact contributed to establish it as the protocol of choice for most data communication networks. The TCP/IP protocol suite consists of four layers (right part of Figure 1.2):

- Application Layer - This layer is provided by the user program that uses TCP/IP for communication. File Transfer Protocol (FTP), Trivial FTP, Telnet, Simple Mail Transfer Protocol (SMTP) are examples of application layer protocols.
- Transport Layer - This layer provides the end-to-end data transfer, being responsible for the reliable exchange of information. The transport layer protocols are TCP, which provides a connection-oriented service, and UDP, which provides a connectionless service. This means that applications using UDP as the transport protocol have to provide their own end-to-end flow control. Usually, UDP is used by applications that need a faster transport mechanism.
- Internetwork Layer - This layer separates the physical network from the layers that are located above it. The Internet Protocol (IP) is the most important protocol in this layer. It is a connectionless protocol that does not assume reliability from the lower layers, does not provide reliability, flow control or error recovery. These functions must be provided at a higher level: the transport layer, if TCP is used, or the application layer, if using UDP. The basic unit of information transmitted across TCP/IP networks is called an IP datagram. IP also provides routing functions for distributing datagrams to the correct destination. Other internetwork layer protocols are the Internet Control Message Protocol (ICMP), the Internet Group Management Protocol (IGMP) and the Address Resolution Protocol (ARP)/Reverse ARP (RARP).

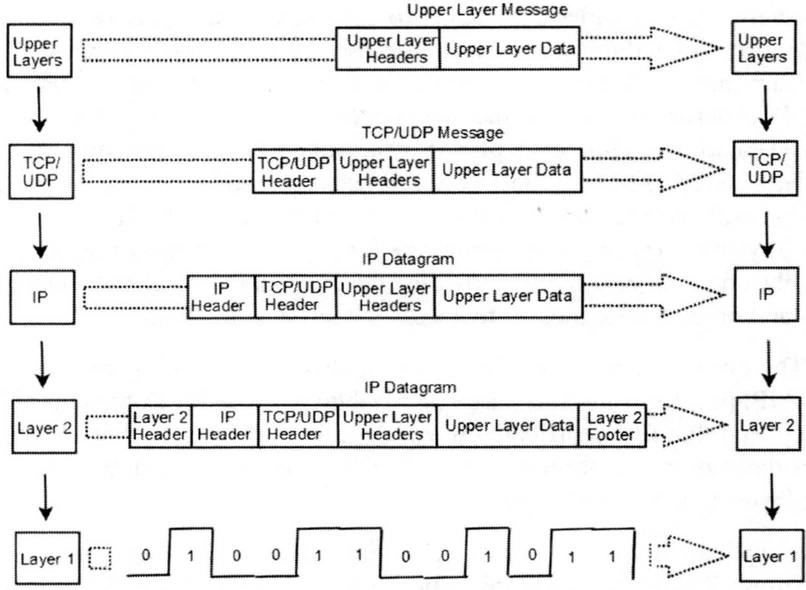

Figure 1.3 IP datagram encapsulation process ([63]).

- Network Interface Layer - This layer is the interface to the actual network hardware and does not guarantee reliable delivery, which is left to the higher layers. The network interface layer may be packet or stream oriented. TCP/IP does not specify any particular protocol for this layer. Examples of supported network interface protocols are IEEE 802.2, X.25 or Asynchronous Transfer Mode (ATM).

When data has to be sent from one host on a network to another host, the process of encapsulation is used to add a header in front of the data at each layer of the protocol stack in descending order. The header must contain a data field that indicates the type of data encapsulated at the layer immediately above the current layer. As the packet ascends the protocol stack on the receiving side of the network, each encapsulation header is removed in the reverse order. So, the term encapsulation refers to the process of adding headers to data at each layer of a particular protocol stack (Figure 1.3).

1.3 Network Design Principles

In order to correctly design a communications network, network designers and technicians must identify network requirements and select the best solutions to meet the needs of the specific business. The following steps are required to design a good network:

- Verify the business goals and technical requirements;
- Determine the features and functions that are required to meet the needs identified in the previous step;
- Make a network-readiness assessment;
- Create a solution and a site acceptance test plan;
- Create a project plan.

1.3.1 Network Requirements

Besides some potential business specificities that may be considered, the most usual network requirements are the following:

- The network should be operational all the time, even in the event of failed links, equipment failure and overloaded conditions.
- The network should reliably deliver applications and provide reasonable response times between hosts.
- The network should be secure, protecting the data that is transmitted and the data that is stored on the different devices.
- The network should be easy to modify in order to adapt to general business changes or network growth.
- Network troubleshooting should be easy, allowing to easily and timely find and fix network problems.

1.3.2 Network Design Considerations

There are some important considerations that must be taken into account when designing a communications network:

- Standards - The whole design and the different components that build the network should be based on open standards. With open standards, there is a higher flexibility when it is necessary to interconnect different devices from different vendors.
- Scalability - The introduction of new hosts, servers or segments to the network should not require a complete redesign of the network topology.

So, the chosen topology should be able to accommodate expansion due to business requirements.

- Availability and Reliability - The mean time between failures (MTBF) and the mean time to repair (MTTR) are metrics that must be considered when designing a network. These goals are achieved by designing logical and physical redundancy in the network.
- Modularity - A complex system (like a network) should be divided into smaller and more manageable sub-systems, making implementation much easier to handle. Besides, modularity also ensures that a failure at a certain part of the network can be isolated so that it will not bring down the entire network.
- Security - This aspect is very important and should be considered in the early stages of the network design; otherwise, we can leave the network open to security attacks until all security holes are closed.
- Network Management - Management provides a way to monitor the health of the network, to ascertain operating conditions, to isolate faults and configure devices to apply changes. A management framework should be integrated into the design of the network from the beginning.
- Performance - Performance metrics (like throughput, delay, response time, among others) need to be taken into account when designing the network.
- Economics - The network design should include a compromise between cost and meeting performance requirements.

1.4 Hierarchical versus Flat Network Design

In order to meet the business and technical goals for a network design, the final network topology will certainly consist of many interconnected components. The most successful large networks are hierarchically designed, presenting a layered structure [77]. Layering creates separate problem domains, focusing the design of each layer on a single goal or set of goals. Layers must be in line with their design goals: trying to add too much functionality into one layer generally leads to a confusing solution that is difficult to document and maintain. Note that the OSI and TCP/IP models also follow this principle, dividing the process of communication between computers into layers, each with different design goals and criteria.

A hierarchical network design methodology allow us to plan a modular topology that limits the number of communicating equipments and helps minimize costs. Using this approach, it is possible to purchase the appro-

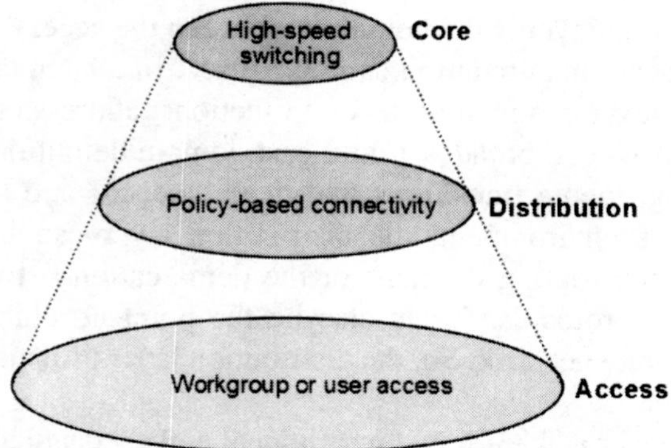

Figure 1.4 Hierarchical network design model.

priate devices for each layer of the hierarchy, avoiding the acquisition of unnecessary features for each layer. Besides, the hierarchical network design allows an accurate capacity planning for each layer and the distribution of the network management responsibilities by the different layers. The hierarchical design also facilitates changes of network components and makes network scalability easier.

So, hierarchical network design has several advantages over flat design. By dividing a flat network into smaller and more manageable hierarchical blocks, it is possible to keep local traffic as local; only traffic destined to other networks is moved to a higher layer. In a flat network, Layer 2 devices are unable to control broadcasts or filter undesirable traffic. As more devices and applications are added to the flat network, response times degrade until the network becomes unusable.

A hierarchical network design includes three layers:

- The backbone (core) layer - provides optimal transport between sites;
- The distribution layer - provides policy-based connectivity;
- The access layer - provides workgroup/user access to the network.

Figure 1.4 shows a high-level view of a hierarchical network design, illustrating the three layers and their main functionalities.

The core layer is usually a high-speed switching backbone and should be designed to switch packets as fast as possible. This layer should not perform any packet manipulation, such as the implementation of access lists and filtering, that would slow down the switching of packets.

The distribution layer is the boundary between the access and core layers and is where packet manipulation can take place. In a campus environment, the distribution layer can include several functions: address or area aggregation, workgroup access, broadcast/multicast domain definition, Virtual LAN (VLAN) routing, media transitions that need to occur and security issues. In a non-campus environment, the distribution layer can be a redistribution point between routing domains or the demarcation between static and dynamic routing protocols. It can also be the point at which remote sites access the corporate network. So, the distribution layer provides policy-based connectivity.

The access layer is the point at which local end users are allowed into the network. In a campus environment, access-layer functions can include shared bandwidth, switched bandwidth, Media Access Control (MAC) layer filtering and micro-segmentation. In a non-campus environment, the access layer can give remote sites access to the corporate network through some wide-area technology, such as Frame Relay, Integrated Services Digital Network (ISDN), Virtual Private Networks (VPNs), leased lines, etc.

Layers instantiation

Layers are defined to aid successful network design and to represent the different functionalities that must exist in a network. The instantiation of each layer can be in distinct routers or switches, can be represented by a physical media, can be combined in a single device, or can be omitted altogether. The way layers are implemented depends on the needs of the network being designed.

1.5 Network Redundancy

A single point of failure is any device, interface on a device, or link that can isolate users from the services they depend on if it fails [86]. Networks that follow a strong, hierarchical model tend to have many single points of failure because of the emphasis on summarization points and clean points of entry between the network layers. In the hierarchical network depicted in Figure 1.5, every device and every link is a single point of failure.

Redundancy provides alternate paths between these failure points. However, if not properly designed and implemented, redundancy can cause a lot of troubles: in fact, each redundant link and each redundant connection point in a network weakens the hierarchy and reduces stability. So, when adding

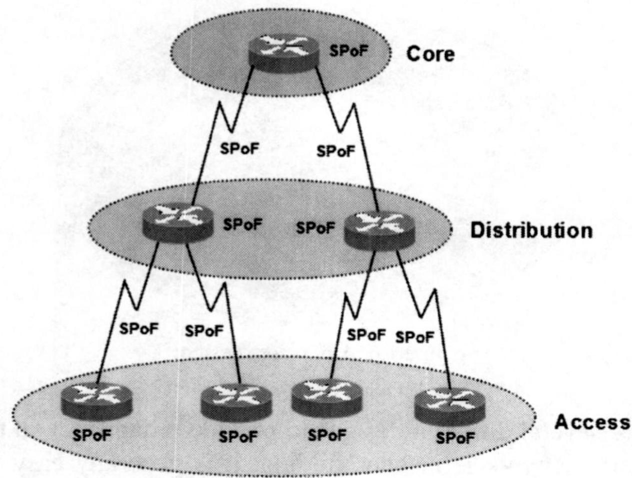

Figure 1.5 Hierarchical network design where every device and link is a single point of failure.

redundancy to a hierarchical design, some important issues should be taken into account:

- Redundant paths should be used only when the normal path is broken, unless the paths are carefully engineered for load balancing.
- Traffic should not pass through devices or links that are not designed to handle traffic. Normal traffic should only use backup paths when normal paths are unavailable.

1.5.1 Redundancy at the Core Layer

The core layer of an enterprise LAN may connect multiple buildings or multiple sites and includes one or more links to the devices at the enterprise edge to support Internet, Virtual Private Networks (VPN) or WAN access. The most common technologies and methodologies used at the core layer include routers or multilayer switches that combine routing and switching in the same device, redundancy and load balancing, high-speed and aggregate links and routing protocols that scale well and converge quickly. Single all core devices usually have complete routing information, that is, full reachability, core redundancy design is generally simple and there is a small chance of occurring routing loops.

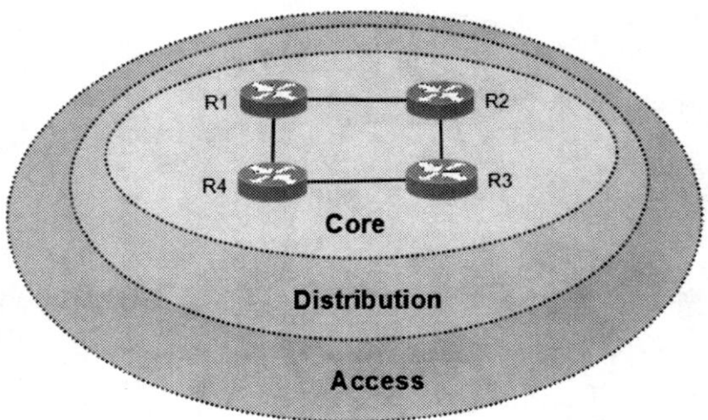

Figure 1.6 A ring core design.

There are several different designs to provide redundancy in the core. If the entire core network is in one building, it is generally easy to connect each router to two high speed LANs, such a high speed Ethernet. If the core routers are not all in one building (or on one campus), the options available become more limited and expensive. Core designs based on ring topologies are relatively common since they are easy to design and maintain. A ring core is usually formed by using multiple point-to-point links to interconnect multiple routers (Figure 1.6) but there are some designs that rely on a ring at the physical layer. Some ring technologies were designed to include redundancy in an implicit way, such as Synchronous Optical Network (SONET) or Synchronous Digital Hierarchy (SDH), and Fiber Distributed Data Interface (FDDI). These technologies provide redundancy at the Layer 2 of the OSI model, resolving many of the issues that arise when providing redundancy at the network layer. However, these methods do not provide redundancy for the devices on the core; they only provide redundancy for the links between devices.

Ring core designs are appropriate to reduce the number of available paths, while still providing redundancy. However, they have some drawbacks: the number of possible routes through the network is low during normal operation, while the number of hops a packet may have to cross with a single link down is very high; the redundancy level is not perfect, because losing any two links on the core automatically isolates some part of the network.

Full mesh designs, where every core router has a connection to every other core router, provide the most redundancy possible, as shown in Figure 1.7. This topology provides a large number of alternate paths to any destination, a two hop path to any destination under normal use, a four hop

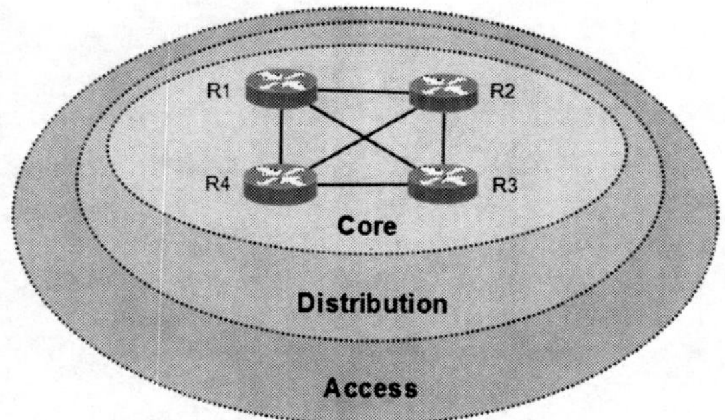

Figure 1.7 A full mesh core design.

maximum path in the worst case scenario (multiple links down with full connectivity) and exceptional redundancy, because every router has a link to every other router. This network would have to lose at least three links before any destination became unreachable.

Unfortunately, full mesh designs can provide too much redundancy in larger networks, forcing a core router to choose between a large number of paths to any destination, which increases convergence times. Besides, full mesh networks with n nodes will have $(n(n-1))/2$ links (the division by 2 is necessary to count links Router X-to-Router Y and Router Y-to-Router X as only one link). Full mesh networks can be expensive because of the number of links required and also need a lot of configuration management. So, it is difficult to perform traffic engineering on a full mesh network since the path that traffic normally takes can be confusing and quite difficult to predict.

Partial mesh cores tend to be a good compromise in hop count, redundancy, and the number of paths through the network. In Figure 1.8, there are four paths between any two points on the network. There is a clear difference in the lengths of the four paths available and no more than three hops will be required to traverse the network during normal operation; if any single link fails, the maximum number of hops to traverse the network will increase to four. These hop counts tend to stay low as a partial mesh core grows. The redundancy provided by a partial mesh design is good, as well: the network provides full connectivity with three links down as long as no single router loses both of its connections to the mesh.

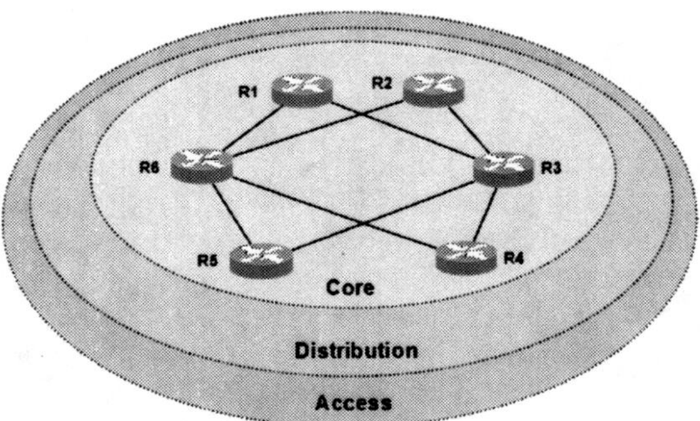

Figure 1.8 A partial mesh core design.

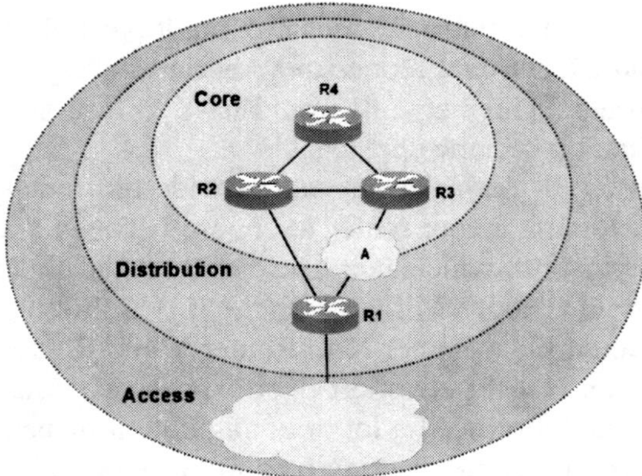

Figure 1.9 Dual homing in the distribution layer.

1.5.2 Redundancy at the Distribution Layer

The distribution layer is usually built using Layer 3 devices. Routers or multilayer switches located at the distribution layer provide many critical functions: filtering and managing traffic flows; enforcing access control policies; summarizing routes before advertising the routes to the core; isolating the core from access layer failures or disruptions; routing between access layer VLANs. Distribution layer devices are also used to manage queues and prioritize traffic before transmission through the campus core.

The two most common methods for providing redundancy at the distribution layer are dual homing (Figure 1.9) and backup links to other distribution layer routers (Figure 1.10).

In Figure 1.9, Router R1 has two connections to the core through separate routers. This provides very good redundancy but it can also create some problems. If Router R1 was connected to only one core router, Router R4 would have two paths to network A; with Router R1 dual-homed to the core, Router R4 has four paths to this destination. So, dual homing Router R1 to the core effectively doubles the number of paths available to network A from the core. This doubling of possible routes for every dual-homed distribution layer router slows network convergence. It is sometimes possible to force the metric or cost of one of the two paths to be worse so that traffic will normally flow over only one link. The number of paths is still doubled, so this is not a very effective solution for advanced routing protocols. A better solution would be to only advertise network A over one link, unless that link becomes unusable.

Dual homing also presents another problem: if the link between Router R2 and Router R3 goes down, Router R1 could be effectively drawn into a core role, passing transit traffic between Router R2 and Router R3. This may be a valid design if it is anticipated and planned for, but it is generally not. The easiest way to prevent this from occurring is to configure Router R4 so that it does not advertise routes learned from Router R3 back to Router R2 and does not advertise routes learned from Router R2 back to Router R3.

Installing a link between distribution layer routers to provide redundancy (Figure 1.10) can have the following consequences:

- The routing table size of the core will double.
- The redundant path can be used by traffic transiting the core - Routers R1 and R2 can play a core routing role if the link between Router R3 and Router R4 fails, because Router R3 can begin forwarding traffic that is destined to some network beyond R4 to Router R1, rather than forwarding it to Router R5.
- The redundant link can be used preferentially by core paths - This can happen if distribution layer routers prefer the redundant path through the distribution layer rather than the path through the core.
- Incomplete routing information - Routers in one branch of the distribution layer will need to advertise the destinations in the other branch through the redundant link.

Trunk links are often configured between access and distribution layer networking devices. Trunks are used to carry traffic that belongs to multiple VLANs between devices over the same link. The network designer considers the overall VLAN strategy and network traffic patterns when designing the

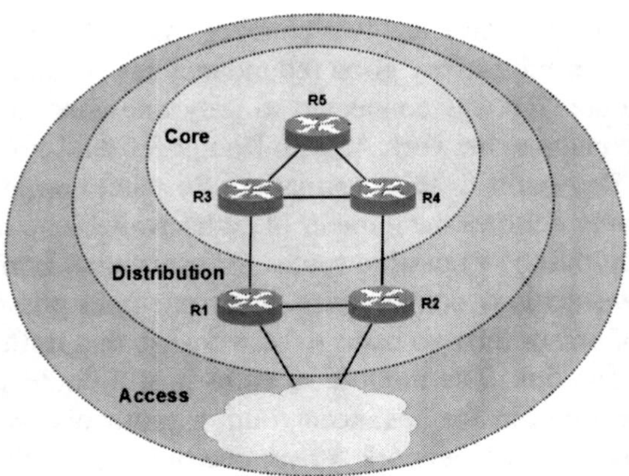

Figure 1.10 Redundant links in the distribution layer.

trunk links. When using Layer 3 devices at the distribution layer, every router functions as a gateway for a limited number of access layer users.

Usually, devices at the distribution layer have redundant connections to switches at the access layer and to devices at the core layer. If a link or device fails, these connections provide alternate paths. Using an appropriate routing protocol at the distribution layer, Layer 3 devices react quickly to link failures so that they do not impact network operations. Providing multiple connections to Layer 2 switches requires the use of the Spanning Tree Protocol (STP). STP guarantees that only one path is active between two devices and, if one of the links fails, the switch recalculates the spanning tree topology and automatically begins using the alternate link. This protocol will be addressed in detail in Chapter 2.

1.5.3 Redundancy at the Access Layer

Access layer devices are located inside each building of a campus, each remote site, the servers room or data centers and at the enterprise edge. The access layer infrastructure uses Layer 2 switching technology to provide access into the network. The access can be made through a permanent wired infrastructure or through wireless access points.

Wiring closets act as the termination point for infrastructure cabling within buildings or within floors of a building. The placement and physical size of the wiring closets depends on network size and expansion plans. Many access layer switches have Power-over-Ethernet (PoE) functionality. Inside a server room or a data center the access layer devices are typically redundant multilayer switches that combine the functionality of both routing and switching. Multilayer switches can provide firewall and intrusion protection features, as well as Layer 3 functions.

Power over Ethernet

Power over Ethernet (PoE) technology passes electrical power safely, along with data, on Ethernet cabling. This minimizes the number of wires that must be strung in order to install the network, which results in lower cost, less downtime, easier maintenance and greater installation flexibility.

Many different devices can connect to an IP network, like IP telephones, video cameras and videoconferencing systems. All these services can be converged onto a single physical access layer infrastructure. However, the logical network design to support them becomes more complex because of considerations such as quality of service (QoS), traffic segregation, and filtering. These new types of end devices, and the associated applications and services, change the requirements for scalability, availability, security, and manageability at the access layer.

Access layer management is crucial because of the increase in the number and types of devices connecting at the access layer and the introduction of wireless access points into the LAN. Besides providing basic connectivity, the designer needs to consider naming structures, VLAN architecture, traffic patterns and prioritization strategies.

Dual homing access layer devices are the most common way of providing redundancy to remote locations, but it is also possible to interconnect access layer devices to provide redundancy. In Figure 1.11, the access layer routers R7 and R8 are dual-homed to different branches of the distribution layer. If these redundant links are actually constantly up and carrying traffic, the number of paths between networks A and B is excessive and with each addition of a dual-homed access layer router, things get worse. The huge number of paths causes major problems in the core, since the size of the routing table will explode. So, if the redundant link crosses the boundary of a distribution layer branch, it should not be advertised as a normal path.

Another option to provide access layer redundancy is to provide links between the access layer routers themselves (Figure 1.12). This saves one link and reduces the number of paths between networks A and B down to two. In this case, it is important to provide enough bandwidth to handle the traffic from both remote sites.

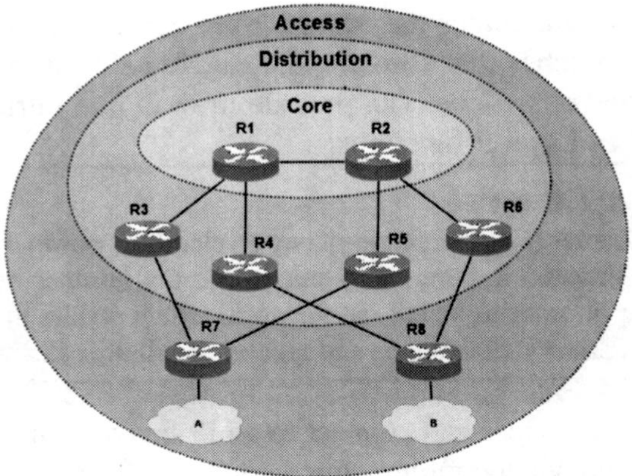

Figure 1.11 Dual homing though different distribution branches.

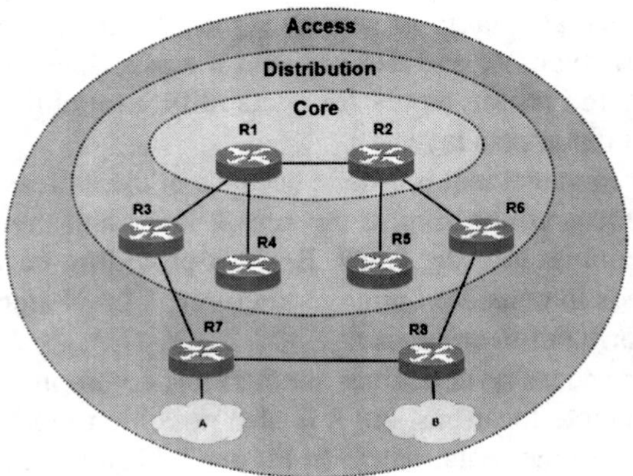

Figure 1.12 Redundancy through interconnected access layer devices.

These last two solutions work quite well as long as the redundant route is not advertised until needed, so traffic will not normally flow across the redundant link.

It is possible to design load sharing and redundancy within the access layer, as illustrated in Figure 1.13. In this case, both links to Routers R7 or R8 are connected to routers within the same distribution layer branch. In this topology, it is still possible for packets traveling from Router R3 to Router R4 to pass through Router R7, but this can be avoided using route filtering. Routers R7 and R8 should only advertise the networks below them in the hierarchy.

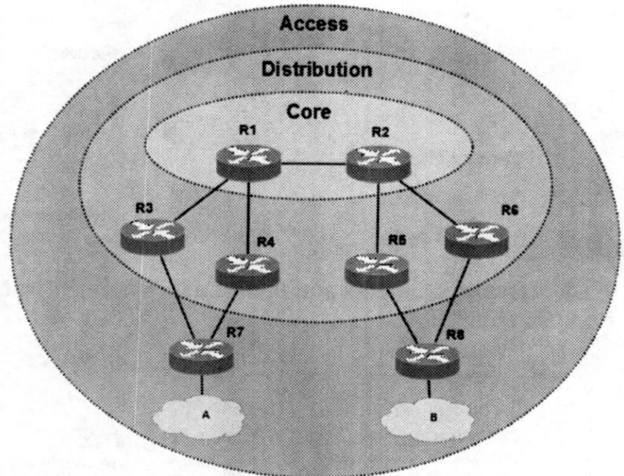

Figure 1.13 Redundancy and load sharing through interconnected access layer devices.

1.6 Network Devices

There are different network devices with the ability to forward packets: repeaters, bridges/switches, and routers. A repeater is a device, typically equipped with two ports, that simply copies what it receives on one port to the other and vice versa. It copies data bit by bit; it does not have any knowledge of protocols and, therefore, cannot distinguish among different frames or packets. Multiport repeaters are called hubs. These devices are not used on modern network architectures because the price of switches has dropped significantly.

Unlike a repeater, a bridge understands link layer (layer 2) protocols and therefore copies data frame by frame, instead of bit by bit. Most LANs are implemented with bridges (that are commonly called switches). The terms bridge and switch can be used to refer to the same device. Historically, the term bridge was used to designate a device equipped with only two ports, so a switch is indeed a multiport bridge.

In order to control broadcast propagation throughout the network, it is important to reduce the amount of overhead associated with these frames. Routers, which operate at layer 3, provide broadcast domain segmentation for each interface, forwarding ingress packets based on a routing table.

Any device that has upper-layer capabilities is considered a host. Figure 1.14 illustrate the different network elements and the corresponding symbols that will be used throughout this book.

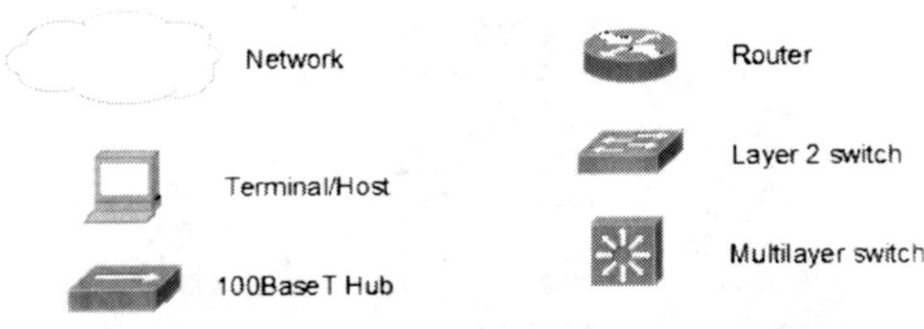

Figure 1.14 Network elements and their corresponding symbols.

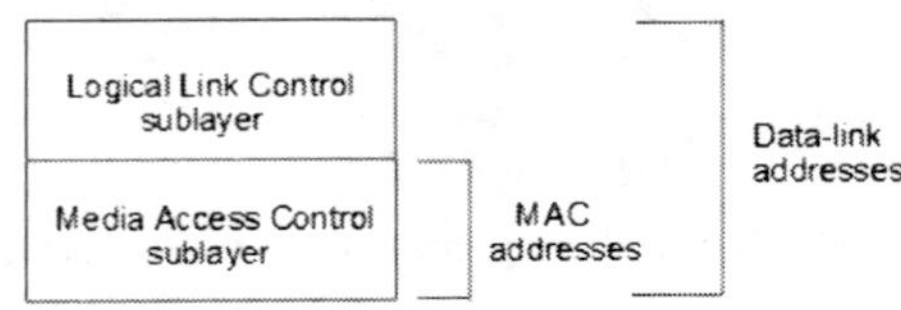

Figure 1.15 Relationship between MAC and Data-Link Addresses.

1.7 Data Link Layer Addressing

A data link layer address uniquely identifies each physical network connection of a network device, so data-link addresses are sometimes referred to as physical or hardware addresses.

End hosts usually have only one physical network connection and thus only one data-link address. Routers and other internetworking devices typically have multiple physical network connections and therefore have multiple data-link addresses.

Media Access Control (MAC) addresses consist of a subset of data link layer addresses. MAC addresses are unique for each LAN interface and identify network entities in local networks that implement the IEEE MAC addresses of the data link layer. Figure 1.15 illustrates the relationship between MAC addresses, data-link addresses, and the IEEE sublayers of the data link layer.

MAC addresses are 48 bits in length and are expressed as 12 hexadecimal digits. The first 6 hexadecimal digits, which are administered by the IEEE, identify the manufacturer or vendor and include the Organizationally Unique Identifier (OUI). The last 6 hexadecimal digits comprise the interface serial number, or another value administered by the specific vendor. MAC addresses are burned into read-only memory (ROM) and are copied into random-access memory (RAM) when the interface card initializes.

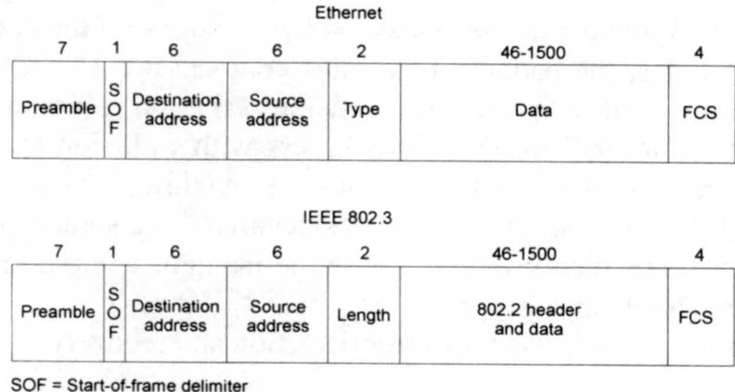

Figure 1.16 Ethernet and IEEE 802.3 Frame Formats.

> **MAC addresses**
> MAC addresses are 48 bits in length and are expressed as 12 hexadecimal digits.

Over the last 30 years, Ethernet became the dominant local area network technology. Nowadays, almost all companies and domestic users use Ethernet to connect their computers to each other. Due to historical reasons, there are two different standards: Ethernet and IEEE 802.3. The differences are subtle: Ethernet provides services corresponding to Layers 1 and 2 of the OSI reference model, while IEEE 802.3 specifies the physical layer and the channel-access portion of the link layer, but does not define a logical link control protocol [99, 94, 64]. Ethernet and IEEE 802.3 frame formats are shown in Figure 1.16, consisting of the following fields:

- Preamble - This consists of seven bytes, all of the form "10101010". This allows the receiver's clock to be synchronized with the sender's.
- Start of Frame Delimiter - This is a single byte ("10101011") used to indicate the start of a frame.
- Destination Address - This is the address of the intended recipient of the frame.
- Source Address - This is the address of the source.
- Length - This is the length of the data in the Ethernet frame, which can be anything from 0 to 1500 bytes.
- Type - In Ethernet frames, this field specifies the upper-layer protocol to receive the data after Ethernet processing is complete.
- Data - This is the information being sent by the frame.

- Pad - 802.3 frame must be at least 64 bytes long, so if the data is shorter than 46 bytes, the padding field must compensate. The reason for the minimum length lies with the collision detection mechanism. In CSMA/CD (Carrier Sense Multiple Access with Collision Detection) the sender must wait at least two times the maximum propagation delay before it knows that no collision has occurred. If a station sends a very short message, then it might release the medium without knowing that the frame has been corrupted.
- Checksum - This is used for error detection and recovery.

1.8 Network Link Layer Addressing

Every device that communicates over a network has also a logical address, called network address. The relationship between a network address and a device is logical and unfixed; typically, it is based either on physical network characteristics (the device is located on a particular network segment) or on a group logic that has no physical basis. End systems require one network layer address for each network layer protocol they support, assuming that the device has only one physical network connection. Routers and other internetworking devices require one network layer address per physical network connection for each network layer protocol supported.

On the Internet, the Internet Protocol (IP) is the network layer protocol, so every network interface has an IP address. Since there are two versions of the IP protocol currently in use, two different addressing schemes exist.

1.8.1 The Internet Protocol, version 4

The Internet Protocol (IP) is a network-layer protocol that contains addressing and control information to enable packets to be routed. IP has two primary goals: providing connectionless, best-effort delivery of datagrams through an internetwork and providing fragmentation and reassembly of datagrams to support data links with different maximum transmission unit (MTU) sizes.

The format of the IP version 4 (IPv4) packet is illustrated in Figure 1.17. The different fields of the packet have the following meaning [94, 64]:

- Version - Indicates the version of IP currently used.
- IP Header Length (IHL) - Indicates the datagram header length in 32-bit words.

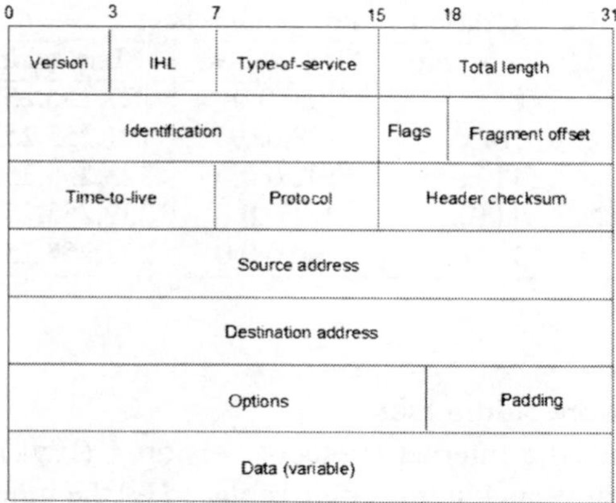

Figure 1.17 Format of the IPv4 packet.

- Type-of-Service - Specifies how an upper-layer protocol would like a current datagram to be handled and assigns various levels of importance to datagrams.
- Total Length - Specifies the length (in bytes) of the entire IP packet, including the data and header.
- Identification - An integer that identifies the current datagram (used in the fragmentation process).
- Flags - A 3 bit field, where the low-order bit specifies whether the packet can be fragmented, the middle bit specifies whether the packet is the last fragment in a series of fragmented packets (these two low-order bits control the fragmentation process), the high-order bit is not used.
- Fragment Offset - Indicates the position of the fragment's data relative to the beginning of the data in the original datagram, which allows the destination IP process to properly reconstruct the original datagram.
- Time-to-Live - Maintains a counter that gradually decrements down to zero, at which point the datagram is discarded. This keeps packets from looping endlessly.
- Protocol - Indicates which upper-layer protocol receives incoming packets after IP processing is complete.
- Header Checksum - Helps ensure IP header integrity.
- Source Address - Specifies the sending node.
- Destination Address - Specifies the receiving node.
- Options - Allows IP to support various options, such as security.
- Data - Contains upper-layer information.

Table 1.1 IPv4 address classes

Class	Leftmost bits	First address	Last address
A	0xxx	0.0.0.0	127.255.255.255
B	10xx	128.0.0.0	191.255.255.255
C	110x	192.0.0.0	223.255.255.255
D	1110	224.0.0.0	239.255.255.255
E	1111	240.0.0.0	255.255.255.255

1.8.1.1 IPv4 Network Addresses

As previously said, the Internet Protocol version 4 (IPv4) was the defined standard for several years but it is being replaced by the more advanced IPv6, mainly in order to solve the address exhaustion problem.

An IPv4 address consists of four bytes (32 bits) that are known as octets. For readability purposes, addresses are written in dotted decimal, that is, periods are placed between each of the four numbers (octets) that comprise an IP address. Because each byte contains 8 bits, each octet in an IP address ranges in value from a minimum of 0 to a maximum of 255. Therefore, the full range of IP addresses is from 0.0.0.0 through 255.255.255.255. That represents a total of 4,294,967,296 possible IP addresses.

The IPv4 address space was originally subdivided into 5 classes (A, B, C, D and E), each one consisting of a contiguous subset of the overall IPv4 address range. Table 1.1 represents the address range of each class.

Classes A, B and C are the three classes of addresses used on IP networks in common practice. Class A addresses have the first bit set to 0 and the next 7 bits are used for the network number. This gives a possibility of 128 networks. However, there are two cases, the all bits 0 number and the all bits 1 number, which have special significance: these special case addresses are reserved, which results in only 126 (128-2) networks of Class A. The remaining 24 bits of a Class A address are used for the host number. Once again, the two special cases apply to the host number part of an IP address. Each Class A network can therefore have a total of 16,777,214 hosts. Class A addresses are assigned only to networks with very large numbers of hosts (large corporations).

The Class B address is more suited to medium-sized networks. The first two bits of the address are predefined as 10. The next 14 bits are used for the network number and the remaining 16 bits identify the host number. This gives a possibility of 16,382 networks, each containing up to 65,534 hosts.

The Class C address offers a maximum of 254 hosts per network and is therefore suited to smaller networks. However, with the first three bits of the address predefined to 110, the next 21 bits provide for a maximum of 2,097,150 such networks.

The IPv4 networking standard defines Class D addresses as reserved for multicast. Multicast is a mechanism for defining groups of nodes and sending IP messages to that group rather than to every node on the LAN (broadcast) or just one other node (unicast). Class D addresses should not be used by ordinary nodes on the Internet.

Class E addresses are reserved, so that they should not be used on IP networks. A special type of IP address is the limited broadcast address 255.255.255.255. When a sender directs an IP broadcast to 255.255.255.255, all other nodes on the local network (LAN) should read that message. This broadcast is "limited" because it does not reach every node on the Internet, only nodes on the LAN.

There are several reserved addresses, with special meanings. When the network part of the address has all bits set to 0, the address corresponds to a source IP address and can be used to identify this host on this network (both network and host number parts set to all bits 0) or a particular host on this network - <network part>, <host part= *whatever* >. Both cases are related to situations where the source IP address appears as part of an initialization procedure when a host is trying to determine its own IP address.

The all bits 1 value is used for broadcast messages and, again, may appear in several combinations. However, it is used only as a destination address. When both the network number and host number parts of an IP address are set to the all bits 1 value, the IP protocol will issue a limited broadcast to all hosts on the network. This is restricted to networks that support broadcasting and will appear only on the local segment. The broadcast will never be forwarded by any routers. If the network number is set to a valid network address while the host number remains set to all bits 1 then a directed broadcast will be sent to all hosts on the specified network.

Of all the broadcast addresses there is one with special significance: 127.0.0.0. This is used as a loopback address and, if implemented, must be used correctly to point back at the originating host itself. In many implementations, this address is used to provide test functions.

The IP standard defines specific address ranges within Class A, Class B, and Class C reserved for use by private networks (intranets). Table 1.2 lists these reserved ranges of the IP address space.

Table 1.2 Private addresses

Class	Private start address	Private finish address
A	10.0.0.0	10.255.255.255
B	172.16.0.0	172.31.255.255
C	192.168.0.0	192.168.255.255

Use of private addresses

Nodes are effectively free to use addresses in the private ranges if they are not connected to the Internet, or if they reside behind firewalls or other gateways that use Network Address Translation (NAT). We will discuss private addressing in more detail in Section 1.13.

1.8.1.2 Subnetting in IPv4

Breaking down the host number part of an IP address provides an extra level of addressability. The host number can be subdivided into a subnet number and a host number to provide a second logical network within the first. This second network is known as the subnetwork or subnet. A subnetted address now has three parts: <network number><subnet number><host number>. The subnet number is transparent to remote networks. Remote hosts still regard the local part of the address (the subnet number and the host number) as a host number. Only those hosts within the network that are configured to use subnets are aware that subnetting is in effect.

A subnet is created by the use of a subnet mask. This is a 32-bit number just like the IP address itself and has bits relating to the network number, subnet number and host number. The bit positions in the subnet mask that identify the network number are set to 1s to maintain the original routing. In the remaining local part of the address, bits set to 1 indicate the subnet number and bits set to zero indicate the host number. In order for a host or router to apply the mask, it performs a logical AND of the mask with the IP address it is trying to route. IP addresses are commonly represented as <IP address>/<network mask>.

The use case that will be solved at the end of this chapter will explain subnetting in detail, considering specific examples.

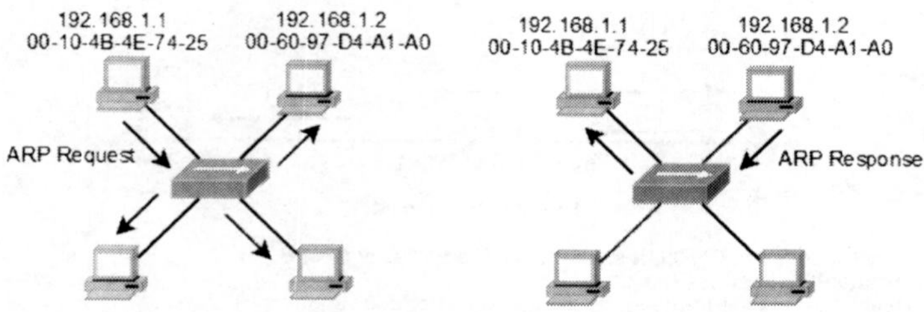

Figure 1.18 The ARP protocol.

1.8.1.3 The Address Resolution Protocol

Because internetworks generally use network addresses to route traffic, there is a need to map network addresses to MAC addresses. When the network layer has determined the destination station's network address, it must forward the information over a physical network using a MAC address. In the TCP/IP suite, the Address Resolution Protocol (ARP) is used to perform this mapping [82].

When a network device needs to send data to another device on the same network, it knows the source and destination network addresses for the data transfer. It must somehow map the destination address to a MAC address before forwarding the data. First, the sending station will check its ARP table to see if it has already discovered this destination station's MAC address. If it has not, it will send a broadcast on the network with the destination station's IP address contained in the broadcast. Every station on the network receives the broadcast and compares the embedded IP address to its own. Only the station with the matching IP address replies to the sending station with a packet containing the MAC address for the station. The first station then adds this information to its ARP table for future reference and proceeds to transfer the data. This process is illustrated in Figure 1.18.

When the destination device lies on a remote network, one beyond a router, the process is the same except that the sending station sends the ARP request for the MAC address of its default gateway. It then forwards the information to that device. The default gateway will then forward the information over whatever networks necessary to deliver the packet to the network on which the destination device resides. The router on the destination device's network then uses ARP to obtain the MAC of the actual destination device and delivers the packet. So, the default gateway of a host is the device that passes traffic from the local subnet to devices on other subnets and is typically one interface of one of the routers that are connected to the local subnet.

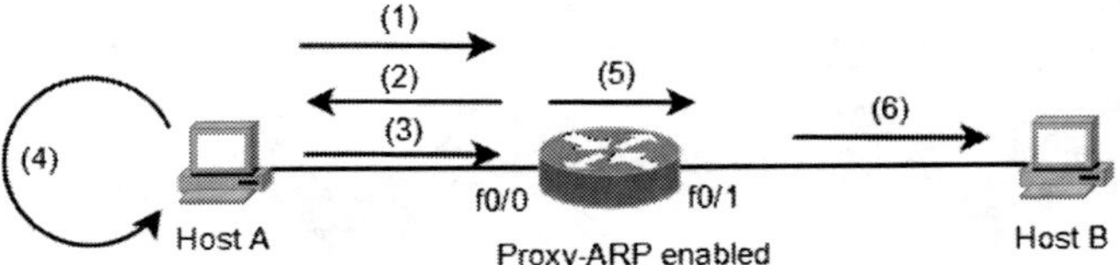

(1) Host A sends ARP Request for the MAC address of Host B
(2) Router f0/0 interface replies for Host B
(3) Host A sends the frame to the Router f0/0 interface
(4) Host A inserts the entry "Router f0/0 MAC address - Host B IP address" in its ARP table
(5) Router forwards the frame to interface f0/1
(6) Router f0/1 interface sends the frame to Host B

Figure 1.19 Illustration of the proxy ARP mechanism.

The Reverse Address Resolution Protocol (RARP) is an obsolete protocol used by a host computer to request its IPv4 address from an administrative host, when it has available its link layer or hardware address. RARP is described in the Internet Engineering Task Force (IETF) RFC 903 [27] and has been rendered obsolete by DHCP, which supports a much greater feature set than RARP. RARP requires one or more server hosts to maintain a database of mappings of link layer addresses to their respective protocol addresses.

The proxy ARP mechanism, illustrated in Figure 1.19, is used to hide physical networks from each other. In this case, a router answers ARP requests targeted for a host.

Gratuitous ARP occurs when a host sends an ARP request resolving its own IP address. This usually happens when the interface is configured at bootstrap time. The interface uses gratuitous ARP to determine if there are other hosts using the same IP address. The sender's IP and MAC address are broadcast, and other hosts will insert this mapping into their ARP tables.

Exercise - The ARP Operation

Let us suppose that a host tries to contact a remote host on the same LAN using the "ping" utility. It is assumed that no previous IP datagrams have been received from this host and, therefore, ARP must first be used to identify the MAC address of the remote host.

The ARP request message, illustrated in Figure 1.20, is sent to the Ethernet broadcast address, and the Ethernet protocol is set to 0x806. Since it is broadcast, this message is received by all systems in the same collision domain. If the target of the query is connected to the network, it will receive a copy of the query. Only this system responds, while other systems discard the packet.

```
No.  Time       Source            Destination       Protocol  Info
  9  27.549000  c4:00:1b:d0:00:   Broadcast         ARP       Who has 192.168.1.2?  Tell 192.168.1.1
 10  27.577000  c4:01:1b:d0:00:   c4:00:1b:d0:00:00 ARP       192.168.1.2 is at c4:01:1b:d0:00:00
 11  29.547000  192.168.1.1       192.168.1.2       ICMP      Echo (ping) request  (id=0x0000, seq(be/le)=1/256, ttl=255)
 12  29.587000  192.168.1.2       192.168.1.1       ICMP      Echo (ping) reply    (id=0x0000, seq(be/le)=1/256, ttl=255)

Frame 9: 60 bytes on wire (480 bits), 60 bytes captured (480 bits)
Ethernet II, Src: c4:00:1b:d0:00:00 (c4:00:1b:d0:00:00), Dst: Broadcast (ff:ff:ff:ff:ff:ff)
Address Resolution Protocol (request)
    Hardware type: Ethernet (0x0001)
    Protocol type: IP (0x0800)
    Hardware size: 6
    Protocol size: 4
    Opcode: request (0x0001)
    [Is gratuitous: False]
    Sender MAC address: c4:00:1b:d0:00:00 (c4:00:1b:d0:00:00)
    Sender IP address: 192.168.1.1 (192.168.1.1)
    Target MAC address: 00:00:00_00:00:00 (00:00:00:00:00:00)
    Target IP address: 192.168.1.2 (192.168.1.2)
```

Figure 1.20 The ARP Request message.

```
No.  Time       Source            Destination       Protocol  Info
  9  27.549000  c4:00:1b:d0:00:   Broadcast         ARP       Who has 192.168.1.2?  Tell 192.168.1.1
 10  27.577000  c4:01:1b:d0:00:   c4:00:1b:d0:00:00 ARP       192.168.1.2 is at c4:01:1b:d0:00:00
 11  29.547000  192.168.1.1       192.168.1.2       ICMP      Echo (ping) request  (id=0x0000, seq(be/le)=1/256, ttl=255)
 12  29.587000  192.168.1.2       192.168.1.1       ICMP      Echo (ping) reply    (id=0x0000, seq(be/le)=1/256, ttl=255)

Frame 10: 60 bytes on wire (480 bits), 60 bytes captured (480 bits)
Ethernet II, Src: c4:01:1b:d0:00:00 (c4:01:1b:d0:00:00), Dst: c4:00:1b:d0:00:00 (c4:00:1b:d0:00:00)
Address Resolution Protocol (reply)
    Hardware type: Ethernet (0x0001)
    Protocol type: IP (0x0800)
    Hardware size: 6
    Protocol size: 4
    Opcode: reply (0x0002)
    [Is gratuitous: False]
    Sender MAC address: c4:01:1b:d0:00:00 (c4:01:1b:d0:00:00)
    Sender IP address: 192.168.1.2 (192.168.1.2)
    Target MAC address: c4:00:1b:d0:00:00 (c4:00:1b:d0:00:00)
    Target IP address: 192.168.1.1 (192.168.1.1)
```

Figure 1.21 The ARP Reply message.

In this case, the target system (with IP address 192.168.1.2/24) forms an ARP response. This packet is unicast to the address of the host sending the query (Figure 1.21). Since the original ARP request also included the hardware address (Ethernet source address) of the requesting host, this is already known and does not require another ARP message to find it.

1.8.1.4 The Internet Control Message Protocol, version 4

The Internet Control Message Protocol (ICMP) is a network-layer Internet protocol that provides messages to report errors and other information regarding IP packet processing back to the source. ICMP includes several messages, like for example Destination Unreachable, Echo Request and Reply, Redirect, Time Exceeded, and Router Advertisement and Router Solicitation. Table 1.3 presents some of the most relevant ICMP message types.

Hierarchically, ICMP lies just above IP, that is, ICMP messages are carried as IP payload. ICMP messages have a type and a code field, and contain the header and the first 8 bytes of the IP datagram that caused the ICMP message to be generated (so that the sender can determine the datagram that caused the error). Let us now explain examples of the most common ICMP message exchanges.

When an ICMP destination-unreachable message is sent by a router, it means that the router is unable to send the packet to its final destination, so it will discard the original packet. There are two reasons for this situation: the source host has specified a nonexistent address or the router does not have a route to the destination. Destination-unreachable messages include four basic types: network unreachable, host unreachable, protocol unreachable, and port unreachable. Network-unreachable messages usually mean that a failure has occurred in the routing or addressing of a packet. Host-unreachable messages usually indicate delivery failure, such as a wrong subnet mask. Protocol-unreachable messages generally mean that the destination does not support the upper-layer protocol specified in the packet. Port-unreachable messages imply that the UDP socket or port is not available.

An ICMP echo-request message, which is generated by the ping command, is sent by any host to test node reachability across an internetwork. The ICMP echo-reply message indicates that the node can be successfully reached. An ICMP Redirect message is sent by the router to the source host on the same LAN when there is a more efficient routing path. The router still forwards the original packet to the destination. ICMP redirects allow host routing tables to remain small because it is necessary to know the address of only one router, even if that router does not provide the best path. Even after receiving an ICMP Redirect message, some devices might continue using the less-efficient route.

An ICMP Time-exceeded message is sent by the router if an IP packet's Time-to-Live field reaches zero. The Time-to-Live field prevents packets from continuously circulating if the internetwork contains a routing loop. The router then discards the original packet.

The traceroute utility allows to trace a route from a source host to any other host in the world and is implemented with ICMP messages. The source sends several (usually three) ICMP Echo Request messages with a TTL of 1 to the destination. The source also starts timers for each of the datagrams. When the datagram arrives to the first router, it decrements the TTL value and observes that it has just expired. According to the rules of the IP protocol, the router discards the datagram and sends an ICMP warning message to the source (a TTL Expired message). This warning message includes the name of the first router and its IP address, so when this message arrives back at the source, the source obtains the round-trip time from the timer and the name and IP address of the first router from the message. This process continues

Table 1.3 Some ICMP message types

ICMP Type	Code	Description
0	0	Echo Reply
3	0	Destination Network Unreachable
3	1	Destination Host Unreachable
3	2	Destination Protocol Unreachable
3	3	Destination Port Unreachable
8	0	Echo Request
9	0	Router Advertisement
11	0	TTL Expired

with the source host sending new ICMP Echo Request messages with a TTL of 2 to the destination. These messages will pass through the first router and reach the second router, which will decrement the TTL value to zero and trigger the same actions that were described above. In this way, the source host will obtain the round-trip time from the different timers and the name and IP address of the second router from the TTL Expired message. This process will continue until the datagrams reach the destination, which will then reply with ICMP Echo Reply messages. In this manner, the source host learns the identities of the routers that lie between it and the destination host and the round-trip time between the two hosts.

Exercise - The ICMP Protocol

Let us suppose that a host with IP address 192.168.1.1/24 executes a *ping* command to another host with IP address 192.168.1.2/24. First, an ICMP Request message is sent from IP address 192.168.1.1 to IP address 192.168.1.2, as illustrated in Figure 1.22. The ICMP protocol is identified in the IP header; the ICMP message has a type of 8, corresponding to a Request. The data field of the ICMP packet, with a length of 72 bytes, has no meaning. The called host replies to the calling counterpart with an ICMP Reply message, illustrated in Figure 1.23.

Now, let us exemplify the operation of the traceroute utility, using the network setup that is depicted in Figure 1.24. By executing command

```
traceroute 192.168.3.2
```

at the PC, we can see from the capture of Figure 1.25 that a first ICMP Echo Request message is encapsulated in an IP packet with TTL equal to 1. As routers decrement by 1 the TTL value of every IP packet that traverses

```
No.  Time      Source        Destination   Protocol  Info
  3 0.244000  192.168.1.1   192.168.1.2   ICMP      Echo (ping) request  (id=0x0002, seq(be/le)=0/0, ttl=255)
  4 0.273000  192.168.1.2   192.168.1.1   ICMP      Echo (ping) reply    (id=0x0002, seq(be/le)=0/0, ttl=255)

Frame 3: 114 bytes on wire (912 bits), 114 bytes captured (912 bits)
Ethernet II, Src: c4:00:1b:d0:00:00 (c4:00:1b:d0:00:00), Dst: c4:01:1b:d0:00:00 (c4:01:1b:d0:00:00)
Internet Protocol, Src: 192.168.1.1 (192.168.1.1), Dst: 192.168.1.2 (192.168.1.2)
    Version: 4
    Header length: 20 bytes
  Differentiated Services Field: 0x00 (DSCP 0x00: Default; ECN: 0x00)
    Total Length: 100
    Identification: 0x000a (10)
  Flags: 0x00
    Fragment offset: 0
    Time to live: 255
    Protocol: ICMP (1)
  Header checksum: 0x383b [correct]
    Source: 192.168.1.1 (192.168.1.1)
    Destination: 192.168.1.2 (192.168.1.2)
Internet Control Message Protocol
    Type: 8 (Echo (ping) request)
    Code: 0
    Checksum: 0x861d [correct]
    Identifier: 0x0002
    Sequence number: 0 (0x0000)
    Sequence number (LE): 0 (0x0000)
  Data (72 bytes)
      Data: 0000000000c2f768abcdabcdabcdabcdabcdabcdabcdabcd...
      [Length: 72]
```

Figure 1.22 The ICMP Request message.

them, the first router will discard the datagram and send an ICMP warning message (a *TTL Exceeded in Transit* message) to the source 192.168.1.1/24. This process is repeated three times in order to obtain reliable round trip time statistics. Then, another ICMP Echo Request message is generated and sent with a TTL of 2 to the same destination (packet number 7). Since this TTL value is enough for the IP datagram to reach the second router, this equipment will answer with the ICMP *TTL Exceeded in Transit* message. In this way, the source becomes aware of the second hop in the way from source to destination. Once again, this process is repeated three times. Finally, the source generates an ICMP Echo Request message with a TTL of 3 (packet number 13). Since this TTL value is enough for the IP datagram to reach the destination, the answer (which is a normal ICMP Echo Reply message) now comes from IP address 192.168.3.2. As always, the process is repeat three times.

1.8.1.5 Fragmentation in IPv4

IP uses the fragmentation technique to solve the problem of heterogeneous Maximum Transmission Units (MTUs). When a datagram is larger than the MTU of the network over which it must be sent, it is divided into smaller fragments which are sent separately. This process is illustrated in Figure 1.26 and was first presented in [75].

To fragment a datagram, a host or router uses the MTU and the datagram header size to calculate how many fragments are required (they must be in

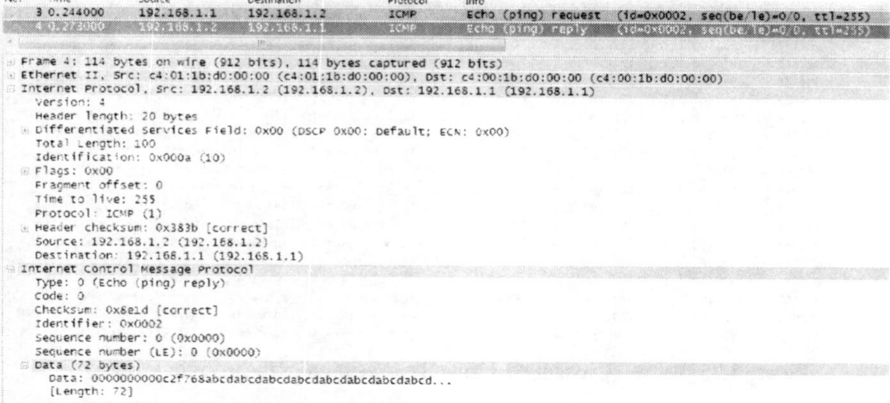

Figure 1.23 The ICMP Reply message.

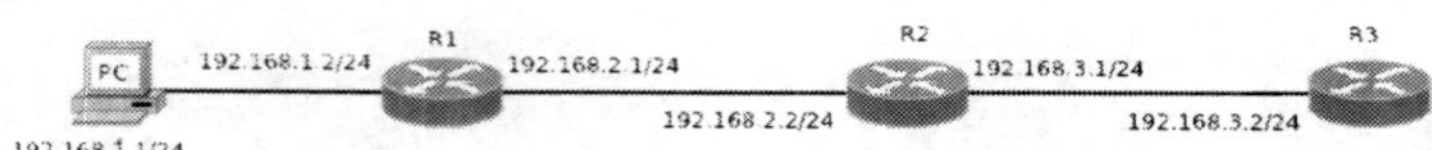

Figure 1.24 Setup used to illustrate the traceroute utility.

multiples of 8 octets). Then the header of the original datagram is copied into the headers of each one of the fragments. The following IPv4 fields have to change: Total Length, to reflect the shorter size; More flag in all but the last fragment; Fragment Offset to reflect the position of the fragment within the original datagram and the Header checksum. Each fragment becomes its own datagram and is routed independently of any other datagrams. This makes it possible for the fragments of the original datagram to arrive at the final destination out of order. Further fragmentation could occur.

At the final destination, the process of re-constructing the original datagram is called reassembly and was first presented in [11]. The unique Identification field is used to group fragments together, even from the same source. The Fragment Offset field tells the receiver how to order the fragments, while the absence of the More flag signals the last fragment. In the example of Figure 1.26, If host A sends a 1500 octet datagram (20-octet header and 1480 octets of data) to host B, and the network between Routers A and B have a MTU of 620 bytes, router R1 will fragment the datagram into three fragments:

- The first fragment will contain a 20-octet header (Fragment Offset is zero and the More flag is set) and 600 octets of data;

No.	Time	Source	Destination	Protocol	Info
1	0.000000	192.168.1.1	192.168.3.2	ICMP	Echo (ping) request (id=0x04ab, seq(be/le)=1/256, ttl=1)
2	0.002127	192.168.1.2	192.168.1.1	ICMP	Time-to-live exceeded (Time to live exceeded in transit)
3	0.002271	192.168.1.1	192.168.3.2	ICMP	Echo (ping) request (id=0x04ab, seq(be/le)=2/512, ttl=1)
4	0.004349	192.168.1.2	192.168.1.1	ICMP	Time-to-live exceeded (Time to live exceeded in transit)
5	0.004422	192.168.1.1	192.168.3.2	ICMP	Echo (ping) request (id=0x04ab, seq(be/le)=3/768, ttl=1)
6	0.006499	192.168.1.2	192.168.1.1	ICMP	Time-to-live exceeded (Time to live exceeded in transit)
7	0.006589	192.168.1.1	192.168.3.2	ICMP	Echo (ping) request (id=0x04ab, seq(be/le)=4/1024, ttl=2)
8	0.012867	192.168.2.2	192.168.1.1	ICMP	Time-to-live exceeded (Time to live exceeded in transit)
9	0.012994	192.168.1.1	192.168.3.2	ICMP	Echo (ping) request (id=0x04ab, seq(be/le)=5/1280, ttl=2)
10	0.019264	192.168.2.2	192.168.1.1	ICMP	Time-to-live exceeded (Time to live exceeded in transit)
11	0.019349	192.168.1.1	192.168.3.2	ICMP	Echo (ping) request (id=0x04ab, seq(be/le)=6/1536, ttl=2)
12	0.025594	192.168.2.2	192.168.1.1	ICMP	Time-to-live exceeded (Time to live exceeded in transit)
13	0.025682	192.168.1.1	192.168.3.2	ICMP	Echo (ping) request (id=0x04ab, seq(be/le)=7/1792, ttl=3)
14	0.038181	192.168.3.2	192.168.1.1	ICMP	Echo (ping) reply (id=0x04ab, seq(be/le)=7/1792, ttl=253)
15	0.038321	192.168.1.1	192.168.3.2	ICMP	Echo (ping) request (id=0x04ab, seq(be/le)=8/2048, ttl=3)
16	0.050559	192.168.3.2	192.168.1.1	ICMP	Echo (ping) reply (id=0x04ab, seq(be/le)=8/2048, ttl=253)
17	0.052663	192.168.1.1	192.168.3.2	ICMP	Echo (ping) request (id=0x04ab, seq(be/le)=9/2304, ttl=3)
18	0.063093	192.168.3.2	192.168.1.1	ICMP	Echo (ping) reply (id=0x04ab, seq(be/le)=9/2304, ttl=253)

Figure 1.25 Packets exchanged during the execution of the traceroute utility.

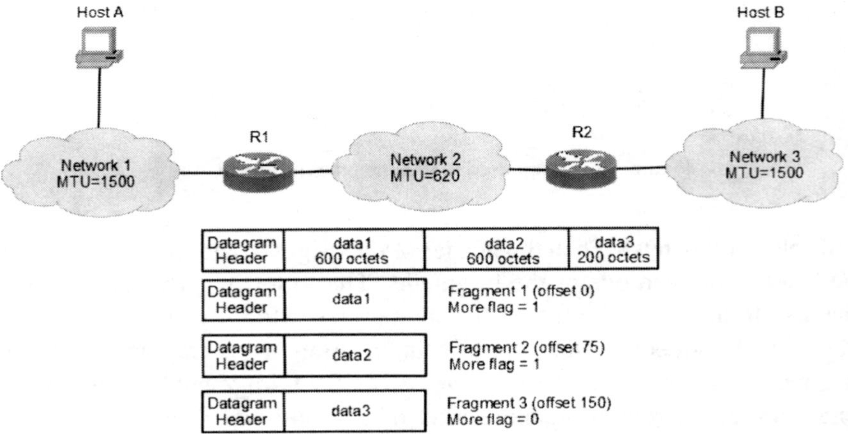

Figure 1.26 IPv4 fragmentation process.

- The second fragment will contain a 20-octet header (Fragment Offset is 600/8 = 75 and the More flag is set) and 600 octets of data;
- The third fragment will contain a 20-octet header (Fragment Offset is 150 and the More flag is clear) and the remaining 300 octets of data.

Exercise - Fragmentation Process

Now, suppose that command

```
ping -l 2000 192.168.1.2
```

is executed at the PC of Figure 1.24 forcing it to send a total amount of 2000 bytes of data. Since the MTU of the Ethernet links is equal to 1500 bytes,

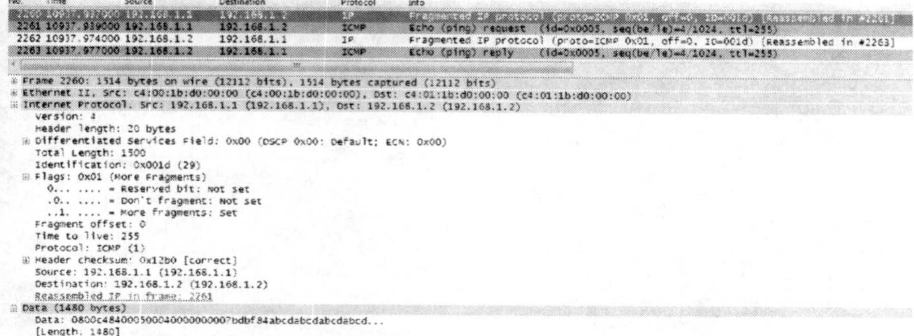

Figure 1.27 The first fragment of the message.

fragmentation will occur, splitting each message (both ICMP Request and Reply) in two fragments. Figure 1.27 shows the first fragment: it is identified by wireshark as an IP packet (since it does not include any ICMP header), has a payload of 1480 byes (adding the 20 bytes of the IP header completes the amount of 1500 bytes that is allowed for the MTU), the *More Fragments* flag is set and the *Fragment Offset* field is set to zero, indicating that this is the first fragment. Figure 1.28 illustrates the second segment, where we can see that the payload has a length of 520 bytes (actually, wireshark identifies 528 bytes of payload, but 8 byes compose the ICMP packet header), the *More Fragments* flag is set to zero and the *Fragment Offset* field is set to 1480, indicating that when the receiver performs the reassembly of the original data it should place the data of the second fragment after the 1480 bytes of data corresponding to the first segment. Note that both fragments have the same *Identification* value, since they belong to the same message.

A completely analogous situation occurs in the ICMP Reply message.

1.8.1.6 The Dynamic Host Configuration Protocol

In small networks, it is often more practical to define static IP addresses, rather than set up and install a server dedicated to assigning IP addresses. Although statically assigning IP addresses is simple, this option presents some problems because maintaining this type of network can be expensive. In these cases, a dynamic protocol like the Dynamic Host Configuration Protocol (DHCP) should be used. DHCP was presented in [21] and supports the following functions:

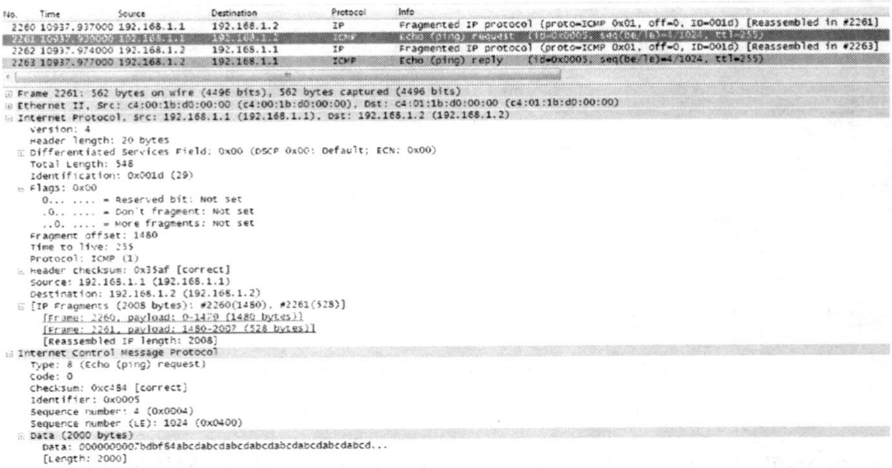

```
No.   Time           Source        Destination   Protocol  Info
2260 10937.937000 192.168.1.1   192.168.1.2   IP        Fragmented IP protocol (proto=ICMP 0x01, off=0, ID=001d) [Reassembled in #2261]
2261 10937.[illegible] 192.168.1.1   192.168.1.2   ICMP      Echo (ping) request  (id=0x0005, seq(be/le)=4/1024, ttl=255)
2262 10937.974000 192.168.1.2   192.168.1.1   IP        Fragmented IP protocol (proto=ICMP 0x01, off=0, ID=001d) [Reassembled in #2263]
2263 10937.977000 192.168.1.2   192.168.1.1   ICMP      Echo (ping) reply    (id=0x0005, seq(be/le)=4/1024, ttl=255)

Frame 2261: 562 bytes on wire (4496 bits), 562 bytes captured (4496 bits)
Ethernet II, Src: c4:00:1b:d0:00:00 (c4:00:1b:d0:00:00), Dst: c4:01:1b:d0:00:00 (c4:01:1b:d0:00:00)
Internet Protocol, Src: 192.168.1.1 (192.168.1.1), Dst: 192.168.1.2 (192.168.1.2)
  Version: 4
  Header length: 20 bytes
  Differentiated Services Field: 0x00 (DSCP 0x00: Default; ECN: 0x00)
  Total Length: 548
  Identification: 0x001d (29)
  Flags: 0x00
    0... .... = Reserved bit: Not set
    .0.. .... = Don't fragment: Not set
    ..0. .... = More fragments: Not set
  Fragment offset: 1480
  Time to live: 255
  Protocol: ICMP (1)
  Header checksum: 0x35af [correct]
  Source: 192.168.1.1 (192.168.1.1)
  Destination: 192.168.1.2 (192.168.1.2)
  [IP Fragments (2008 bytes): #2260(1480), #2261(528)]
    [Frame: 2260, payload: 0-1479 (1480 bytes)]
    [Frame: 2261, payload: 1480-2007 (528 bytes)]
    [Reassembled IP length: 2008]
Internet Control Message Protocol
  Type: 8 (Echo (ping) request)
  Code: 0
  Checksum: 0xc484 [correct]
  Identifier: 0x0005
  Sequence number: 4 (0x0004)
  Sequence number (LE): 1024 (0x0400)
  Data (2000 bytes)
    Data: 000000007bdbf84abcdabcdabcdabcdabcdabcdabcdabcd...
    [Length: 2000]
```

Figure 1.28 The second fragment of the message.

- Automatic allocation - DHCP assigns a permanent IP address to a device.
- Dynamic allocation - DHCP assigns a leased IP address to the device for a limited period of time (*lease time*).
- Manual allocation - The devices address is manually configured by the network administrator, and the DHCP is used to inform devices of the assigned address.

We can have only one DHCP server in a LAN or several servers, but in this case the address pools should not overlap. This service is based on a client-server architecture. A client may be configured to broadcast a request for address assignment and will select the most appropriate response from those servers that answer the request. Some important parameters should be conveniently configured at the server: the address range, that is, the set of IP addresses defined by an initial address and a final address; an exclusion range, that is, the sets of IP addresses to be excluded; reserved addresses, which are IP addresses that are attributed in a permanent way to MAC addresses and the lease duration. The DHCP protocol is an extension of the Bootstrap Protocol (BOOTP, RFC 1542) and runs over UDP (with a server port number of 67 and a client port number of 68).

An example of DHCP in operation is shown in Figure 1.29. The basic assignment process consists of four steps: (i) when a new host is added to the network and is initialized, it sends a broadcast message (DHCP Dis-

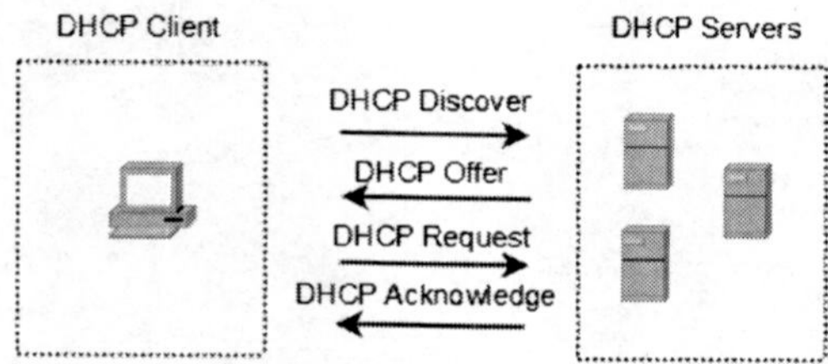

Figure 1.29 Steps of the DHCP negotiation process.

cover message) to the network that will be received by any DHCP server (optionally, the client may also indicate the IP address he wants to lease); (ii) all available servers respond to the broadcast with a DHCP Offer message that is encapsulated in a BootP Reply packet (each server indicates an IP address to be leased and, if possible, servers respect the client preferences); (iii) the client selects the most appropriate server/offer and sends a DHCP Request message (encapsulated in a BootP Request packet); (iv) the selected server will send the client the necessary configuration parameters in a DHCP Acknowledgment message (encapsulated in a BootP Reply packet).

Besides the four basic messages that were explained, the DHCP service includes some additional messages:

- DHCP Decline: the client rejects the offer that was made and restarts the address leasing process;
- DHCP Nack: the server informs the client that it cannot satisfy its demand, that was made using the DHCP Request message;
- DHCP Release: the client informs that it intends to finish the address leasing;
- DHCP Inform: the client asks for a limited set of parameters (in this case, the client already has an IP address but wants to ask for, for example, the address of a DNS server).

Exercise - The DHCP Operation

In the first experiment, whose setup is depicted in Figure 1.30, the DHCP server will assign the basic IP configuration to PC 1. Note that both elements belong to the same IP network. The server configuration is illustrated in Box 1.3: we can see that the default lease time is equal to 10 minutes, there are two DNS servers to assign and the IP addresses that will be assigned belong to one of two different ranges, 10.1.1.100-10.1.1.200 (as stated in command line 5) and 10.2.2.100-10.2.2.200 (as stated in command line 9).

```
default-lease-time 600;
max-lease-time 7200;
option domain-name-servers 10.0.0.1, 10.0.0.2;
subnet 10.1.1.0 netmask 255.255.255.0 {
  range 10.1.1.100 10.1.1.200;
}

subnet 10.2.2.0 netmask 255.255.255.0 {
  range 10.2.2.100 10.2.2.200;
}
```

Box 1.3 DHCP Server configuration

The four steps of the assignment process are illustrated in the capture of Figure 1.31, being important to notice that all exchanged messages are sent to a broadcast address, both at the MAC and IP levels. This guarantees that all DHCP servers that can be present in the LAN are aware of the negotiation that is taking place.

The DHCP Discover message, illustrated in detail in Figure 1.31, is encapsulated in a BootP Request message and includes the MAC address of the DHCP client. The Offer message is encapsulated in a BootP reply packet and includes the IP address and subnet mask that the server is offering to the client, the lease time value and the IP address of the DHCP server (Figure 1.32).

DHCP Offer message

Note that more than one DHCP server, if they exist, can offer addresses and configurations to any client.

The Request message is also encapsulated in a BootP Request packet and includes the configuration parameters that the client is requesting from one of the servers (Figure 1.33). Finally, the DHCP Acknowledgment message is encapsulated in a BootP Reply message and confirms the assignment of the requested parameters (Figure 1.34).

Note that, besides the IP address leasing, the DHCP service also allows the dynamic allocation of other important configuration parameters: the subnet mask, the default gateway, DNS servers, WINS servers and the DNS domain, for example. The Renewal Time (T1) and the Rebinding Time (T2) values are equal to 50% and 85% of the Lease Time, respectively. The host will try to renew the network configurations at T1 and, if it doesn't succeed, it will try again at T2.

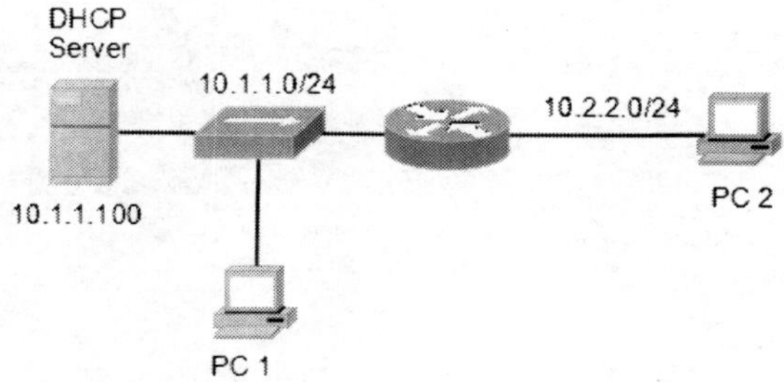

Figure 1.30 Network set up used to study the DHCP operation.

```
interface FastEthernet0/0
 ip address 10.1.1.1 255.255.255.0

interface FastEthernet0/1
 ip address 10.2.2.1 255.255.255.0
 ip helper-address 10.1.1.100
```

Box 1.4 DHCP Relay configuration

DHCP transaction ID

During all the DHCP negotiation phase, the transaction ID is always the same (0xa5b42413 in this case).

Figure 1.35 illustrates the message that is exchanged when the client releases the IP configuration that was previously allocated. Now, the source and destination MAC and IP addresses are filled with the information corresponding to the DHCP Client and Server, respectively.

Now, suppose that PC 2 in Figure 1.30, which is located on a different IP network, is switched on and requests its basic IP configuration using DHCP. By default, the router does not forward broadcast packets, so a BootP Relay Agent must be activated on the router in order to forward the Discover message to the IP network where the server is located. The configuration of the router should be similar to the one illustrated in Box 1.4.

Using this mechanism, the router will forward the BootP Request packet to the IP network of the DHCP server, filling the *Relay Agent IP Address* field with the IP address of the interface that received the DHCP Discover message from the client (Figure 1.36). In this way, the DHCP server knows from which pool of addresses it should pick the address to lease. All subsequent DHCP negotiation packets are now correctly routed to their destinations.

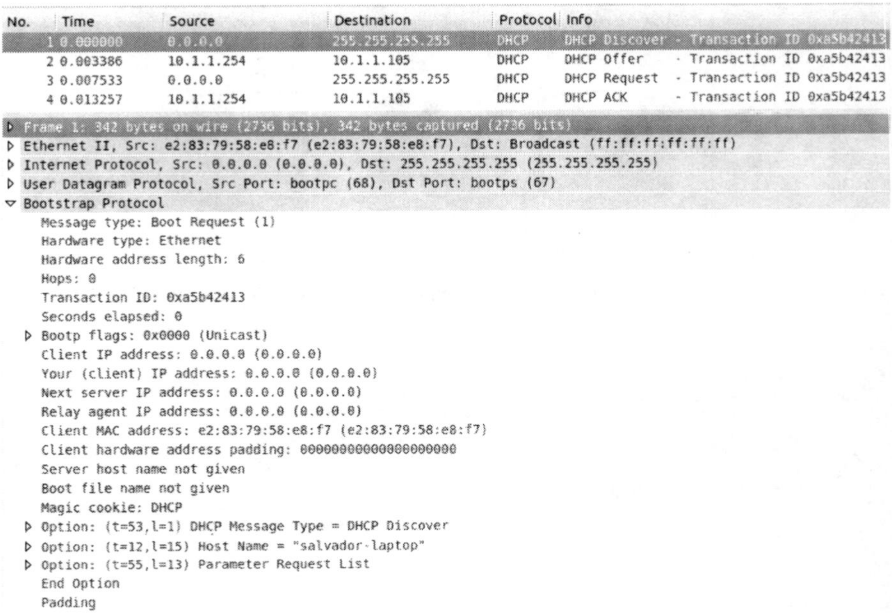

No.	Time	Source	Destination	Protocol	Info
1	0.000000	0.0.0.0	255.255.255.255	DHCP	DHCP Discover - Transaction ID 0xa5b42413
2	0.003386	10.1.1.254	10.1.1.105	DHCP	DHCP Offer - Transaction ID 0xa5b42413
3	0.007533	0.0.0.0	255.255.255.255	DHCP	DHCP Request - Transaction ID 0xa5b42413
4	0.013257	10.1.1.254	10.1.1.105	DHCP	DHCP ACK - Transaction ID 0xa5b42413

```
▷ Frame 1: 342 bytes on wire (2736 bits), 342 bytes captured (2736 bits)
▷ Ethernet II, Src: e2:83:79:58:e8:f7 (e2:83:79:58:e8:f7), Dst: Broadcast (ff:ff:ff:ff:ff:ff)
▷ Internet Protocol, Src: 0.0.0.0 (0.0.0.0), Dst: 255.255.255.255 (255.255.255.255)
▷ User Datagram Protocol, Src Port: bootpc (68), Dst Port: bootps (67)
▽ Bootstrap Protocol
    Message type: Boot Request (1)
    Hardware type: Ethernet
    Hardware address length: 6
    Hops: 0
    Transaction ID: 0xa5b42413
    Seconds elapsed: 0
  ▷ Bootp flags: 0x0000 (Unicast)
    Client IP address: 0.0.0.0 (0.0.0.0)
    Your (client) IP address: 0.0.0.0 (0.0.0.0)
    Next server IP address: 0.0.0.0 (0.0.0.0)
    Relay agent IP address: 0.0.0.0 (0.0.0.0)
    Client MAC address: e2:83:79:58:e8:f7 (e2:83:79:58:e8:f7)
    Client hardware address padding: 00000000000000000000
    Server host name not given
    Boot file name not given
    Magic cookie: DHCP
  ▷ Option: (t=53,l=1) DHCP Message Type = DHCP Discover
  ▷ Option: (t=12,l=15) Host Name = "salvador-laptop"
  ▷ Option: (t=55,l=13) Parameter Request List
    End Option
    Padding
```

Figure 1.31 The DHCP negotiation process: details of the Discover message.

1.8.2 The Internet Protocol, version 6

IPv6 was developed to replace IPv4 mainly to the exhaustion of IPv4 addresses and is described in Internet standard document RFC 2460 [16]. Figure 1.37 illustrates the format of the IPv6 packet header, where it is possible to notice some differences and similarities when compared to IPv4.

Basically, there is a header format simplification, where some fields have been dropped and adapted to make the header size constant, thus reducing the processing cost. Besides, in IPv6 header extensions are intensively used. The main changes can be summarized as follows:

- The alignment changed from 32 bit to 64 bit multiples;
- Header length was eliminated and replaced by the Payload Length field;
- Size of source and destination addresses changed to 16 octets;
- Fragmentation information moved out of fixed fields in base header to extension header;
- Time-to-Live field changed to Hop Limit;
- Type-of-Service field was renamed to Traffic Class and extended with the Flow Label field;

No.	Time	Source	Destination	Protocol	Info
1	0.000000	0.0.0.0	255.255.255.255	DHCP	DHCP Discover - Transaction ID 0xa5b42413
2	0.003386	10.1.1.254	10.1.1.105	DHCP	DHCP Offer - Transaction ID 0xa5b42413
3	0.007533	0.0.0.0	255.255.255.255	DHCP	DHCP Request - Transaction ID 0xa5b42413
4	0.013257	10.1.1.254	10.1.1.105	DHCP	DHCP ACK - Transaction ID 0xa5b42413

```
▷ Frame 2: 342 bytes on wire (2736 bits), 342 bytes captured (2736 bits)
▷ Ethernet II, Src: CadmusCo_de:98:9e (08:00:27:de:98:9e), Dst: e2:83:79:58:e8:f7 (e2:83:79:58:e8:f7)
▷ Internet Protocol, Src: 10.1.1.254 (10.1.1.254), Dst: 10.1.1.105 (10.1.1.105)
▷ User Datagram Protocol, Src Port: bootps (67), Dst Port: bootpc (68)
▽ Bootstrap Protocol
    Message type: Boot Reply (2)
    Hardware type: Ethernet
    Hardware address length: 6
    Hops: 0
    Transaction ID: 0xa5b42413
    Seconds elapsed: 0
  ▷ Bootp flags: 0x0000 (Unicast)
    Client IP address: 0.0.0.0 (0.0.0.0)
    Your (client) IP address: 10.1.1.105 (10.1.1.105)
    Next server IP address: 0.0.0.0 (0.0.0.0)
    Relay agent IP address: 0.0.0.0 (0.0.0.0)
    Client MAC address: e2:83:79:58:e8:f7 (e2:83:79:58:e8:f7)
    Client hardware address padding: 00000000000000000000
    Server host name not given
    Boot file name not given
    Magic cookie: DHCP
  ▷ Option: (t=53,l=1) DHCP Message Type = DHCP Offer
  ▷ Option: (t=54,l=4) DHCP Server Identifier = 10.1.1.254
  ▷ Option: (t=51,l=4) IP Address Lease Time = 10 minutes
  ▷ Option: (t=1,l=4) Subnet Mask = 255.255.255.0
  ▷ Option: (t=6,l=8) Domain Name Server
    End Option
    Padding
```

Figure 1.32 The DHCP negotiation process: details of the Offer message.

- The Protocol field was replaced by a field that specifies the type of next header;
- The Payload Length is the length of all extension headers plus data, that is, total length minus 40 octets (the Base Header);
- IPv6 datagram can now contain up to 64K octets of data.

Figure 1.38 shows the generic format of an IPv6 datagram. Following the Base Header, a series of optional Extension Headers can precede the Data field.

1.8.2.1 IPv6 Addressing

IP addresses changed significantly with IPv6. IPv6 addresses are 16 bytes (128 bits) long rather than four bytes (32 bits). This larger size means that IPv6 supports more than $3 * 10^{38}$ possible addresses, allowing every consumer electronic device to connect to the Internet. In IPv6 addresses, pairs of bytes are separated by a colon and each byte in turns is represented as a pair of hexadecimal numbers, like for ex-

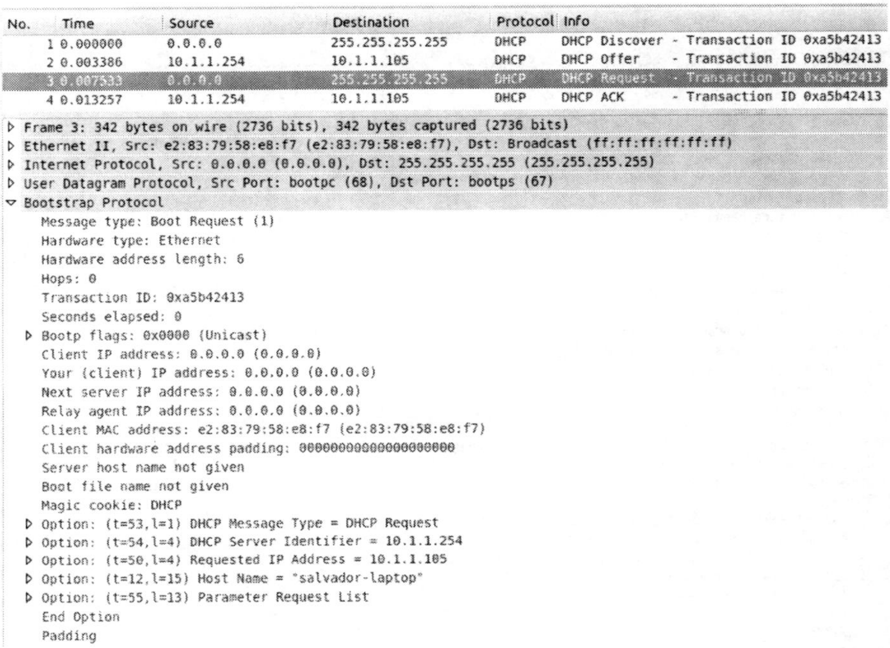

No.	Time	Source	Destination	Protocol	Info
1	0.000000	0.0.0.0	255.255.255.255	DHCP	DHCP Discover - Transaction ID 0xa5b42413
2	0.003386	10.1.1.254	10.1.1.105	DHCP	DHCP Offer - Transaction ID 0xa5b42413
3	0.007533	0.0.0.0	255.255.255.255	DHCP	DHCP Request - Transaction ID 0xa5b42413
4	0.013257	10.1.1.254	10.1.1.105	DHCP	DHCP ACK - Transaction ID 0xa5b42413

```
▷ Frame 3: 342 bytes on wire (2736 bits), 342 bytes captured (2736 bits)
▷ Ethernet II, Src: e2:83:79:58:e8:f7 (e2:83:79:58:e8:f7), Dst: Broadcast (ff:ff:ff:ff:ff:ff)
▷ Internet Protocol, Src: 0.0.0.0 (0.0.0.0), Dst: 255.255.255.255 (255.255.255.255)
▷ User Datagram Protocol, Src Port: bootpc (68), Dst Port: bootps (67)
▽ Bootstrap Protocol
    Message type: Boot Request (1)
    Hardware type: Ethernet
    Hardware address length: 6
    Hops: 0
    Transaction ID: 0xa5b42413
    Seconds elapsed: 0
  ▷ Bootp flags: 0x0000 (Unicast)
    Client IP address: 0.0.0.0 (0.0.0.0)
    Your (client) IP address: 0.0.0.0 (0.0.0.0)
    Next server IP address: 0.0.0.0 (0.0.0.0)
    Relay agent IP address: 0.0.0.0 (0.0.0.0)
    Client MAC address: e2:83:79:58:e8:f7 (e2:83:79:58:e8:f7)
    Client hardware address padding: 00000000000000000000
    Server host name not given
    Boot file name not given
    Magic cookie: DHCP
  ▷ Option: (t=53,l=1) DHCP Message Type = DHCP Request
  ▷ Option: (t=54,l=4) DHCP Server Identifier = 10.1.1.254
  ▷ Option: (t=50,l=4) Requested IP Address = 10.1.1.105
  ▷ Option: (t=12,l=15) Host Name = "salvador-laptop"
  ▷ Option: (t=55,l=13) Parameter Request List
    End Option
    Padding
```

Figure 1.33 The DHCP negotiation process: details of the Request message.

ample: 2001:0db8:0000:130F:0000:0000:087C:140B. Since IPv6 addresses commonly contain many bytes with a zero value, a shorthand notation in IPv6 represents leading zeros in contiguous block by a double colon (2001:0db8:0:130F::87C:140B), which should only appear once in any address.

IPv6 does not use classes and supports the following three IP address types: unicast, multicast and anycast. Unicast and multicast messaging in IPv6 are conceptually the same as in IPv4. A unicast address is the address of a single interface. A packet forwarded to a unicast address is delivered only to the interface identified by that address. IPv6 does not support broadcast, but its multicast mechanism accomplishes essentially the same effect. Multicast addresses in IPv6 start with “FF” (255), just like IPv4 addresses, and represent the address of a set of interfaces that typically belong to different nodes. So, a packet forwarded to a multicast address is delivered to all interfaces belonging to the set. Anycast in IPv6 is a variation on multicast. Whereas multicast delivers messages to all nodes in the multicast group, anycast delivers messages to any one node in the multicast group (the nearest to the source

No.	Time	Source	Destination	Protocol	Info
1	0.000000	0.0.0.0	255.255.255.255	DHCP	DHCP Discover - Transaction ID 0xa5b42413
2	0.003386	10.1.1.254	10.1.1.105	DHCP	DHCP Offer - Transaction ID 0xa5b42413
3	0.007533	0.0.0.0	255.255.255.255	DHCP	DHCP Request - Transaction ID 0xa5b42413
4	0.013257	10.1.1.254	10.1.1.105	DHCP	DHCP ACK - Transaction ID 0xa5b42413

```
▷ Frame 4: 342 bytes on wire (2736 bits), 342 bytes captured (2736 bits)
▷ Ethernet II, Src: CadmusCo_de:98:9e (08:00:27:de:98:9e), Dst: e2:83:79:58:e8:f7 (e2:83:79:58:e8:f7)
▷ Internet Protocol, Src: 10.1.1.254 (10.1.1.254), Dst: 10.1.1.105 (10.1.1.105)
▷ User Datagram Protocol, Src Port: bootps (67), Dst Port: bootpc (68)
▽ Bootstrap Protocol
    Message type: Boot Reply (2)
    Hardware type: Ethernet
    Hardware address length: 6
    Hops: 0
    Transaction ID: 0xa5b42413
    Seconds elapsed: 0
  ▷ Bootp flags: 0x0000 (Unicast)
    Client IP address: 0.0.0.0 (0.0.0.0)
    Your (client) IP address: 10.1.1.105 (10.1.1.105)
    Next server IP address: 0.0.0.0 (0.0.0.0)
    Relay agent IP address: 0.0.0.0 (0.0.0.0)
    Client MAC address: e2:83:79:58:e8:f7 (e2:83:79:58:e8:f7)
    Client hardware address padding: 00000000000000000000
    Server host name not given
    Boot file name not given
    Magic cookie: DHCP
  ▷ Option: (t=53,l=1) DHCP Message Type = DHCP ACK
  ▷ Option: (t=54,l=4) DHCP Server Identifier = 10.1.1.254
  ▷ Option: (t=51,l=4) IP Address Lease Time = 10 minutes
  ▷ Option: (t=1,l=4) Subnet Mask = 255.255.255.0
  ▷ Option: (t=6,l=8) Domain Name Server
    End Option
    Padding
```

Figure 1.34 The DHCP negotiation process: details of the Acknowledge message.

node, according to the routing metric). Anycast is an advanced networking concept designed to support the failover and load balancing needs of some applications.

IPv6 reserves just two special addresses: 0:0:0:0:0:0:0:0 is internal to the protocol implementation, so nodes cannot use it for their own communication purposes, and 0:0:0:0:0:0:0:1 is the loopback address, equivalent to 127.0.0.1 in IPv4.

For unicast addresses, the network ID is administratively assigned, and the host ID can be configured manually or auto-configured by any of the following methods: using a randomly generated number, using DHCPv6 or using the Extended Unique Identifier (EUI-64) format. This format expands the device interface 48-bit MAC address to 64 bits by inserting FFFE into the middle 16 bits (as illustrated in Figure 1.39).

As with the IPv4 Classless Inter-Domain Routing (CIDR) network prefix representation, an IPv6 address network prefix is represented the same way (<IP address>/<network mask>).

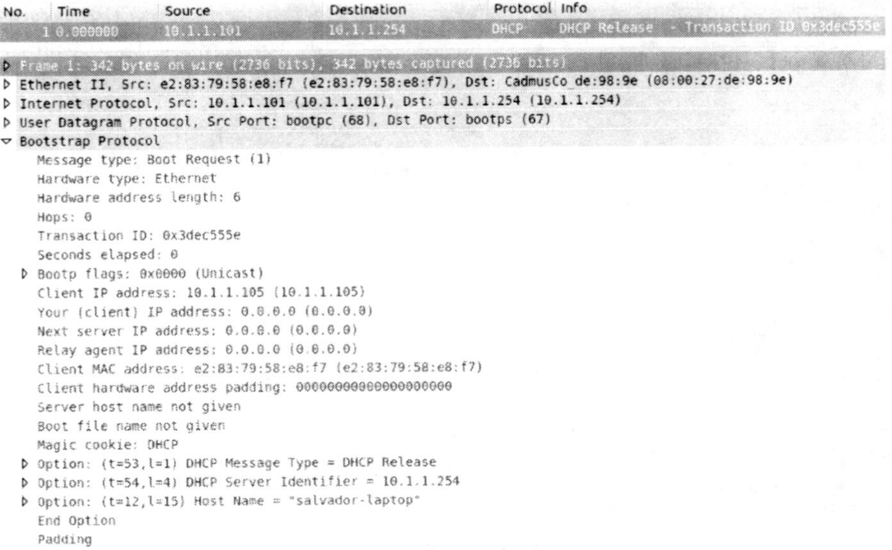
No. Time Source Destination Protocol Info
1 0.000000 10.1.1.101 10.1.1.254 DHCP DHCP Release - Transaction ID 0x3dec555e
Frame 1: 342 bytes on wire (2736 bits), 342 bytes captured (2736 bits)
Ethernet II, Src: e2:83:79:58:e8:f7 (e2:83:79:58:e8:f7), Dst: CadmusCo_de:98:9e (08:00:27:de:98:9e)
Internet Protocol, Src: 10.1.1.101 (10.1.1.101), Dst: 10.1.1.254 (10.1.1.254)
User Datagram Protocol, Src Port: bootpc (68), Dst Port: bootps (67)
Bootstrap Protocol
Message type: Boot Request (1)
Hardware type: Ethernet
Hardware address length: 6
Hops: 0
Transaction ID: 0x3dec555e
Seconds elapsed: 0
Bootp flags: 0x0000 (Unicast)
Client IP address: 10.1.1.105 (10.1.1.105)
Your (client) IP address: 0.0.0.0 (0.0.0.0)
Next server IP address: 0.0.0.0 (0.0.0.0)
Relay agent IP address: 0.0.0.0 (0.0.0.0)
Client MAC address: e2:83:79:58:e8:f7 (e2:83:79:58:e8:f7)
Client hardware address padding: 00000000000000000000
Server host name not given
Boot file name not given
Magic cookie: DHCP
Option: (t=53,l=1) DHCP Message Type = DHCP Release
Option: (t=54,l=4) DHCP Server Identifier = 10.1.1.254
Option: (t=12,l=15) Host Name = "salvador-laptop"
End Option
Padding

Figure 1.35 DHCP release.

Unlike IPv4, IPv6 addresses have different scopes: an address scope defines the region where an address can be defined as a unique identifier of an interface. These scopes or regions are the link, the site network, and the global network, corresponding to link-local, unique local unicast, and global addresses.

Global unicast addresses have the following characteristics: are routable and reachable across the Internet; are IPv6 addresses for widespread generic use; are structured as a hierarchy to allow address aggregation and are identified by their three high-level bits set to 001 (2000::/3). The format of a global unicast address is illustrated in Figure 1.40.

The global routing prefix is assigned to a service provider by the Internet Assigned Numbers Authority (IANA). The site level aggregator (SLA), or subnet ID, is assigned to a customer by their service provider. The LAN ID represents individual networks within the customer site and is administered by the customer. The Host or Interface ID has the same meaning for all unicast addresses. It is 64 bits long and is typically created by using the EUI-64 format. An example of a global unicast address is: 2001:0DB8:BBBB:CCCC:0987:65FF:FE01:2345.

No.	Time	Source	Destination	Protocol	Info
1	0.000000	10.2.2.1	10.1.1.254	DHCP	DHCP Discover - Transaction ID 0xa8ec8346
2	0.003094	10.1.1.254	10.2.2.1	DHCP	DHCP Offer - Transaction ID 0xa8ec8346
3	0.007282	10.2.2.1	10.1.1.254	DHCP	DHCP Request - Transaction ID 0xa8ec8346
4	0.012324	10.1.1.254	10.2.2.1	DHCP	DHCP ACK - Transaction ID 0xa8ec8346

```
▷ Frame 1: 342 bytes on wire (2736 bits), 342 bytes captured (2736 bits)
▷ Ethernet II, Src: c2:00:21:73:00:00 (c2:00:21:73:00:00), Dst: CadmusCo_de:98:9e (08:00:27:de:98:9e)
▷ Internet Protocol, Src: 10.2.2.1 (10.2.2.1), Dst: 10.1.1.254 (10.1.1.254)
▷ User Datagram Protocol, Src Port: bootps (67), Dst Port: bootps (67)
▽ Bootstrap Protocol
    Message type: Boot Request (1)
    Hardware type: Ethernet
    Hardware address length: 6
    Hops: 1
    Transaction ID: 0xa8ec8346
    Seconds elapsed: 0
  ▷ Bootp flags: 0x0000 (Unicast)
    Client IP address: 0.0.0.0 (0.0.0.0)
    Your (client) IP address: 0.0.0.0 (0.0.0.0)
    Next server IP address: 0.0.0.0 (0.0.0.0)
    Relay agent IP address: 10.2.2.1 (10.2.2.1)
    Client MAC address: ea:92:3b:17:53:9d (ea:92:3b:17:53:9d)
    Client hardware address padding: 00000000000000000000
    Server host name not given
    Boot file name not given
    Magic cookie: DHCP
  ▷ Option: (t=53,l=1) DHCP Message Type = DHCP Discover
  ▷ Option: (t=50,l=4) Requested IP Address = 10.2.2.100
  ▷ Option: (t=12,l=15) Host Name = "salvador-laptop"
  ▷ Option: (t=55,l=13) Parameter Request List
    End Option
    Padding
```

Figure 1.36 DHCP Discover when Relay Agent is configured.

Unique local unicast addresses are: analogous to private IPv4 addresses; used for local communications, inter-site VPNs, and so forth; not routable on the Internet (routing would require IPv6 NAT). Figure 1.41 illustrates the format of a unique local unicast address.

Global IDs do not have to be aggregated and are defined by the administrator of the local domain. Subnet IDs are also defined by the administrator of the local domain. Subnet IDs are typically defined using a hierarchical addressing plan to allow for route summarization. The Host or Interface ID has the same meaning for all unicast addresses. It is 64 bits long and is typically created by using the EUI-64 format. Example of a unique local unicast address is: FD00:aaaa:bbbb:CCCC:0987:65FF:FE01:2345.

Link local unicast addresses are: mandatory addresses that are used exclusively for communication between two IPv6 devices on the same link; automatically assigned by device as soon as IPv6 is enabled; not routable addresses, since their scope is link-specific only; identified by the first 10 bits (FE80). Figure 1.42 illustrates the format of a link local unicast address.

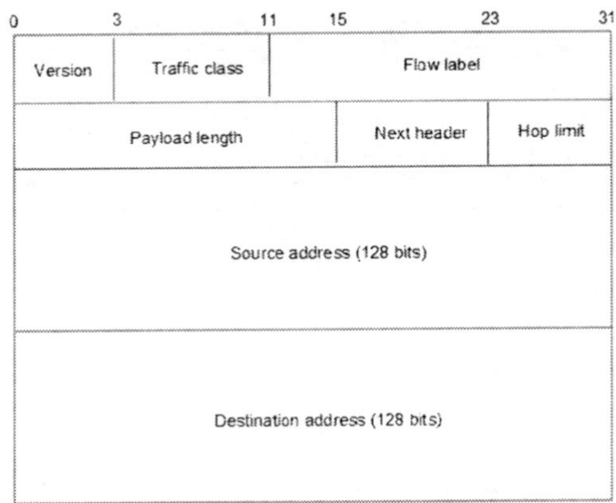

Figure 1.37 Format of the IPv6 packet header.

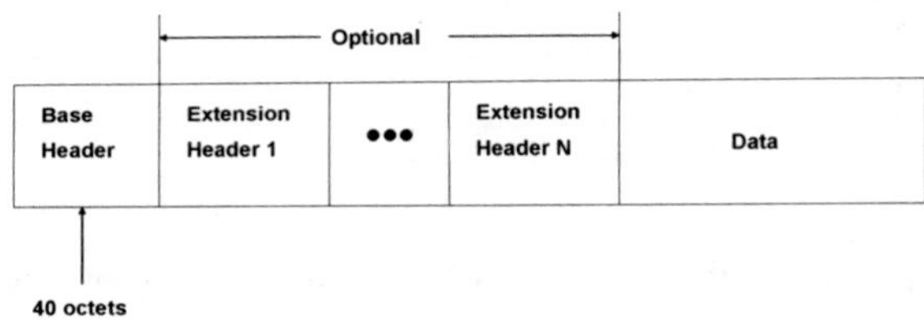

Figure 1.38 Format of the IPv6 datagram.

The remaining 54 bits of the network ID could be zero or any manually configured value. The interface ID has the same meaning for all unicast addresses. It is 64 bits long and is typically created by using the EUI-64 format. Example of a link local unicast address is FE80:0000:0000:0000:0987:65FF:FE01:2345, which is represented in shorthand notation as FE80::987:65FF:FE01:2345.

IPv6 multicast addresses have an 8-bit prefix, FF00::/8 (1111 1111). The second octet defines the lifetime and scope of the multicast address (Figure 1.43). Multicast addresses are always destination addresses. Multicast addresses are used for router solicitations (RS), router advertisements (RA), DHCPv6, multicast applications, and so forth. Table 1.4 lists some well known multicast addresses.

Finally, for each unicast and anycast address configured there is a corresponding solicited-node multicast address (FF02::1:FF:<interface ID's lower 24 bits>) that has link local significance only and is used in Neighbor Solicitation Messages for MAC address resolution and Duplicate Address

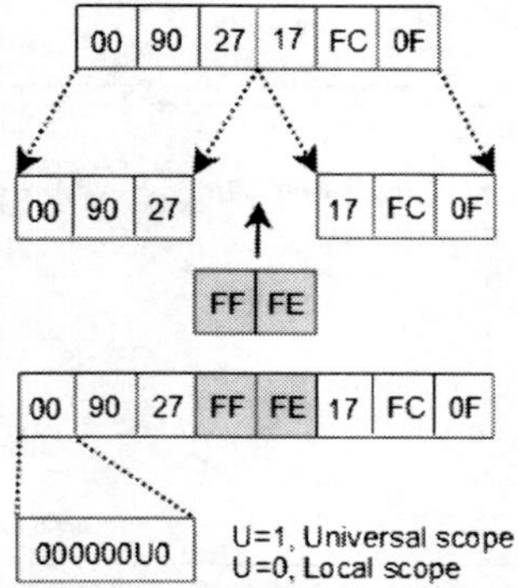

Figure 1.39 Conversion of EUI-64 MAC Address to IPv6 Host Address format.

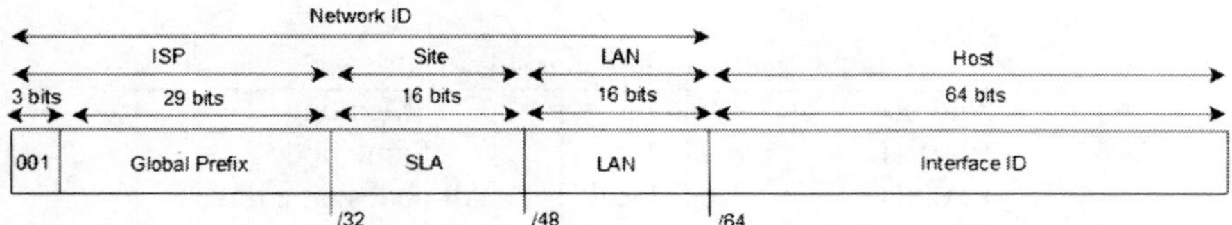

Figure 1.40 Global Unicast Address format.

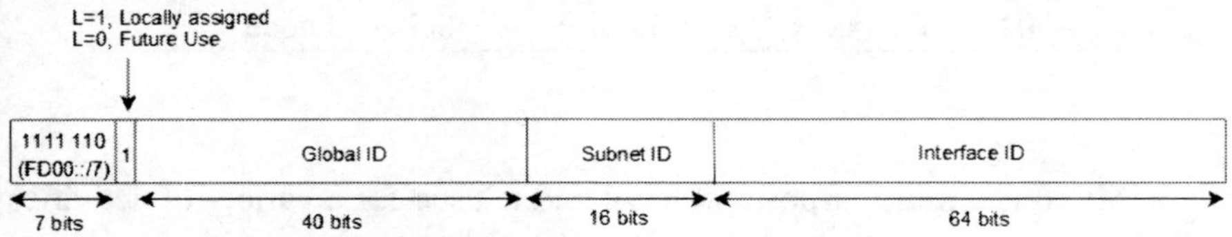

Figure 1.41 Unique Local Unicast Address format.

Detection (DAD) (note that random or assigned interface IDs may result in equal global/link addresses). Figure 1.44 illustrates the solicited node multicast address format.

1.8.2.2 The Internet Control Message Protocol, version 6

IPv6 includes ICMP Version 6 (ICMPv6) intrinsically in its architecture. So, ICMPv6 messages are transported within an IPv6 packet that may include IPv6 extension headers. ICMPv6 offers a comprehensive solution by offering the different functions that were earlier subdivided among different protocols such as ICMP, ARP (Address Resolution Protocol) and IGMP. The general format of the ICMPv6 packet is illustrated in Figure 1.45.

Figure 1.42 Link Local Unicast Address format.

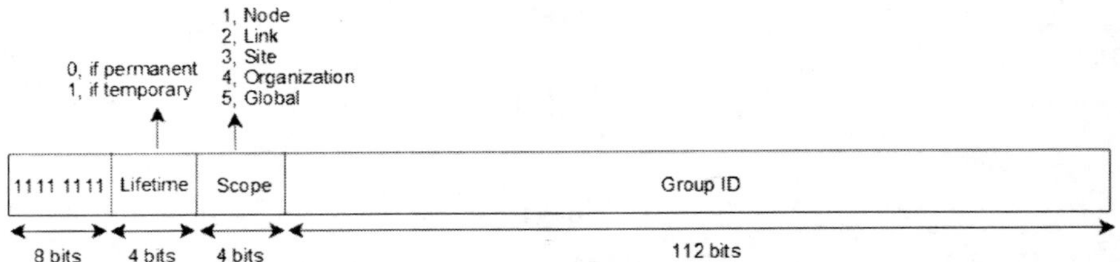

Figure 1.43 Multicast Address format.

Table 1.4 Well known IPv6 multicast addresses

Address	Scope	Meaning
FF01::1	Node-local	Same node
FF02::1	Link-local	All nodes on a link
FF01::2	Node-local	Same router
FF02::2	Link-local	All routers on a link
FF05::2	Site-local	All routers on the Internet
FF02::1:FFxx:xxxx	Link-local	Solicited node

ICMPv6 is a multipurpose protocol and is used for a variety of activities including error reporting in packet processing, diagnostic activities, neighbor discovery processes and IPv6 multicast membership reporting. To perform these activities, ICMPv6 messages are subdivided into two classes:

- Error Messages, which belong to four different categories: Destination Unreachable, Time Exceeded, Packet Too Big and Parameter Problems.
- Information Messages, which belong to three groups: diagnostic messages, Neighbor Discovery messages and messages for the management of multicast groups.

Neighbor discovery messages are originated from the node link local address with an hop limit of 255, and consist of IPv6 header, ICMPv6 header, neighbor discovery header and neighbor discovery options. There are five neighbor discovery messages: Router solicitation (ICMPv6 type 133), Router advertisement (ICMPv6 type 134), Neighbor solicitation (ICMPv6 type 135),

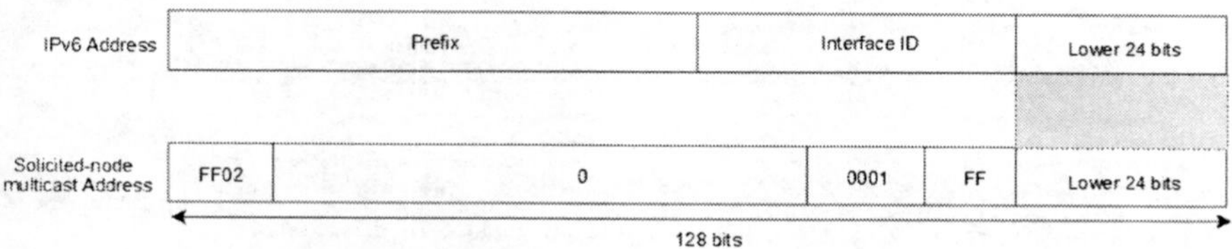

Figure 1.44 Solicited Node Multicast Address format.

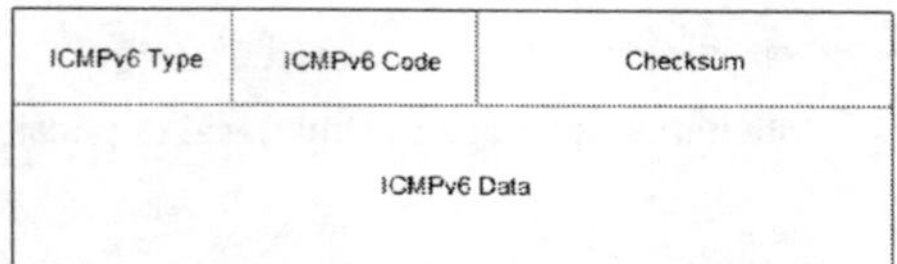

Figure 1.45 ICMPv6 packet format.

Neighbor advertisement (ICMPv6 type 136) and Redirect (ICMPV6 type 137):

- Router Solicitation is sent by a host to inquire about the presence of a router on the link; it is sent to the all routers multicast address of FF02::2 and the source IP address is either the link local address or an unspecified IPv6 address.
- Router advertisement is sent out by routers periodically or in response to a router solicitation. It includes auto-configuration information, a "preference level" for each advertised router address and a "lifetime" field.
- Neighbor Solicitation is sent to discover the link layer address of an IPv6 node. The source address of the IPv6 header is set to the unicast address of the sending node or :: for Duplicate Address Detection (DAD) and the destination address is set to the unicast address for reachability purposes and to the solicited node multicast for address resolution and DAD.
- Neighbor Advertisement is sent in response to a neighbor solicitation message or to inform about the change of a link layer address.
- The Redirect message is used by a router to signal the reroute of a packet to a better router.

1.8.2.3 Address Assignment in IPv6

An IPv6 address can be configured statically by a human operator. This can be an appropriate method of assigning addresses for router interfaces and static network elements and resources. However, manual assignment is open

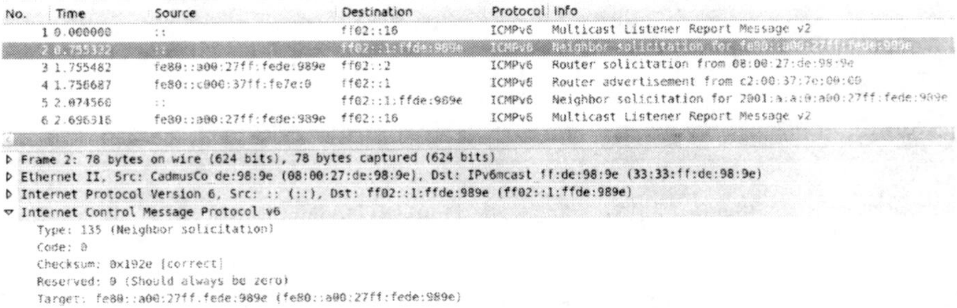

No.	Time	Source	Destination	Protocol	Info
1	0.000000	::	ff02::16	ICMPv6	Multicast Listener Report Message v2
2	0.755332	::	ff02::1:ffde:989e	ICMPv6	Neighbor solicitation for fe80::a00:27ff:fede:989e
3	1.755482	fe80::a00:27ff:fede:989e	ff02::2	ICMPv6	Router solicitation from 08:00:27:de:98:9e
4	1.756687	fe80::c000:37ff:fe7e:0	ff02::1	ICMPv6	Router advertisement from c2:00:37:7e:00:00
5	2.074560	::	ff02::1:ffde:989e	ICMPv6	Neighbor solicitation for 2001:a:a:0:a00:27ff:fede:989e
6	2.696316	fe80::a00:27ff:fede:989e	ff02::16	ICMPv6	Multicast Listener Report Message v2

```
▷ Frame 2: 78 bytes on wire (624 bits), 78 bytes captured (624 bits)
▷ Ethernet II, Src: CadmusCo de:98:9e (08:00:27:de:98:9e), Dst: IPv6mcast ff:de:98:9e (33:33:ff:de:98:9e)
▷ Internet Protocol Version 6, Src: :: (::), Dst: ff02::1:ffde:989e (ff02::1:ffde:989e)
▽ Internet Control Message Protocol v6
    Type: 135 (Neighbor solicitation)
    Code: 0
    Checksum: 0x192e [correct]
    Reserved: 0 (Should always be zero)
    Target: fe80::a00:27ff.fede:989e (fe80::a00:27ff:fede:989e)
```

Figure 1.46 Neighbor Solicitation message for link local duplicate address detection.

to errors and operational overhead due to the 128-bit length and hexadecimal attributes of the addresses.

Stateless address auto-configuration provides a convenient method to assign IP addresses to IPv6 nodes. This method does not require any human intervention from an IPv6 user. In order to use this method on an IPv6 node, it is important to connect that IPv6 node to a network with at least one IPv6 router. This router is configured by the network administrator and sends out Router Advertisement (RA) announcements onto the link. These announcements can allow the on-link connected IPv6 nodes to configure themselves with an IPv6 address and routing parameters, without further human intervention. The node uses the IPv6 network prefix advertised in the link-local router's RAs and creates the IPv6 host ID by using its MAC address and the EUI-64 format for host IDs.

So, when a IPv6 node is connected, it sends a Neighbor Solicitation message to the multicast solicited-node address to detect possible duplicate addresses of its own link local address (Figure 1.46).

Then, a Router Solicitation message is sent from the link-local address of the host to the *all routers* multicast address (ff02::2), trying to look for the presence of a router on the local network. This packet is shown in Figure 1.47. The connected router sends a Router Advertisement message (Figure 1.48), using the link local address as source address and the multicast *All Hosts* address (ff02::1) as destination. Note that the global prefix 2001:a:a::/64 is announced by the router, since this equipment is configured with a global address from this network.

Finally, another Neighbor Solicitation message is sent by the host to the multicast solicited-node address to detect possible duplicate addresses of its own global address (Figure 1.46).

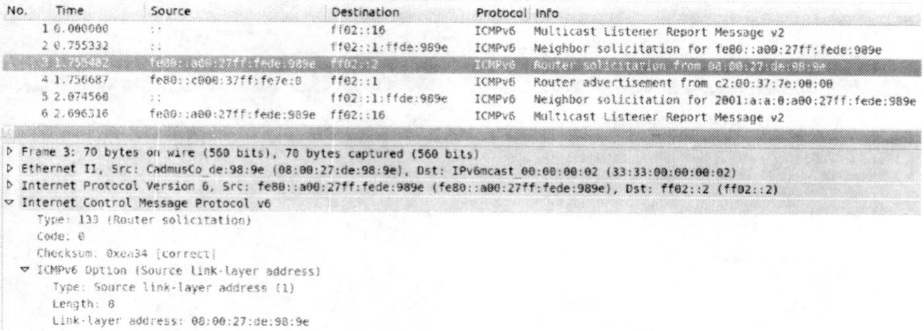

No.	Time	Source	Destination	Protocol	Info
1	0.000000	::	ff02::16	ICMPv6	Multicast Listener Report Message v2
2	0.755332	::	ff02::1:ffde:989e	ICMPv6	Neighbor solicitation for fe80::a00:27ff:fede:989e
3	1.755482	fe80::a00:27ff:fede:989e	ff02::2	ICMPv6	Router solicitation from 08:00:27:de:98:9e
4	1.756687	fe80::c000:37ff:fe7e:0	ff02::1	ICMPv6	Router advertisement from c2:00:37:7e:00:00
5	2.074560	::	ff02::1:ffde:989e	ICMPv6	Neighbor solicitation for 2001:a:a:0:a00:27ff:fede:989e
6	2.696316	fe80::a00:27ff:fede:989e	ff02::16	ICMPv6	Multicast Listener Report Message v2

```
▷ Frame 3: 70 bytes on wire (560 bits), 70 bytes captured (560 bits)
▷ Ethernet II, Src: CadmusCo_de:98:9e (08:00:27:de:98:9e), Dst: IPv6mcast_00:00:00:02 (33:33:00:00:00:02)
▷ Internet Protocol Version 6, Src: fe80::a00:27ff:fede:989e (fe80::a00:27ff:fede:989e), Dst: ff02::2 (ff02::2)
▽ Internet Control Message Protocol v6
    Type: 133 (Router solicitation)
    Code: 0
    Checksum: 0xea34 [correct]
  ▽ ICMPv6 Option (Source link-layer address)
      Type: Source link-layer address (1)
      Length: 8
      Link-layer address: 08:00:27:de:98:9e
```

Figure 1.47 Router Solicitation message.

Two messages of the Multicast Listener Discovery Protocol are also visible in the previous captures. Among other utilities, this protocol is used by routers and hosts to report interest in the respective Solicited-Node Multicast Addresses, but its operation details will be explained in Chapter 3.

IPv6 devices use multicast to acquire IP addresses and to find DHCPv6 servers. The basic DHCPv6 client-server concept is similar to DHCP for IPv4. If a client wishes to receive configuration parameters, it will send out a request on the attached local network to detect available DHCPv6 servers (Figure 1.50). This is done through the Solicit and Advertise messages. Well-known DHCPv6 multicast addresses are used for this process. Next, the DHCPv6 client will Request parameters from an available server, which will respond with the requested information in a Reply message. Like DHCPv4, DHCPv6 uses an architectural concept of "options" to carry additional parameters and information within DHCPv6 messages.

The DHCPv6 client knows whether to use DHCPv6 based upon the instruction from a router on its link-local network. The default gateway has two configurable bits in its Router Advertisement (RA) available for this purpose:

- O bit - When this bit is set, the client can use DHCPv6 to retrieve other configuration parameters (for example, TFTP server address or DNS server address) but not the client's IP address.
- M bit - When this bit is set, the client can use DHCPv6 to retrieve a managed IPv6 address and other configuration parameters from a DHCPv6 server.

Stateless DHCPv6 is a combination of stateless address autoconfiguration and DHCP for IPv6 [45]. When a router sends a RA with the O bit set but does not set the M bit, the client can use stateless address auto-

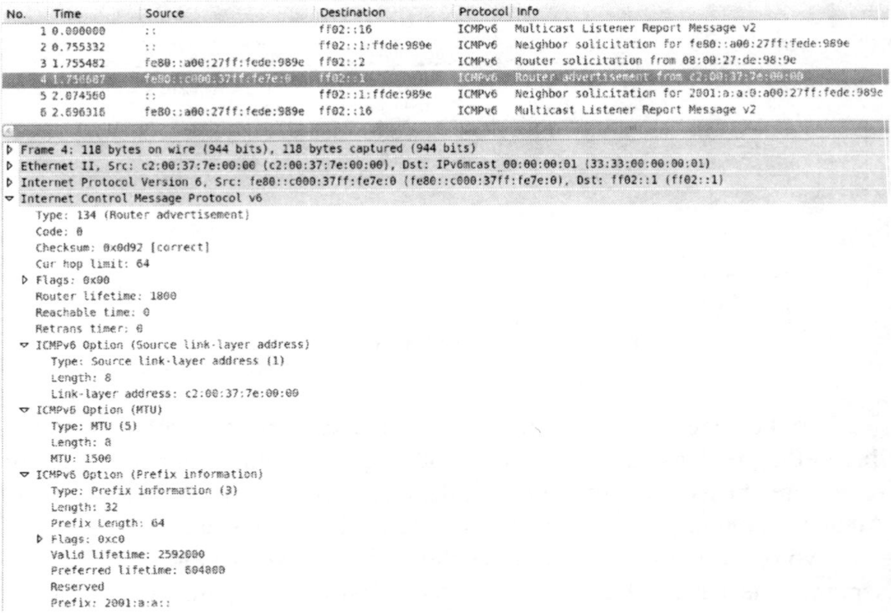

No.	Time	Source	Destination	Protocol	Info
1	0.000000	::	ff02::16	ICMPv6	Multicast Listener Report Message v2
2	0.755332	::	ff02::1:ffde:989e	ICMPv6	Neighbor solicitation for fe80::a00:27ff:fede:989e
3	1.755482	fe80::a00:27ff:fede:989e	ff02::2	ICMPv6	Router solicitation from 08:00:27:de:98:9e
4	1.758687	fe80::c000:37ff:fe7e:0	ff02::1	ICMPv6	Router advertisement from c2:00:37:7e:00:00
5	2.074560	::	ff02::1:ffde:989e	ICMPv6	Neighbor solicitation for 2001:a:a:0:a00:27ff:fede:989e
6	2.696316	fe80::a00:27ff:fede:989e	ff02::16	ICMPv6	Multicast Listener Report Message v2

```
▷ Frame 4: 118 bytes on wire (944 bits), 118 bytes captured (944 bits)
▷ Ethernet II, Src: c2:00:37:7e:00:00 (c2:00:37:7e:00:00), Dst: IPv6mcast_00:00:00:01 (33:33:00:00:00:01)
▷ Internet Protocol Version 6, Src: fe80::c000:37ff:fe7e:0 (fe80::c000:37ff:fe7e:0), Dst: ff02::1 (ff02::1)
▽ Internet Control Message Protocol v6
    Type: 134 (Router advertisement)
    Code: 0
    Checksum: 0x0d92 [correct]
    Cur hop limit: 64
  ▷ Flags: 0x90
    Router lifetime: 1800
    Reachable time: 0
    Retrans timer: 0
  ▽ ICMPv6 Option (Source link-layer address)
      Type: Source link-layer address (1)
      Length: 8
      Link-layer address: c2:00:37:7e:00:00
  ▽ ICMPv6 Option (MTU)
      Type: MTU (5)
      Length: 8
      MTU: 1500
  ▽ ICMPv6 Option (Prefix information)
      Type: Prefix information (3)
      Length: 32
      Prefix Length: 64
    ▷ Flags: 0xc0
      Valid lifetime: 2592000
      Preferred lifetime: 604800
      Reserved
      Prefix: 2001:a:a::
```

Figure 1.48 Router Advertisement message.

configuration to obtain its IPv6 address and use DHCPv6 to obtain additional information (such as the TFTP server address or the DNS server address). This mechanism is known as Stateless DHCPv6 because the DHCPv6 server does not have to keep track of the client address bindings.

The DHCP for IPv6 has been standardized by the IETF through RFC 3315 [44]. When a router sends a RA with the M bit set, this indicates that clients should use DHCP to obtain their IP addresses. When the M bit is set, the setting of the O bit is irrelevant because the DHCP server will also return "other" configuration information together with the addresses. This mechanism is known as Stateful DHCPv6 because the DHCPv6 server does keep track of the client address bindings.

Figure 1.51 presents the format of the DHCPv6 messages. DHCPv6 uses options to carry additional parameters and information within DHCPv6 messages. These options are aligned in a Type-Length-Value (TLV) structure. Each Type and Length field has a length of 16 bits, with a variable length available for the Value field.

No.	Time	Source	Destination	Protocol	Info
1	0.000000	::	ff02::16	ICMPv6	Multicast Listener Report Message v2
2	0.755332	::	ff02::1:ffde:989e	ICMPv6	Neighbor solicitation for fe80::a00:27ff:fede:989e
3	1.755482	fe80::a00:27ff:fede:989e	ff02::2	ICMPv6	Router solicitation from 08:00:27:de:98:9e
4	1.756687	fe80::c000:37ff:fe7e:0	ff02::1	ICMPv6	Router advertisement from c2:00:37:7e:00:00
5	2.074500	::	ff02::1:ffde:989e	ICMPv6	Neighbor solicitation for 2001:a:a:0:a00:27ff:fede:989e
6	2.696316	fe80::a00:27ff:fede:989e	ff02::16	ICMPv6	Multicast Listener Report Message v2

```
▷ Frame 5: 78 bytes on wire (624 bits), 78 bytes captured (624 bits)
▷ Ethernet II, Src: CadmusCo_de:98:9e (08:00:27:de:98:9e), Dst: IPv6mcast_ff:de:98:9e (33:33:ff:de:98:9e)
▷ Internet Protocol Version 6, Src: :: (::), Dst: ff02::1:ffde:989e (ff02::1:ffde:989e)
▽ Internet Control Message Protocol v6
    Type: 135 (Neighbor solicitation)
    Code: 0
    Checksum: 0xf799 [correct]
    Reserved: 0 (Should always be zero)
    Target: 2001:a:a:0:a00:27ff:fede:989e (2001:a:a:0:a00:27ff:fede:989e)
```

Figure 1.49 Neighbor Solicitation message for global duplicate address detection.

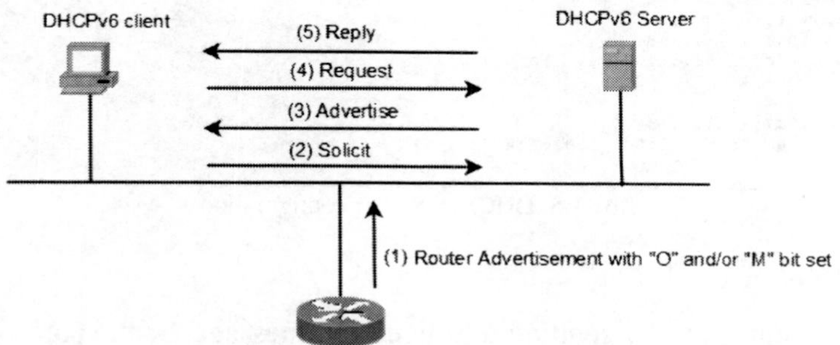

Figure 1.50 DHCP operation in IPv6.

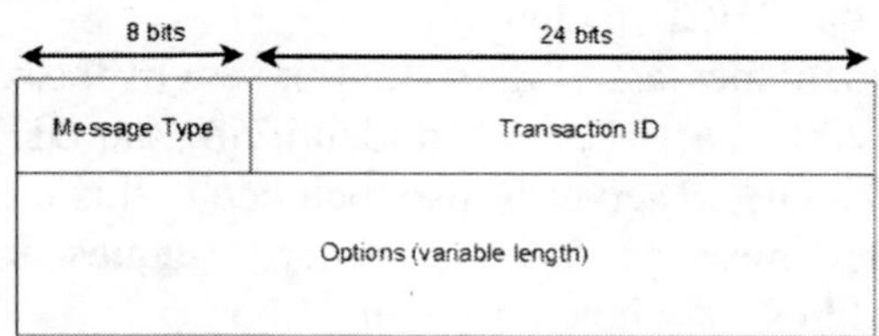

Figure 1.51 Format of the DHCPv6 message.

Exercise - DHCPv6 Operation

In the first experiment, whose setup is depicted in Figure 1.52, the DHCP server will assign a basic IP configuration to PC 1. Note that both elements belong to the same IP network. The server configuration is illustrated in Box 1.5: we can see that the default lease time is equal to 10 minutes and the IP addresses that will be assigned belong to one of two different ranges, 2001:a:a:1::1000-2001:a:a:1::1fff (as stated in command line 4) and 2001:a:a:2::1000-2001:a:a:2::1fff (as stated in command line 8).

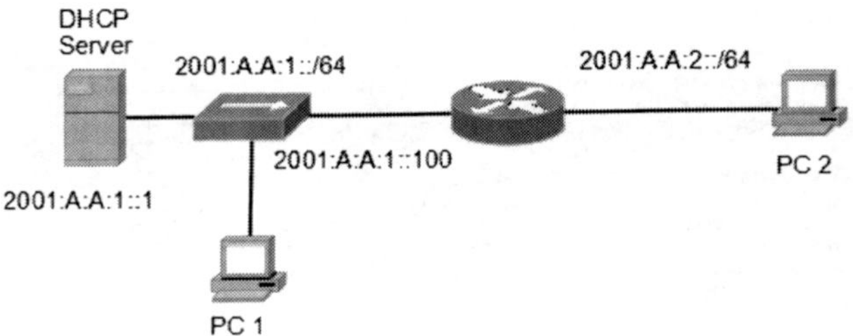

Figure 1.52 Network set up to study the DHCPv6 operation.

```
default-lease-time 600;
max-lease-time 7200;
subnet6 2001:a:a:1::/64 {
  range6 2001:a:a:1::1000 2001:a:a:1::1fff;
}

subnet6 2001:a:a:2::/64 {
  range6 2001:a:a:2::1000 2001:a:a:2::1fff;
}
```

Box 1.5 DHCPv6 Server configuration

The client starts by sending a Solicitation message (with source address equal to its link local address) to a multicast address. This message is shown in Figure 1.53, where we can see that a DHCP Unique Identifier is used to uniquely identify the DHCPv6 client.

An Advertisement message (Figure 1.54) is sent by the server announcing an IPv6 address (2001:a:a:1::1296) and identifying the DHCPv6 client. The identification of the DHCP server is also included in this message.

Then, a Request message is sent in order to request for the previously advertised IPv6 address, as shown in Figure 1.55.

Finally, a DHCPv6 Reply message is sent to confirm the assignment of the previously requested IP address (Figure 1.56).

Now, suppose that PC 2, which is located on a different IPv6 network, asks for an IPv6 configuration using the DHCPv6 mechanism. Similarly to IPv4, the router will have to relay DHCPv6 messages between the client and server IPv6 subnets, which is only possible by activating the DHCP Relay service. This can be done by including the configuration shown in Box 1.6 in the router.

Now, the DHCPv6 negotiation process proceeds exactly in the same way that was explained before but the exchanged packets are encapsulated inside DHCPv6 Relay Forward and Relay Reply messages. The Solicit and Request

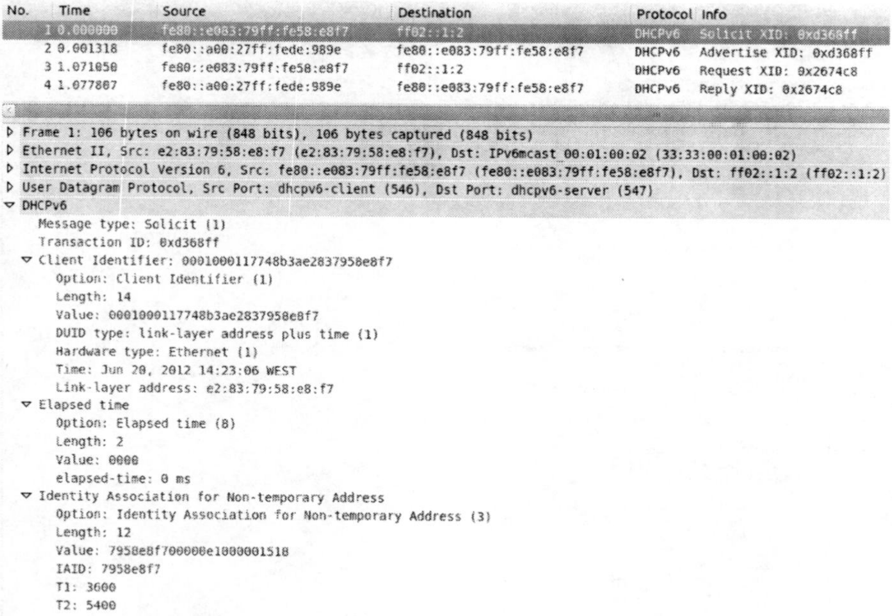

No.	Time	Source	Destination	Protocol	Info
1	0.000000	fe80::e083:79ff:fe58:e8f7	ff02::1:2	DHCPv6	Solicit XID: 0xd368ff
2	0.001318	fe80::a00:27ff:fede:989e	fe80::e083:79ff:fe58:e8f7	DHCPv6	Advertise XID: 0xd368ff
3	1.071050	fe80::e083:79ff:fe58:e8f7	ff02::1:2	DHCPv6	Request XID: 0x2674c8
4	1.077867	fe80::a00:27ff:fede:989e	fe80::e083:79ff:fe58:e8f7	DHCPv6	Reply XID: 0x2674c8

```
▷ Frame 1: 106 bytes on wire (848 bits), 106 bytes captured (848 bits)
▷ Ethernet II, Src: e2:83:79:58:e8:f7 (e2:83:79:58:e8:f7), Dst: IPv6mcast_00:01:00:02 (33:33:00:01:00:02)
▷ Internet Protocol Version 6, Src: fe80::e083:79ff:fe58:e8f7 (fe80::e083:79ff:fe58:e8f7), Dst: ff02::1:2 (ff02::1:2)
▷ User Datagram Protocol, Src Port: dhcpv6-client (546), Dst Port: dhcpv6-server (547)
▽ DHCPv6
    Message type: Solicit (1)
    Transaction ID: 0xd368ff
  ▽ Client Identifier: 0001000117748b3ae2837958e8f7
      Option: Client Identifier (1)
      Length: 14
      Value: 0001000117748b3ae2837958e8f7
      DUID type: link-layer address plus time (1)
      Hardware type: Ethernet (1)
      Time: Jun 20, 2012 14:23:06 WEST
      Link-layer address: e2:83:79:58:e8:f7
  ▽ Elapsed time
      Option: Elapsed time (8)
      Length: 2
      Value: 0000
      elapsed-time: 0 ms
  ▽ Identity Association for Non-temporary Address
      Option: Identity Association for Non-temporary Address (3)
      Length: 12
      Value: 7958e8f700000e1000001518
      IAID: 7958e8f7
      T1: 3600
      T2: 5400
```

Figure 1.53 DHCPv6 Solicitation message.

messages are encapsulated in Relay Forward packets, while the Advertise and Reply messages are encapsulated in Relay Reply packets. Figures 1.57 and 1.58 illustrate both types of relay messages, which were captured in the IPv6 network where the DHCPv6 server is located.

Note that these messages are sent in unicast mode, between the IPv6 addresses of the router interface and the DHCPv6 server.

1.8.2.4 Address Resolution in IPv6

In IPv6, resolution is still dynamic and is based on the use of a cache table that maintains pairings of IPv6 addresses and hardware addresses. Each device on a physical network keeps track of this information for its neighbors. When a source device needs to send an IPv6 datagram to a local network neighbor but does not have its hardware address, it initiates the resolution process. Instead of sending an ARP Request message, device A (that is trying to send to device B) creates an Neighbor Solicitation message. The Neighbor Solicitation message is sent to the solicited-node address of the device whose

```
No. Time      Source                     Destination                Protocol Info
  1 0.000000  fe80::e083:79ff:fe58:e8f7  ff02::1:2                  DHCPv6   Solicit XID: 0xd368ff
  2 0.001318  fe80::a00:27ff:fede:989e   fe80::e083:79ff:fe58:e8f7  DHCPv6   Advertise XID: 0xd368ff
  3 1.071050  fe80::e083:79ff:fe58:e8f7  ff02::1:2                  DHCPv6   Request XID: 0x2674c8
  4 1.077807  fe80::a00:27ff:fede:989e   fe80::e083:79ff:fe58:e8f7  DHCPv6   Reply XID: 0x2674c8

▷ Frame 2: 146 bytes on wire (1168 bits), 146 bytes captured (1168 bits)
▷ Ethernet II, Src: CadmusCo_de:98:9e (08:00:27:de:98:9e), Dst: e2:83:79:58:e8:f7 (e2:83:79:58:e8:f7)
▷ Internet Protocol Version 6, Src: fe80::a00:27ff:fede:989e (fe80::a00:27ff:fede:989e), Dst: fe80::e083:79ff:fe58:e8f7 (fe80::e083:79ff:fe58:e8f7)
▷ User Datagram Protocol, Src Port: dhcpv6-server (547), Dst Port: dhcpv6-client (546)
▽ DHCPv6
    Message type: Advertise (2)
    Transaction ID: 0xd368ff
  ▽ Identity Association for Non-temporary Address
      Option: Identity Association for Non-temporary Address (3)
      Length: 40
      Value: 7958e8f70000000000000000000500182001000a000a0001...
      IAID: 7958e8f7
      T1: 0
      T2: 0
    ▽ IA Address:  2001:a:a:1::1296
        Option: IA Address (5)
        Length: 24
        Value: 2001000a000a000100000000000012960000017700000258
        IPv6 address: 2001:a:a:1::1296
        Preferred lifetime: 375
        Valid lifetime: 600
  ▽ Client Identifier: 0001000117748b3ae2837958e8f7
      Option: Client Identifier (1)
      Length: 14
      Value: 0001000117748b3ae2837958e8f7
      DUID type: link-layer address plus time (1)
      Hardware type: Ethernet (1)
      Time: Jun 20, 2012 14:23:06 WEST
      Link-layer address: e2:83:79:58:e8:f7
  ▽ Server Identifier: 0001000117748450080027de989e
      Option: Server Identifier (2)
      Length: 14
      Value: 0001000117748450080027de989e
      DUID type: link-layer address plus time (1)
      Hardware type: Ethernet (1)
      Time: Jun 20, 2012 13:53:36 WEST
      Link-layer address: 08:00:27:de:98:9e
```

Figure 1.54 DHCPv6 Advertisement message.

IPv6 address we are trying to resolve (Figure 1.59). So, A will not broadcast the message, it will multicast it to device B's solicited-node multicast address.

Device B will receive the Neighbor Solicitation and answers back to device A with a Neighbor Advertisement (Figure 1.60). This message contains the link layer address of host B.

This is analogous to the ARP Reply and tells device A the physical address of B. Device A then adds device B's information to its neighbor cache. For efficiency, cross-resolution is supported as in IPv4 address resolution. This is done by letting device A include its own layer two address in the Neighbor Solicitation, assuming it knows it. Device B will record this along with A's IP address in B's neighbor cache.

1.8.2.5 IPv6 Fragmentation

As with IPv4, IPv6 arranges for destination to perform re-assembly. In IPv6, however, changes were made that avoid fragmentation by routers: IPv4 requires intermediate routers to fragment any datagram that is too large for the MTU of the network over which it must travel, while IPv6 fragmentation is

```
No.  Time        Source                      Destination                 Protocol  Info
  1  0.000000    fe80::e083:79ff:fe58:e8f7   ff02::1:2                   DHCPv6    Solicit XID: 0xd368ff
  2  0.001318    fe80::a00:27ff:fede:989e    fe80::e083:79ff:fe58:e8f7   DHCPv6    Advertise XID: 0xd368ff
  3  1.071050    fe80::e083:79ff:fe58:e8f7   ff02::1:2                   DHCPv6    Request XID: 0x2674c8
  4  1.077807    fe80::a00:27ff:fede:989e    fe80::e083:79ff:fe58:e8f7   DHCPv6    Reply XID: 0x2674c8

▷ Frame 3: 152 bytes on wire (1216 bits), 152 bytes captured (1216 bits)
▷ Ethernet II, Src: e2:83:79:58:e8:f7 (e2:83:79:58:e8:f7), Dst: IPv6mcast_00:01:00:02 (33:33:00:01:00:02)
▷ Internet Protocol Version 6, Src: fe80::e083:79ff:fe58:e8f7 (fe80::e083:79ff:fe58:e8f7), Dst: ff02::1:2 (ff02::1:2)
▷ User Datagram Protocol, Src Port: dhcpv6-client (546), Dst Port: dhcpv6-server (547)
▽ DHCPv6
    Message type: Request (3)
    Transaction ID: 0x2674c8
  ▽ Client Identifier: 0001000117748b3ae2837958e8f7
      Option: Client Identifier (1)
      Length: 14
      Value: 0001000117748b3ae2837958e8f7
      DUID type: link-layer address plus time (1)
      Hardware type: Ethernet (1)
      Time: Jun 20, 2012 14:23:06 WEST
      Link-layer address: e2:83:79:58:e8:f7
  ▽ Server Identifier: 0001000117748450080027de989e
      Option: Server Identifier (2)
      Length: 14
      Value: 0001000117748450080027de989e
      DUID type: link-layer address plus time (1)
      Hardware type: Ethernet (1)
      Time: Jun 20, 2012 13:53:36 WEST
      Link-layer address: 08:00:27:de:98:9e
  ▷ Elapsed time
  ▽ Identity Association for Non-temporary Address
      Option: Identity Association for Non-temporary Address (3)
      Length: 40
      Value: 7958e8f700000e1000001518000500182001000a000a0001...
      IAID: 7958e8f7
      T1: 3600
      T2: 5400
    ▽ IA Address:  2001:a:a:1::1296
        Option: IA Address (5)
        Length: 24
        Value: 2001000a000a0001000000000000129600001c2000001d4c
        IPv6 address: 2001:a:a:1::1296
        Preferred lifetime: 7200
        Valid lifetime: 7500
```

Figure 1.55 DHCPv6 Request message.

end-to-end. No fragmentation is done on intermediate routers and the source that is responsible for fragmentation has two choices: use guaranteed minimum MTU (1280 octets) or perform Path MTU discovery, which identifies minimum MTU along the path to the destination. In either case, the source fragments data and the IPv6 fragmentation process inserts a small extension header, after the base header, in each fragment (Figure 1.61).

Let us look at an illustrative example: suppose we have an IPv6 datagram exactly 2160 bytes wide, consisting of a 40 byte IP header, four 30 byte extension headers, and 2000 bytes of data. Two of the extension headers are unfragmentable, while two are fragmentable. Suppose we need to send this over a link with an MTU of 1500 bytes. The fragmentation process is illus-

```
No.  Time        Source                       Destination                  Protocol  Info
  1  0.000000    fe80::e083:79ff:fe58:e8f7    ff02::1:2                    DHCPv6    Solicit XID: 0xd368ff
  2  0.001318    fe80::a00:27ff:fede:989e     fe80::e083:79ff:fe58:e8f7    DHCPv6    Advertise XID: 0xd368ff
  3  1.071050    fe80::e083:79ff:fe58:e8f7    ff02::1:2                    DHCPv6    Request XID: 0x2674c8
  4  1.077807    fe80::a00:27ff:fede:989e     fe80::e083:79ff:fe58:e8f7    DHCPv6    Reply XID: 0x2674c8

▷ Frame 4: 146 bytes on wire (1168 bits), 146 bytes captured (1168 bits)
▷ Ethernet II, Src: CadmusCo_de:98:9e (08:00:27:de:98:9e), Dst: e2:83:79:58:e8:f7 (e2:83:79:58:e8:f7)
▷ Internet Protocol Version 6, Src: fe80::a00:27ff:fede:989e (fe80::a00:27ff:fede:989e), Dst: fe80::e083:79ff:fe58:e8f7 (fe80::e083:79ff:fe58:e8f7)
▷ User Datagram Protocol, Src Port: dhcpv6-server (547), Dst Port: dhcpv6-client (546)
▽ DHCPv6
    Message type: Reply (7)
    Transaction ID: 0x2674c8
  ▽ Identity Association for Non-temporary Address
      Option: Identity Association for Non-temporary Address (3)
      Length: 48
      Value: 7958e8f7000000000000000000050018200100 0a000a0001...
      IAID: 7958e8f7
      T1: 0
      T2: 0
    ▽ IA Address:  2001:a:a:1::1296
        Option: IA Address (5)
        Length: 24
        Value: 2001000a000a000100000000000012960000017700000258
        IPv6 address: 2001:a:a:1::1296
        Preferred lifetime: 375
        Valid lifetime. 600
  ▽ Client Identifier: 0001000117748b3ae2837958e8f7
      Option: Client Identifier (1)
      Length: 14
      Value: 0001000117748b3ae2837958e8f7
      DUID type: link-layer address plus time (1)
      Hardware type: Ethernet (1)
      Time: Jun 20, 2012 14:23:06 WEST
      Link-layer address: e2:83:79:58:e8:f7
  ▽ Server Identifier: 0001000117748450080027de989e
      Option: Server Identifier (2)
      Length: 14
      Value: 0001000117748450080027de989e
      DUID type: link-layer address plus time (1)
      Hardware type: Ethernet (1)
      Time: Jun 20, 2012 13:53:36 WEST
      Link-layer address: 08.00:27:de:98:9e
```

Figure 1.56 DHCPv6 Reply message.

trated in Figure 1.62: since we have to put the two 30 byte unfragmentable extension headers in each fragment, fragments should be structured like this:

- First fragment - This fragment would consist of the 100 byte unfragmentable part, followed by an 8 byte Fragment Header and the first 1400 bytes of the fragmentable part of the original datagram (the two fragmentable extension headers and the first 1332 bytes of data). This leaves 668 bytes of data to send.
- Second fragment - This would also contain the 100 byte unfragmentable part, followed by a Fragment Header and 668 bytes of data

As already explained for IPv4, the More Fragments flag would be set to one in the first fragment and zero in the second, and the Fragment Offset values would be set appropriately set.

```
ipv6 unicast-routing

interface FastEthernet0/0
 ipv6 address 2001:A:A:1::1/64
 ipv6 nd managed-config-flag
 ipv6 nd other-config-flag
!
interface FastEthernet0/1
 ipv6 address 2001:A:A:2::1/64
 ipv6 nd managed-config-flag
 ipv6 nd other-config-flag
 ipv6 dhcp relay destination 2001:A:A:1::100
```

Box 1.6 DHCPv6 Relay configuration.

No.	Time	Source	Destination	Protocol	Info
1	0.000000	2001:a:a:1::1	2001:a:a:1::100	DHCPv6	Relay-forw L: 2001:a:a:2::1 Solicit XID: 0x39750b
2	0.001158	2001:a:a:1::100	2001:a:a:1::1	DHCPv6	Relay-reply L: 2001:a:a:2::1 Advertise XID: 0x39750b
3	1.030767	2001:a:a:1::1	2001:a:a:1::100	DHCPv6	Relay-forw L: 2001:a:a:2::1 Request XID: 0x4981c1
4	1.032201	2001:a:a:1::100	2001:a:a:1::1	DHCPv6	Relay-reply L: 2001:a:a:2::1 Reply XID: 0x4981c1

```
▷ Frame 1: 152 bytes on wire (1216 bits), 152 bytes captured (1216 bits)
▷ Ethernet II, Src: c2:00:21:73:00:00 (c2:00:21:73:00:00), Dst: CadmusCo_de:98:9e (08:00:27:de:98:9e)
▷ Internet Protocol Version 6, Src: 2001:a:a:1::1 (2001:a:a:1::1), Dst: 2001:a:a:1::100 (2001:a:a:1::100)
▷ User Datagram Protocol, Src Port: dhcpv6-server (547), Dst Port: dhcpv6-server (547)
▽ DHCPv6
    Message type: Relay-forw (12)
    Hopcount: 0
    Link address: 2001:a:a:2::1 (2001:a:a:2::1)
    Peer address: fe80::e892:3bff:fe17:539d (fe80::e892:3bff:fe17:539d)
  ▽ Relay Message
      Option: Relay Message (9)
      Length: 44
      Value: 0139750b0001000e0001000117748b3ae2837958e8f70008...
    ▽ DHCPv6
        Message type: Solicit (1)
        Transaction ID: 0x39750b
      ▷ Client Identifier: 0001000117748b3ae2837958e8f7
      ▷ Elapsed time
      ▷ Identity Association for Non-temporary Address
  ▽ Interface-Id
      Option: Interface-Id (18)
      Length: 4
      Value: 00000005
      Interface-ID:
```

Figure 1.57 Illustration of a DHCPv6 Relay Forward message.

1.9 Transition Mechanisms from IPv4 to IPv6

There are several mechanisms to implement IPv6 or enable the transition from IPv4 to IPv6. One of them is based on the use of dual-stack backbones [32]. In this case, IPv4 and IPv6 applications coexist in a dual IP layer routing backbone and all routers in the network need to be upgraded to be dual-stack (Figure 1.63).

Another important transition mechanism is overlay tunneling [32, 109, 51]. In this case, three possible types of tunneling can be used: manual, semi-automatic and automatic.

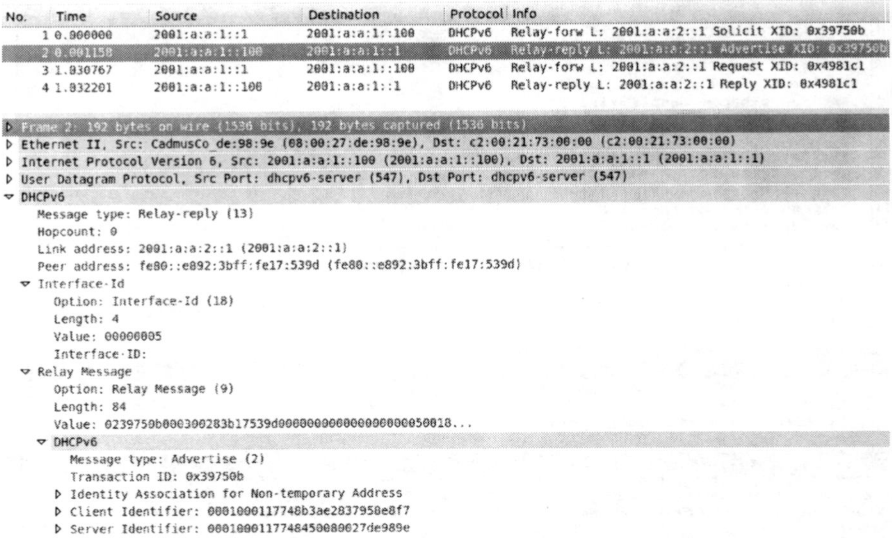

No.	Time	Source	Destination	Protocol	Info
1	0.000000	2001:a:a:1::1	2001:a:a:1::100	DHCPv6	Relay-forw L: 2001:a:a:2::1 Solicit XID: 0x39750b
2	0.001158	2001:a:a:1::100	2001:a:a:1::1	DHCPv6	Relay-reply L: 2001:a:a:2::1 Advertise XID: 0x39750b
3	1.030767	2001:a:a:1::1	2001:a:a:1::100	DHCPv6	Relay-forw L: 2001:a:a:2::1 Request XID: 0x4981c1
4	1.032201	2001:a:a:1::100	2001:a:a:1::1	DHCPv6	Relay-reply L: 2001:a:a:2::1 Reply XID: 0x4981c1

```
▷ Frame 2: 192 bytes on wire (1536 bits), 192 bytes captured (1536 bits)
▷ Ethernet II, Src: CadmusCo_de:98:9e (08:00:27:de:98:9e), Dst: c2:00:21:73:00:00 (c2:00:21:73:00:00)
▷ Internet Protocol Version 6, Src: 2001:a:a:1::100 (2001:a:a:1::100), Dst: 2001:a:a:1::1 (2001:a:a:1::1)
▷ User Datagram Protocol, Src Port: dhcpv6-server (547), Dst Port: dhcpv6-server (547)
▽ DHCPv6
    Message type: Relay-reply (13)
    Hopcount: 0
    Link address: 2001:a:a:2::1 (2001:a:a:2::1)
    Peer address: fe80::e892:3bff:fe17:539d (fe80::e892:3bff:fe17:539d)
  ▽ Interface-Id
      Option: Interface-Id (18)
      Length: 4
      Value: 00000005
      Interface-ID:
  ▽ Relay Message
      Option: Relay Message (9)
      Length: 84
      Value: 0239750b000300283b17539d00000000000000000000050018...
    ▽ DHCPv6
        Message type: Advertise (2)
        Transaction ID: 0x39750b
      ▷ Identity Association for Non-temporary Address
      ▷ Client Identifier: 000100011774 8b3ae2837950e8f7
      ▷ Server Identifier: 0001000117748450080027de989e
```

Figure 1.58 Illustration of a DHCPv6 Relay Reply message.

Regarding manual tunnels, a permanent link can be configured between two IPv6 domains over an IPv4 backbone, as illustrated in Figure 1.64. The primary use of this option is for stable connections that require regular secure communication between two edge routers, an end system and an edge router or for connection to remote IPv6 networks. Basically, this mechanism establishes a tunnel between two points and requires a complex management.

Another possibility is the configuration of a manual GRE (Generic Route Encapsulation) tunnel, ad shown in Figure 1.65. Its primary use is for stable connections that require regular stable communication. Using this standard technique, it is also possible to configure IPv4 over IPv6 tunnels. GRE tunnels are designed to be completely stateless, that is, none of the tunnel end-points keep information about the state or availability of the remote tunnel end-point. So, the local tunnel end-point router does not have the ability to bring the line protocol of the GRE tunnel interface down if the remote end-point is unreachable.

Regarding semi-automatic mechanisms, a tunnel broker service allows IPv6 applications on dual-stack systems access to an IPv6 backbone. Requests from users are processed by a server that automatically creates and configure the server part of the tunnel and provides to the client the information necessary to configure the client side. Tunnel brokers automate the

```
No.  Time      Source                         Destination                    Protocol Info
  1 0.000000   2001:a:a:0:a00:27ff:fede:989e  ff02::1:ff00:1                 ICMPv6   Neighbor solicitation
  2 0.002337   2001:a:a::1                    2001:a:a:0:a00:27ff:fede:989e  ICMPv6   Neighbor advertisement 2001:a:a::1 (rtr, sol, ovr)
  3 0.002649   2001:a:a:0:a00:27ff:fede:989e  2001:a:a::1                    ICMPv6   Echo (ping) request id=0x4e7a, seq=1
  4 0.004458   2001:a:a::1                    2001:a:a:0:a00:27ff:fede:989e  ICMPv6   Echo (ping) reply id=0x4e7a, seq=1
  5 0.999524   2001:a:a:0:a00:27ff:fede:989e  2001:a:a::1                    ICMPv6   Echo (ping) request id=0x4e7a, seq=2
  6 1.001491   2001:a:a::1                    2001:a:a:0:a00:27ff:fede:989e  ICMPv6   Echo (ping) reply id=0x4e7a, seq=2
  7 2.000538   2001:a:a:0:a00:27ff:fede:989e  2001:a:a::1                    ICMPv6   Echo (ping) request id=0x4e7a, seq=3
  8 2.002721   2001:a:a::1                    2001:a:a:0:a00:27ff:fede:989e  ICMPv6   Echo (ping) reply id=0x4e7a, seq=3
  9 3.000977   2001:a:a:0:a00:27ff:fede:989e  2001:a:a::1                    ICMPv6   Echo (ping) request id=0x4e7a, seq=4
 10 3.002055   2001:a:a::1                    2001:a:a:0:a00:27ff:fede:989e  ICMPv6   Echo (ping) reply id=0x4e7a, seq=4
 11 4.998221   fe80::c000:37ff:fe7e:0         2001:a:a:0:a00:27ff:fede:989e  ICMPv6   Neighbor solicitation
 12 4.998790   2001:a:a:0:a00:27ff:fede:989e  fe80::c000:37ff:fe7e:0         ICMPv6   Neighbor advertisement

▷ Frame 1: 86 bytes on wire (688 bits), 86 bytes captured (688 bits)
▷ Ethernet II, Src: CadmusCo_de:98:9e (08:00:27:de:98:9e), Dst: IPv6mcast_ff:00:00:01 (33:33:ff:00:00:01)
▷ Internet Protocol Version 6, Src: 2001:a:a:0:a00:27ff:fede:989e (2001:a:a:0:a00:27ff:fede:989e), Dst: ff02::1:ff00:1 (ff02::1:ff00:1)
▽ Internet Control Message Protocol v6
    Type: 135 (Neighbor solicitation)
    Code: 0
    Checksum: 0xa779 [correct]
    Reserved: 0 (Should always be zero)
    Target: 2001:a:a::1 (2001:a:a::1)
  ▽ ICMPv6 Option (Source link-layer address)
      Type: Source link-layer address (1)
      Length: 8
      Link-layer address: 08:00:27:de:98:9e
```

Figure 1.59 Neighbor Solicitation for the address resolution process.

```
No.  Time      Source                         Destination                    Protocol Info
  1 0.000000   2001:a:a:0:a00:27ff:fede:989e  ff02::1:ff00:1                 ICMPv6   Neighbor solicitation
  2 0.002337   2001:a:a::1                    2001:a:a:0:a00:27ff:fede:989e  ICMPv6   Neighbor advertisement 2001:a:a::1 (rtr, sol, ovr)
  3 0.002649   2001:a:a:0:a00:27ff:fede:989e  2001:a:a::1                    ICMPv6   Echo (ping) request id=0x4e7a, seq=1
  4 0.004458   2001:a:a::1                    2001:a:a:0:a00:27ff:fede:989e  ICMPv6   Echo (ping) reply id=0x4e7a, seq=1
  5 0.999524   2001:a:a:0:a00:27ff:fede:989e  2001:a:a::1                    ICMPv6   Echo (ping) request id=0x4e7a, seq=2
  6 1.001491   2001:a:a::1                    2001:a:a:0:a00:27ff:fede:989e  ICMPv6   Echo (ping) reply id=0x4e7a, seq=2
  7 2.000538   2001:a:a:0:a00:27ff:fede:989e  2001:a:a::1                    ICMPv6   Echo (ping) request id=0x4e7a, seq=3
  8 2.002721   2001:a:a::1                    2001:a:a:0:a00:27ff:fede:989e  ICMPv6   Echo (ping) reply id=0x4e7a, seq=3
  9 3.000977   2001:a:a:0:a00:27ff:fede:989e  2001:a:a::1                    ICMPv6   Echo (ping) request id=0x4e7a, seq=4
 10 3.002055   2001:a:a::1                    2001:a:a:0:a00:27ff:fede:989e  ICMPv6   Echo (ping) reply id=0x4e7a, seq=4
 11 4.998221   fe80::c000:37ff:fe7e:0         2001:a:a:0:a00:27ff:fede:989e  ICMPv6   Neighbor solicitation
 12 4.998790   2001:a:a:0:a00:27ff:fede:989e  fe80::c000:37ff:fe7e:0         ICMPv6   Neighbor advertisement

▷ Frame 2: 86 bytes on wire (688 bits), 86 bytes captured (688 bits)
▷ Ethernet II, Src: c2:00:37:7e:00:00 (c2:00:37:7e:00:00), Dst: CadmusCo_de:98:9e (08:00:27:de:98:9e)
▷ Internet Protocol Version 6, Src: 2001:a:a::1 (2001:a:a::1), Dst: 2001:a:a:0:a00:27ff:fede:989e (2001:a:a:0:a00:27ff:fede:989e)
▽ Internet Control Message Protocol v6
    Type: 136 (Neighbor advertisement)
    Code: 0
    Checksum: 0x7266 [correct]
  ▷ Flags: 0xe0000000
    Target: 2001:a:a::1 (2001:a:a::1)
  ▽ ICMPv6 Option (Target link-layer address)
      Type: Target link-layer address (2)
      Length: 8
      Link-layer address: c2:00:37:7e:00:00
```

Figure 1.60 Neighbor Advertisement for the address resolution process.

management of IPv6 tunnels requests from users [22]. Although this solution automatically manages tunnel requests and configuration, it presents potential security implications since the broker is a single point of failure. Figure 1.66 illustrates this model, showing the configuration of two distinct tunnels. Note that the tunnel broker gets IPv6 addresses from the DNS service (upon the clients' requests) and registers the addresses in the DNS database.

Teredo is an automatic tunneling mechanism intended to provide IPv6 connectivity to IPv4 hosts that are located behind NAT (Figure 1.67). This service employs two entities, a server and a relay. The server is an IPv6/IPv4

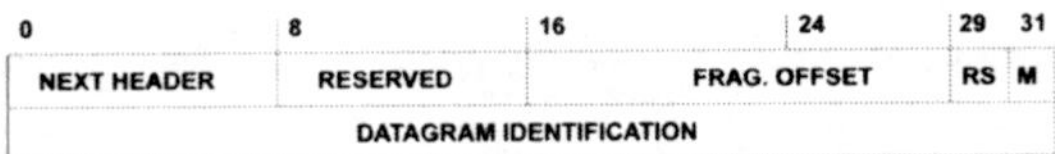

Figure 1.61 IPv6 fragmentation process.

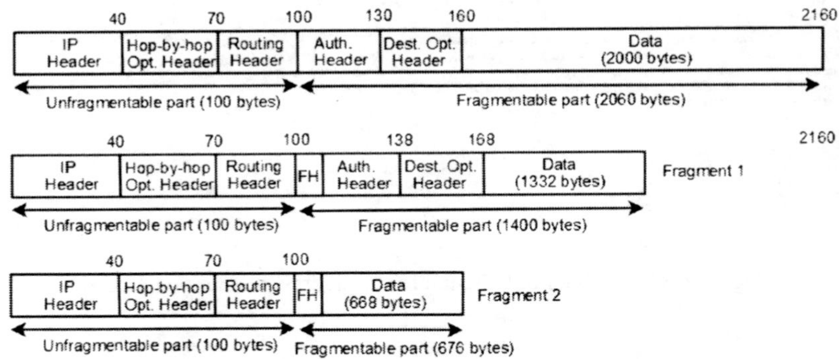

Figure 1.62 IPv6 datagram fragmentation.

node that is connected to both the IPv4 Internet and the IPv6 Internet and assists the initial configuration of Teredo clients and facilitates the initial communication between Teredo clients or Teredo clients and IPv6 hosts. The Teredo relay is an IPv6/IPv4 router that can forward packets between Teredo clients on the IPv4 Internet and IPv6-only hosts on the IPv6 Internet. The Teredo host-specific relay presented in Figure 1.67 is an IPv6/IPv4 node that has an interface and connectivity to both the IPv4 Internet and the IPv6 Internet and can communicate directly. IPv6 packets are encapsulated in IPv4-based UDP messages.

The Dual Stack Transition Mechanism (DSTM), illustrated in Figure 1.68, is used by dual-stack hosts without an assigned IPv4 address and requires a dedicated server that dynamically provides a temporary global IPv4 address. This mechanism uses dynamic tunnels to carry the IPv4 traffic within an IPv6 packet through the IPv6 domain.

In the automatic IPv4 compatible tunnel, the IPv4 tunnel end-point address is embedded within the destination IPv6 address. This tunnel can be configured between edge routers or between an edge router and an end system. The systems must be dual-stack and communication is only possible with other IPv4-compatible sites. This tunneling technique is currently deprecated.

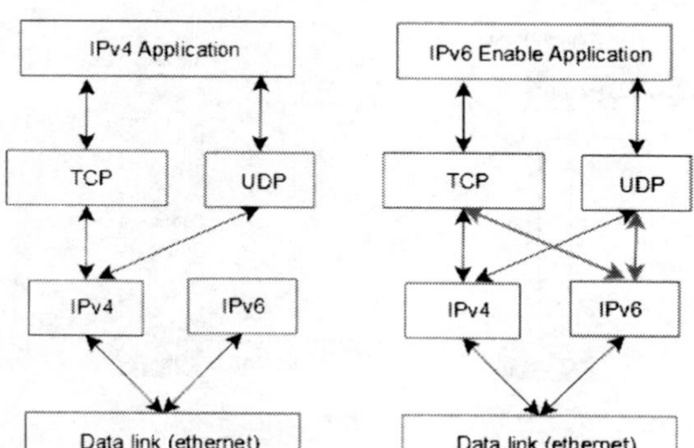

Figure 1.63 Dual stack IPv4 to IPv6 migration.

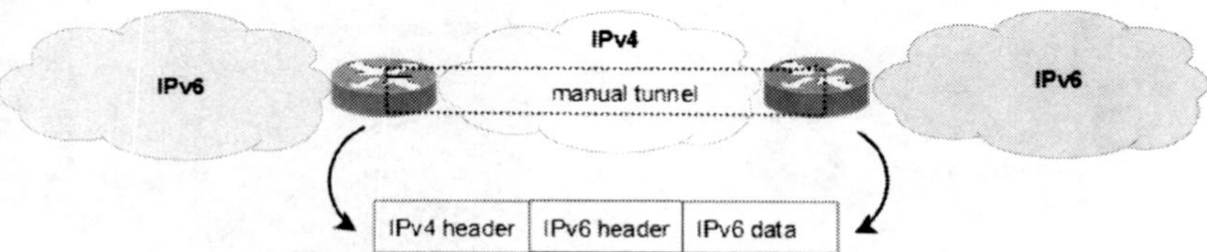

Figure 1.64 IPv6 over IPv4 manual tunnel.

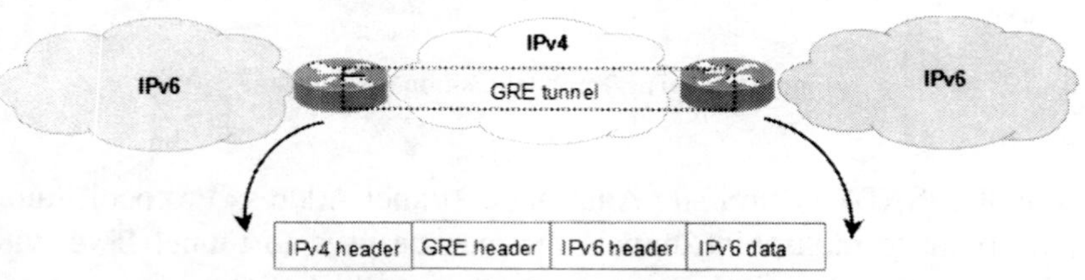

Figure 1.65 IPv6 over IPv4 GRE tunnel.

In automatic 6to4 tunnels an IPv4 tunnel end-point address is embedded within the destination IPv6 address, as shown in Figure 1.69. These tunnels allow isolated IPv6 domains to connect over an IPv4 network and, unlike the manually configured tunnels, they are multipoint tunnels. A 6to4 host/router needs to have a globally addressable IPv4 address, so it cannot be located behind a NAT box. As shown in Figure 1.70, a 6to4 router connects 6to4 hosts from an IPv6 domain and other 6to4 routers or the IPv6 Internet through a 6to4 relay router; a 6to4 relay router connects 6to4 routers on the IPv4 Internet and hosts on the IPv6 Internet.

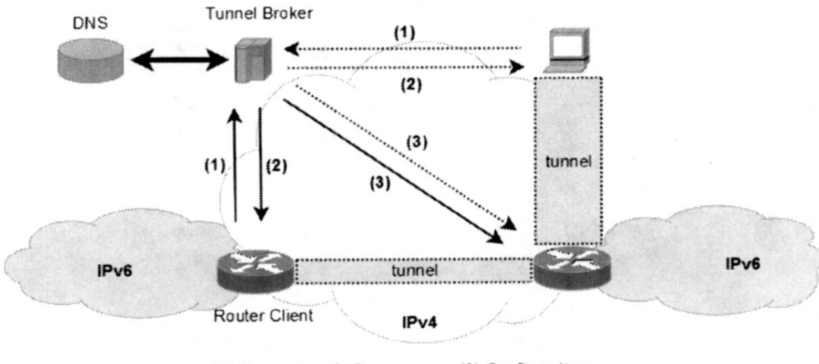

Figure 1.66 The tunnel broker service.

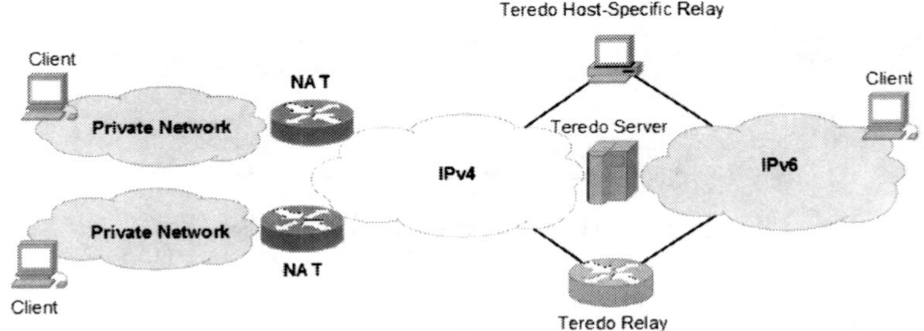

Figure 1.67 The Teredo transition mechanism.

Finally, ISATAP (Intra-site Automatic Tunnel Address Protocol) tunnels rely on point-to-multipoint tunnels that can be used to tunnel IPv4 within as administrative domain in order to create a virtual IPv6 network over an IPv4 network. It is a scalable approach for incremental deployment. The IPv4 address is encoded in IPv6 address within the interface ID (Figure 1.71).

The third transition mechanism is address translation. Several translation solutions have been proposed:

- Stateless IP/ICMP Translator (SIIT) Model - A general mechanism that translates IPv4 headers into IPv6 headers or vice versa [74];
- NAT-PT and NAPT-PT - NAT-PT translates an IPv4 packet into a semantically equivalent IPv6 datagram or vice versa, while NATPT-PT performs network addresses plus port translation plus packet translation (note that DNS Application Level Gateway (DNS-ALG) translates between the IPv4 and IPv6 DNS requests and responses [101, 66];

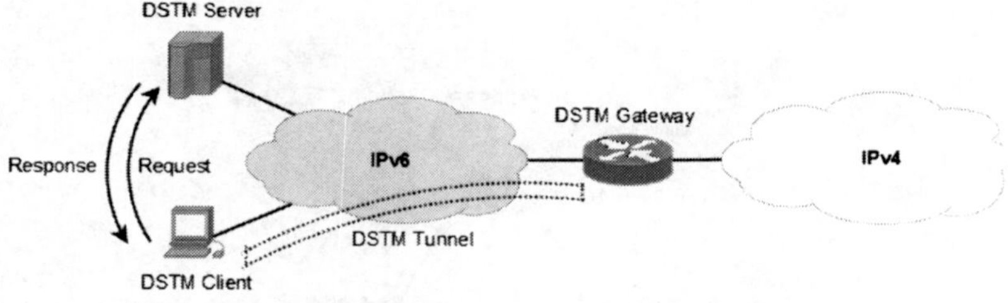

Figure 1.68 The Dual Stack Transition Mechanism.

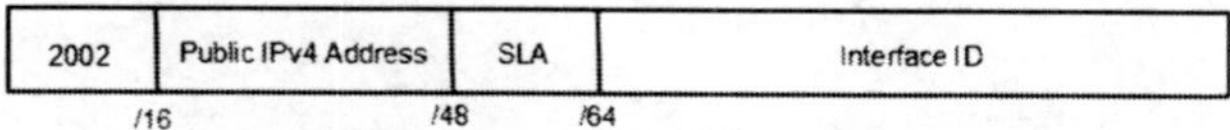

Figure 1.69 Packet header in automatic 6to4 tunnels.

- Bump-In-the-Stack (BIS) - Translation occurs at the protocol stack in each host and is a translation interface between IPv4 applications and the underlying IPv6 network. Three extra layers (name resolver extension, address mapper, and translator) were added to the IPv4 protocol stack [102];
- Bump-In-The-API (BIA) - Very similar to BIS but, instead of translating between IPv4 and IPv6 headers, BIA inserts an API translator between the socket API and the TCP/IP modules of the host stack [67];
- SOCKS-Based IPv6/IPv4 Gateway - Based on SOCKSv5, it permits communication between IPv4-only and IPv6-only hosts. When a client wants to connect to an application server, it sets up a connection to a well known, pre-configured proxy server using a special proxy protocol and then it informs the proxy about the IP address and the port number of the application server it wants to communicate with. The proxy server is responsible to set up a connection to the application server and, after establishing the connection, the proxy relays packets between the client and application server hiding the actual connection [61].
- Transport Relay Translator (TRT) - Enables IPv6-only hosts to exchange traffic with IPv4-only hosts without requiring any modification on hosts [46]. The IPv6 host uses the DNS-ALG service to resolve its DNS queries: it receives an IPv6 address specially constructed from the IPv4 address, consisting of a special network prefix associated with the transport relay and a host ID (the lower 64 bits) that embeds the IPv4 address of the remote host (Figure 1.72).

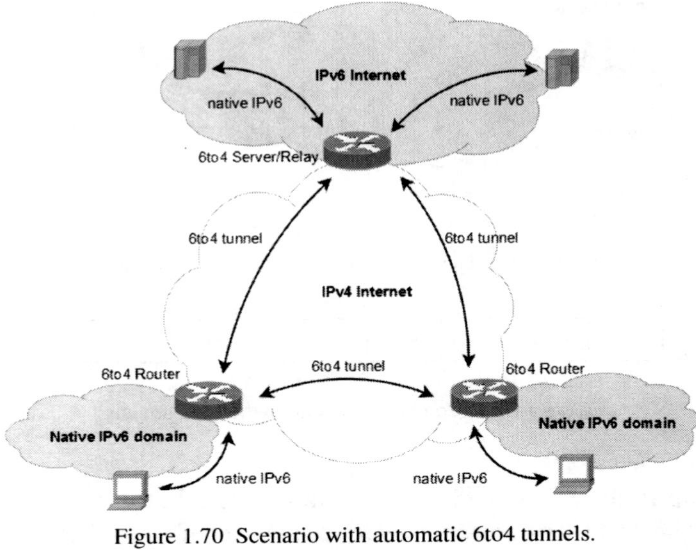

Figure 1.70 Scenario with automatic 6to4 tunnels.

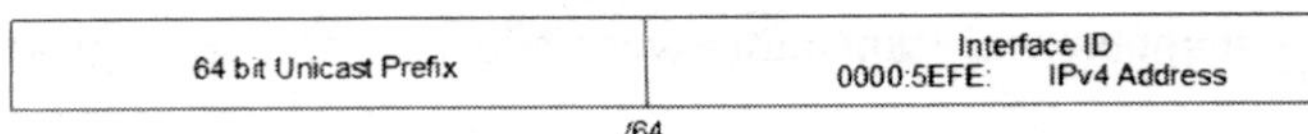

Figure 1.71 ISATAP tunneling.

1.10 The Domain Name System

The Domain Name System is basically a large database that contains the names and IP addresses of various hosts on the Internet and various domains, while the DNS service is the act of querying the database in order to convert human readable names of hosts to IP addresses [70]. The database is divided into sections called zones. The name servers in their respective zones are responsible for answering queries for their zones. A zone is a subtree of the DNS structure and is administered separately. There are multiple name servers for a zone: there is usually one primary name server and one or more secondary name servers. A name server may be authoritative for more than one zone.

Domain names consist of strings of characters separated by dots. The last word in a domain name represents a top-level domain. These top-level domains are controlled by IANA in the Root Zone Database. Some of the most common top-level domains are:

- COM - commercial Web sites, open to everyone;

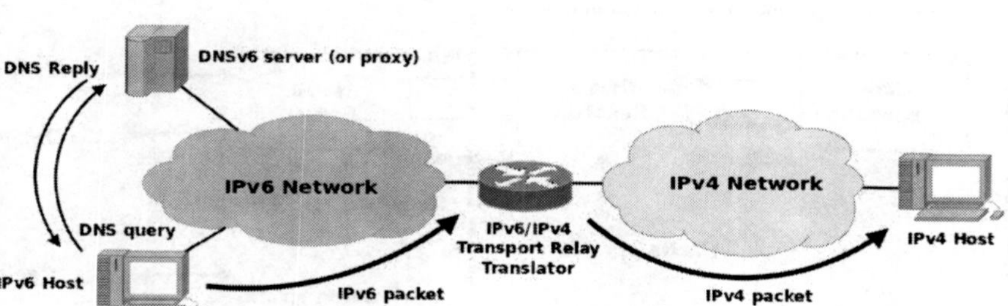

Figure 1.72 The Transport Relay Translator mechanism.

- NET - network Web sites, open to everyone;
- ORG - non-profit organization Web sites, open to everyone;
- EDU - restricted to schools and educational organizations;
- GOV - restricted to the U.S. government;
- Two-letter country codes (like US, UK, PT) - each one is assigned to a domain name authority in the respective country.

In a domain name, each word and dot combination placed before a top-level domain indicates a level in the domain structure. Each level refers to a server or a group of servers that manage that domain level. For example, "metalrd" in the domain name "metalrd.pt" is a second-level domain of the PT top-level domain. An organization may have a hierarchy of sub-domains further organizing its Internet presence, like "xxx.metalrd.pt" which is the *xxx*'s domain under *metalrd*, an additional level created by the domain name authority responsible for the PT country code. The left-most word in the domain name is a host name. A domain can potentially contain millions of host names as long as they are all unique to that domain.

Since all names in a certain domain need to be unique, there has to be some way to control the list and make sure no duplicates arise. A registrar is an authority that can assign domain names directly under one or more top-level domains and register them with InterNIC, a service of ICANN (Internet Corporation for Assigned Names and Numbers) that enforces uniqueness of domain names across the Internet. Each domain registration becomes part of a central domain registration database, known as the whois database.

A DNS server is responsible for two primary tasks: (i) maintain a small database of domain names and IP addresses most often used on its own network and delegate name resolution for all other names to other DNS servers on the Internet; (ii) pair IP addresses with all hosts and sub-domains for which that DNS server has authority. Servers that perform the first task are normally

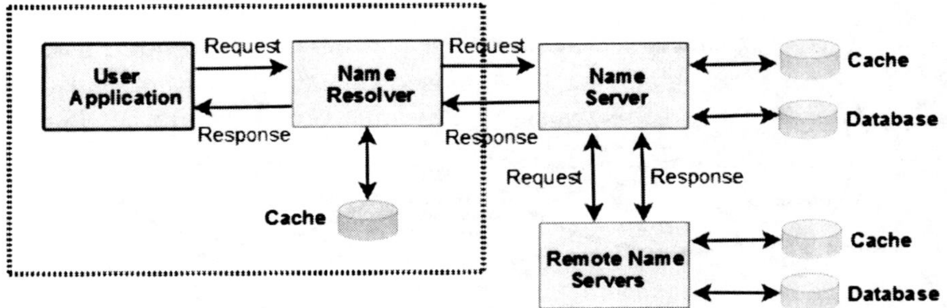

Figure 1.73 The DNS resolution process.

managed by an Internet Service Provider (ISP), residing in its data centers. These servers handle requests as follows:

- If it has the domain name and IP address in its database, it resolves the name itself.
- If it does not have the domain name and IP address in its database, it contacts another DNS server on the Internet; it may have to do this multiple times.
- If it has to contact another DNS server, it caches the lookup results for a limited time so it can quickly resolve subsequent requests to the same domain name.
- If it is not successful in finding the domain name after a reasonable search, it returns an error indicating that the name is invalid or does not exist.

Figure 1.73 illustrates this resolution process. Note that each host also contains a cache that maintains name-IP address pairs for some time (equal to the Time to Live value of the previously obtained response) and is consulted by the host name resolver before contacting the configured name server.

The second category of DNS servers is typically associated with Web, Mail and other Internet domain hosting services. A DNS server that manages a specific domain is called the start of authority (SOA) for that domain. Over time, the results from looking up hosts at the SOA will propagate to other DNS servers, which in turn propagate to other DNS servers, and so on across the Internet. This propagation is a result of each DNS server caching the lookup result for a limited time, known as its TTL, ranging from a few minutes to a few days.

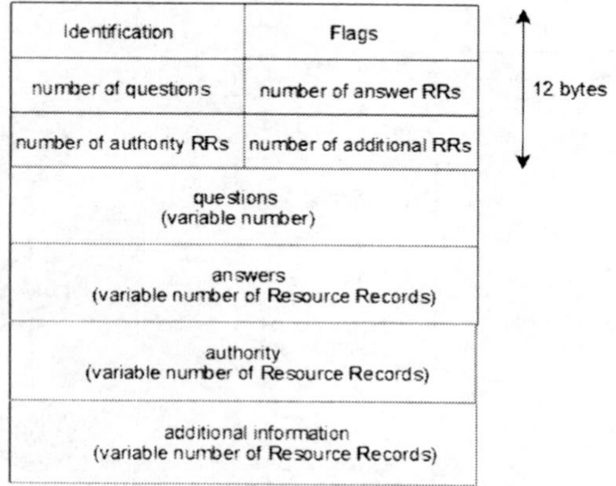

Figure 1.74 Format of the DNS messages.

The format of the query and reply messages is the same and is illustrated in Figure 1.74. The identification number is used to match queries to the corresponding replies, while flags are used to identify if (i) the message belongs to a query or a reply, (ii) recursion is preferred (iii) recursion is available or (iv) the reply is authoritative. The questions field includes the Name and Type of a query; the answers field includes the Resource Records that compose the reply; the authority field includes records for the authoritative servers; the additional information field contains additional information that can be useful.

DNS queries can be recursive or iterative. In a recursive name query, the DNS client requires that the DNS server respond to the client with either the requested resource record or an error message stating that the record or domain name does not exist. The DNS server cannot just refer the DNS client to a different DNS server. Thus, if a DNS server does not have the requested information when it receives a recursive query, it queries other servers until it gets the information, or until the name query fails. Recursive name queries are generally made by a DNS client to a DNS server, or by a DNS server that is configured to pass unresolved name queries to another DNS server, in the case of a DNS server configured to use a forwarder.

An iterative name query is one in which a DNS client allows the DNS server to return the best answer it can give based on its cache or zone data. If the queried DNS server does not have an exact match for the queried name, the best possible information it can return is a referral (that is, a pointer to a DNS server authoritative for a lower level of the domain namespace). The

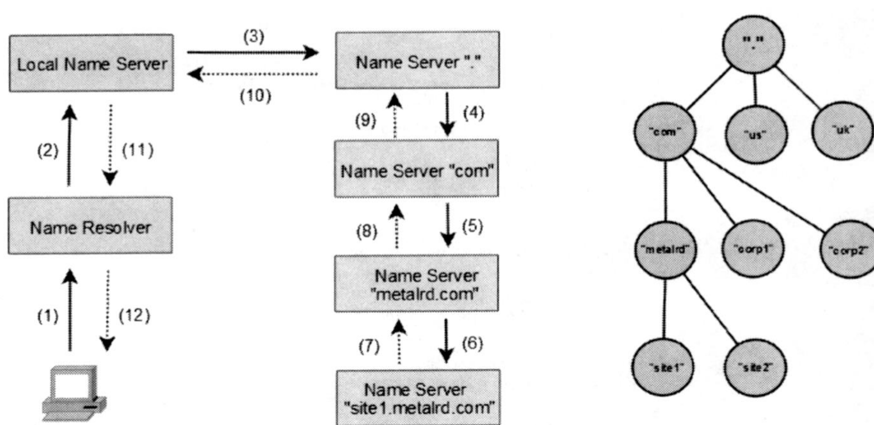

Figure 1.75 Example of a recursive DNS query.

DNS client can then query the DNS server for which it obtained a referral. It continues this process until it locates a DNS server that is authoritative for the queried name, or until an error or time-out condition is met. This type of query is typically initiated by a DNS server that attempts to resolve a recursive name query for a DNS client.

Figures 1.75 and 1.76 show examples of recursive and iterative queries, respectively. The different messages that are exchanged are identified by their order number. The recursive resolution is more efficient since it minimizes the time interval between the DNS query and the corresponding response. However, it penalizes the performance of the DNS servers because each server has a higher average number of simultaneous queries to process. The iterative resolution is less efficient, increasing the average time interval between the DNS query and the corresponding response, but optimizes the performance of the DNS servers, since each server immediately answers to each query.

In order to create a new domain name, the following steps should be taken:

- Use the Whois database to find a unique domain name that is not yet registered. There are several sites that offer free Whois database searches, such as Network Solutions (*www.networksolutions.com*). If the search comes up empty, the domain name is available.
- Register the domain name with a registrar.

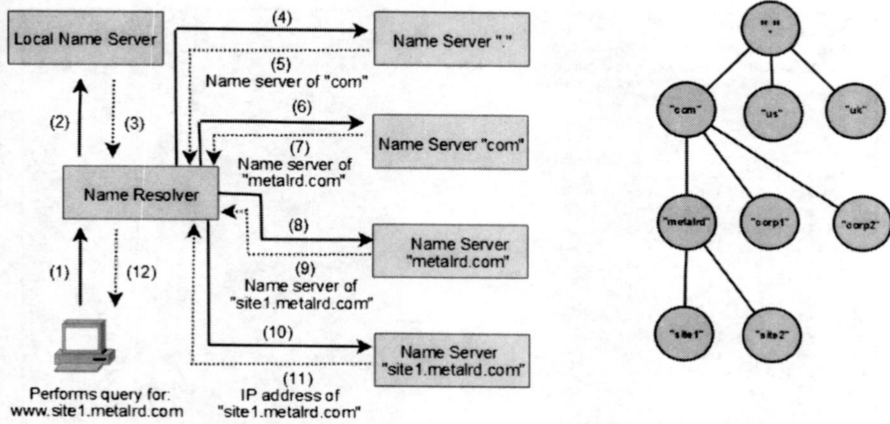

Figure 1.76 Example of an iterative DNS query.

- If we are hosting the domain at a different company than our registrar, we should configure the registrar to point our domain name to the correct host name or IP address for our hosting company.

Using the DNS servers from our registrar or hosting company means that we have a parked domain. This means that someone else owns the computer hardware for the DNS servers, and our domain is just part of that company's larger DNS configuration. Alternatively, if we want to host our own DNS, we can set up our own server, either as a physical or virtual machine. The DNS server (or group of servers) becomes the SOA for our domain.

Whether our SOA is somewhere else or on our own system, we can extend and modify our DNS settings to add sub-domains, redirect e-mail and control other services. This information is kept in a zone file on the DNS server. Each new configuration is called a record, and the following are the most common types of records:

- Host (A) - This is the basic mapping of IP address to host name, the essential component for any domain name.
- Canonical Name (CNAME) - This is an alias for the domain. Anyone accessing that alias will be automatically directed to the server indicated in the A record.
- Mail Exchanger (MX) - This maps e-mail traffic to a specific server. It could indicate another host name or an IP address. For example, people

```
1  $TTL 3600
2  $ORIGIN mydomain.com.
3  @ IN SOA ns1.mydomain.com. hostmaster.mydomain.com. (
4                  20120622 ; serial
5                  172800 ; refresh
6                  3600 ; retry
7                  1728000 ; expire
8                  172800 ) ; neg.TTL
9    IN NS ns1.mydomain.com.
10   IN NS ns2.mydomain.com.
11   IN A 192.168.251.10
12   IN AAAA 2001:A:A::10
13 ns1 IN A 192.168.251.253
14 ns2 IN A 192.168.251.254
15 ns1 IN AAAA 2001:A:A::253
16 ns2 IN AAAA 2001:A:A::254
17 @ IN MX 5 mail.mydomain.com.
18 @ IN MX 10 mail2.mydomain.com.
19 mail IN A 192.168.251.5
20 mail2 IN A 192.168.251.6
21 mail IN AAAA 2001:A:A::5
22 mail2 IN AAAA 2001:A:A::6
23 imapmail IN CNAME mail.mydomain.com.
24 offsite.mydomain.com. IN A 192.168.200.33 www IN A
25 192.168.251.10
26 www IN AAAA 2001:A:A::10 \$ORIGIN subdomain.mydomain.com.
27 www IN A 192.168.251.100
28 www IN AAAA 2001:A:A::100
```

Box 1.7 DNS bind9 sample configuration.

who use Google for the e-mail for their domain will create an MX record that points to *ghs.google.com*.

- Name Server (NS) - This contains the name server information for the zone. When configuring this record, our server will let other DNS servers know that it is the ultimate authority (SOA) for our domain when caching lookup information on oour domain from other DNS servers around the world.
- Start of Authority (SOA) - This is one larger record at the beginning of every zone file with the primary name server for the zone and some other information. If our registrar or hosting company is running our DNS server, we do not need to manage this.

Let us look at the sample domain file illustrated in the *bind9* configuration file of Box 1.7. Box 1.8 is used to define the DNS zone.

The $ORIGIN (line 2 of Box 1.7) simply provides a name to be appended to each name in the file that does not end with a period. For example, "mail" will be taken as "mail.mydomain.com".

The "Start Of Authority" (SOA) (line 3) record includes several interesting items. First, it contains the name of the machine on which this file was

```
// Consider adding the 1918 zones here, if they are not
// used in your organization
//include "/etc/bind/zones.rfc1918";

zone "mydomain.com" {
type master;
file "/etc/bind/zones/db.mydomain.com";
};
```

Box 1.8 DNS bind9 zone definition.

created, followed by the name of the responsible person. Next, enclosed in parentheses, are some parameters and their values:

- Serial is the revision number of this zone file and should be incremented each time the zone file is changed. It is important to increment this value each time a change is made, so that the changes will be distributed to any secondary DNS servers.
- The Refresh Time is the time, in seconds, a secondary DNS server waits before querying the primary DNS server's SOA record to check for changes. When the refresh time expires, the secondary DNS server requests a copy of the current SOA record from the primary. The primary DNS server complies with this request. The secondary DNS server compares the serial number of the primary DNS server's current SOA record and the serial number in it's own SOA record. If they are different, the secondary DNS server will request a zone transfer from the primary DNS server. The default value is 3600.
- The Retry Time is the time, in seconds, a secondary server waits before retrying a failed zone transfer. Normally, the retry time is less than the refresh time. The default value is 600.
- The Expire Time is the time, in seconds, that a secondary server will keep trying to complete a zone transfer. If this time expires prior to a successful zone transfer, the secondary server will expire its zone file. This means the secondary server will stop answering queries, as it considers its data too old to be reliable. The default value is 86,400.
- The Minimum Time-To-Live value applies to all resource records in the zone file. This value is supplied in query responses to inform other servers how long they should keep the data in cache. The default value is 3600.

Next, we have some NS records (lines 9-10) that state which Name Servers exist for this domain. Note that the "name" on each of these records (the first entry) is null, but since it does not end in a period it

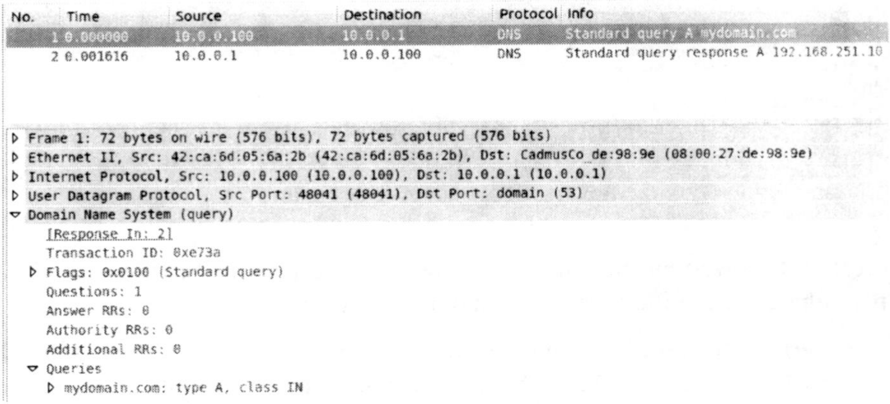

No.	Time	Source	Destination	Protocol	Info
1	0.000000	10.0.0.100	10.0.0.1	DNS	Standard query A mydomain.com
2	0.001616	10.0.0.1	10.0.0.100	DNS	Standard query response A 192.168.251.10

```
▷ Frame 1: 72 bytes on wire (576 bits), 72 bytes captured (576 bits)
▷ Ethernet II, Src: 42:ca:6d:05:6a:2b (42:ca:6d:05:6a:2b), Dst: CadmusCo_de:98:9e (08:00:27:de:98:9e)
▷ Internet Protocol, Src: 10.0.0.100 (10.0.0.100), Dst: 10.0.0.1 (10.0.0.1)
▷ User Datagram Protocol, Src Port: 48041 (48041), Dst Port: domain (53)
▽ Domain Name System (query)
    [Response In: 2]
    Transaction ID: 0xe73a
  ▷ Flags: 0x0100 (Standard query)
    Questions: 1
    Answer RRs: 0
    Authority RRs: 0
    Additional RRs: 0
  ▽ Queries
    ▷ mydomain.com: type A, class IN
```

Figure 1.77 DNS Query message.

has the “mydomain.com” suffix added to it. Alternatively, we could specify “mydomain.com.”.

Now we have “A” address records for IPv4 addresses and “AAA” address records for IPv6 addresses (lines 11-16). Some records start with names, while others do not.

The next record is the MX or Mail exchanger (email server) record (lines 17-18). The number right after MX is the priority of this record. The lower the number, the higher the priority, so that if there are two or more records that would apply to a given name, the lowest number would be tried first. Then, subdomains are defined in lines 19-22 for both IPv4 and IPv6.

The “CNAME” record (line 23) is used to define an alias. This CNAME defines the name “imapmail” within “mydomain.com”.

Finally, there is an example of using $ORIGIN to create a subdomain (line 26). Here, there is only one host defined in the subdomain. Its full name is “www.subdomain.mydomain.com” and it has an address that is still within our class C network. Once again, however, the target address could be anywhere on the Internet.

When a DNS name query is made, a DNS Query message is triggered asking for a type A record, as shown in Figure 1.77. A DNS Query Response message, shown in Figure 1.78 is received with the requested IPv4 address, the name servers with authority over the requested names and the A and AAAA records of those name servers.

No.	Time	Source	Destination	Protocol	Info
1	0.000000	10.0.0.100	10.0.0.1	DNS	Standard query A mydomain.com
2	0.001616	10.0.0.1	10.0.0.100	DNS	Standard query response A 192.168.251.10

```
▷ Frame 2: 212 bytes on wire (1696 bits), 212 bytes captured (1696 bits)
▷ Ethernet II, Src: CadmusCo_de:98:9e (08:00:27:de:98:9e), Dst: 42:ca:6d:05:6a:2b (42:ca:6d:05:6a:2b)
▷ Internet Protocol, Src: 10.0.0.1 (10.0.0.1), Dst: 10.0.0.100 (10.0.0.100)
▷ User Datagram Protocol, Src Port: domain (53), Dst Port: 48041 (48041)
▽ Domain Name System (response)
    [Request In: 1]
    [Time: 0.001616000 seconds]
    Transaction ID: 0xe73a
  ▷ Flags: 0x8580 (Standard query response, No error)
    Questions: 1
    Answer RRs: 1
    Authority RRs: 2
    Additional RRs: 4
  ▽ Queries
    ▷ mydomain.com: type A, class IN
  ▽ Answers
    ▷ mydomain.com: type A, class IN, addr 192.168.251.10
  ▽ Authoritative nameservers
    ▷ mydomain.com: type NS, class IN, ns ns2.mydomain.com
    ▷ mydomain.com: type NS, class IN, ns ns1.mydomain.com
  ▽ Additional records
    ▷ ns1.mydomain.com: type A, class IN, addr 192.168.251.253
    ▷ ns1.mydomain.com: type AAAA, class IN, addr 2001:a:a::253
    ▷ ns2.mydomain.com: type A, class IN, addr 192.168.251.254
    ▷ ns2.mydomain.com: type AAAA, class IN, addr 2001:a:a::254
```

Figure 1.78 DNS Query Response message.

1.11 The Transmission Control Protocol

The Transmission Control Protocol (TCP), which was firstly proposed in [10], provides reliable transmission of data in an IP environment and corresponds to the transport layer of the OSI reference model. Among the services that TCP provides are stream data transfer, reliability, efficient flow control, full-duplex operation and multiplexing. With stream data transfer, TCP delivers an unstructured stream of bytes identified by sequence numbers. This service benefits applications because they do not have to split data into blocks before handing it off to TCP. Instead, TCP groups bytes into segments and passes them to IP for delivery.

TCP offers reliability by providing connection-oriented, end-to-end reliable packet delivery through an internetwork. It does this by sequencing bytes with a forwarding acknowledgment number that indicates to the destination the next byte the source expects to receive. Bytes not acknowledged within a specified time period are retransmitted. The reliability mechanism of TCP allows devices to deal with lost, delayed, duplicate, or misread packets. A timeout mechanism allows devices to detect lost packets and request retransmission.

Efficient flow control means that, when sending acknowledgments back to the source, the receiving TCP process indicates the highest sequence number it can receive without overflowing its internal buffers.

Full-duplex operation means that TCP processes can both send and receive at the same time. Finally, multiplexing means that numerous simultaneous upper-layer conversations can be multiplexed or mixed over a single connection.

The TCP packet format contains several fields, as shown in Figure 1.79, with the following purposes:

- Source Port and Destination Port - Identify points at which upper-layer source and destination processes receive TCP services.
- Sequence Number - Usually specifies the number assigned to the first byte of data in the current message. In the connection-establishment phase, this field also can be used to identify an initial sequence number to be used in an upcoming transmission.
- Acknowledgment Number - Contains the sequence number of the next byte of data the sender of the packet expects to receive.
- Data Offset - Indicates the number of 32-bit words in the TCP header.
- Reserved - Reserved for future use.
- Flags - Carries a variety of control information, including the SYN and ACK bits used for connection establishment, and the FIN bit used for connection termination.
- Window - Specifies the size of the sender's receive window.
- Checksum - Indicates whether the header was damaged in transit.
- Urgent Pointer - Points to the first urgent data byte in the packet.
- Options - Specifies various TCP options.
- Data - Contains upper-layer information.

1.11.1 Connection Establishment Phase

To use reliable transport services, TCP hosts must establish a connection-oriented session with one another. Connection establishment is performed by using a "three-way handshake" mechanism. A three-way handshake synchronizes both ends of a connection by allowing both sides to agree upon initial sequence numbers. This mechanism also guarantees that both sides are ready to transmit data and know that the other side is ready to transmit as well. This

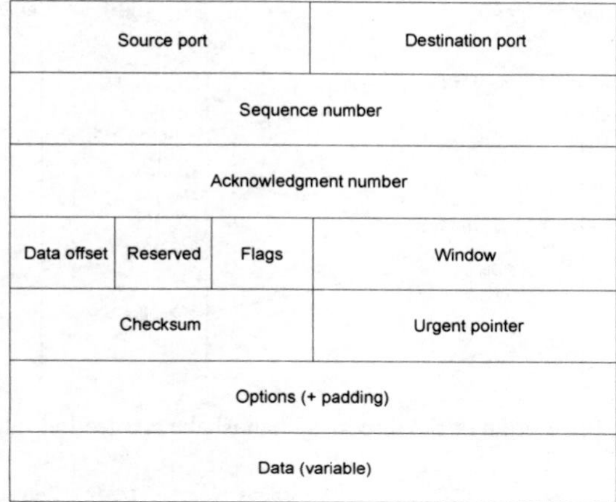

Figure 1.79 TCP packet format.

is necessary so that packets are not transmitted or retransmitted during session establishment or after session termination. Each host randomly chooses a sequence number and the three-way handshake procedure, illustrated in Figure 1.80, proceeds in the following manner:

- the first host (Terminal Application) initiates a connection by sending a packet with the initial sequence number (X) and with the SYN bit set to indicate a connection request;
- the second host (Server Application) receives the SYN, records the sequence number X, and replies by acknowledging the SYN (with an ACK = X + 1). The Server includes its own initial sequence number (SEQ = Y). An ACK = X means the host has received bytes 0 through $X - 1$ and expects byte X next. This technique is called forward acknowledgment;
- The Terminal Application then acknowledges all bytes the Server Application sent with a forward acknowledgment indicating the next byte the Terminal Application expects to receive (ACK = Y + 1). Data transfer then can begin.

If the destination port is not open the connection establishment will not be successful: as illustrated in Figure 1.81, the Server Application replies to the initial packet sent by the Terminal Application with a packet that includes the RESET flag set, acknowledges the received SYN (with an ACK = X + 1) and uses its own initial sequence number (SEQ = Y). The value of this sequence

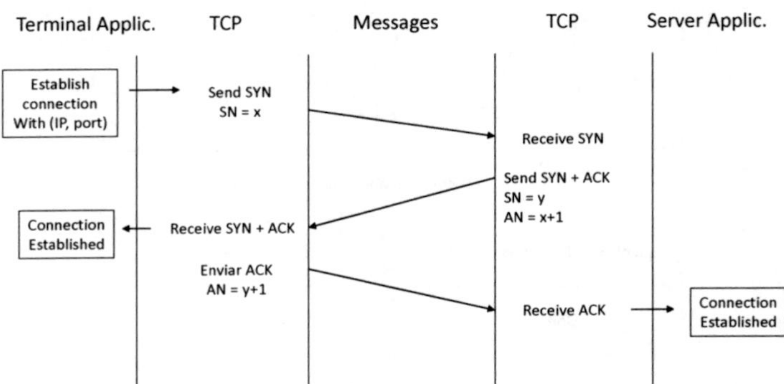

Figure 1.80 Illustration of the three-way handshake connection setup process.

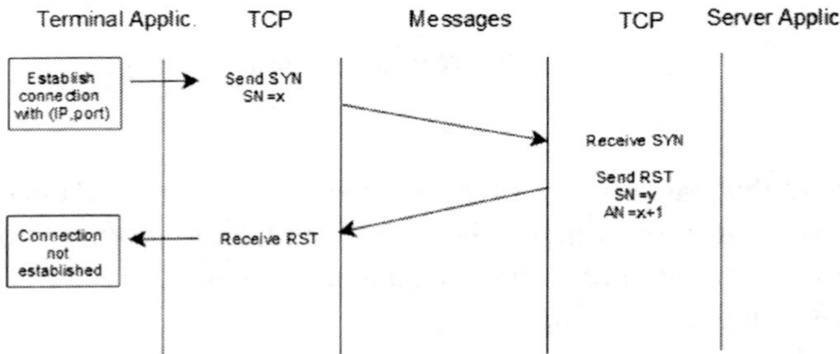

Figure 1.81 Illustration of an unsuccessful connection setup process.

number depends on the operating system, although it is frequently equal to zero.

1.11.2 Flow Control Mechanism

The flow control mechanism is illustrated in the simple example of Figure 1.82, corresponding to the information flow in a single direction (the same occurs in the opposite direction). After establishing the connection, the receiver notifies the sender about its window size (expressed in bytes, because TCP provides a byte-stream connection), representing the number of data bytes that the sender is allowed to send before waiting for an acknowledgment (in this case, 2500 bytes). Initial window sizes are indicated at connection setup, but might vary throughout the data transfer to provide flow control. Then, when the application delivers a set of data bytes to be sent, TCP will fragment it in the most convenient way and sends the maximum number of bytes that the receiver declared to be able to receive. In this example, the first two pack-

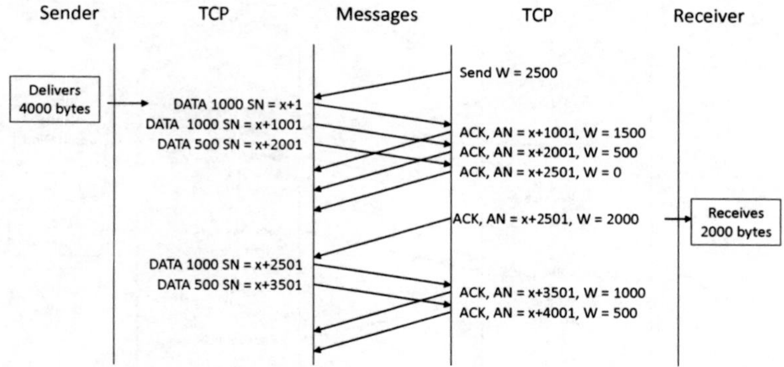

Figure 1.82 Illustration of the TCP flow control mechanism.

ets include 1000 bytes each and the third includes the remaining 500 bytes that are allowed to be sent. The sliding window mechanism enables hosts to send multiple bytes or packets before waiting for an acknowledgment. After that, the sender will wait for the acknowledgment of all sent data packets and for a new window value that will allow him to send more data. Finally, the sender is able to deliver the last 1550 bytes, receiving the corresponding acknowledgment messages. Note that in the sender to receiver direction the Sequence Number identifies the byte in the information flow from where data was sent; in the opposite direction, the Acknowledge Number identifies from which byte it is expecting to receive more data. In all answers, the remaining reception capacity of the receiver is updated. This value decreases every time the receiving host receives data from the network and increases every time it sends information to the receiving application. This mechanism guarantees the reliability of the data transfer, even if the receiver has a lower processing capability, because the sender only sends data after being notified that the receiver is able to receive it. The Sequence and Acknowledge Numbers also allow the sender to verify if all data was completely received and, if it was not, re-send it.

1.11.3 Connection Termination Phase

Unlike the TCP connection establishment, the connection termination phase can be triggered by any one of the end applications. This process is quite similar to the connection establishment process, using now the FIN code bit instead of the SYN bit (Figure 1.83). However, there is a fundamental difference between both phases: after receiving the first message, the TCP stack

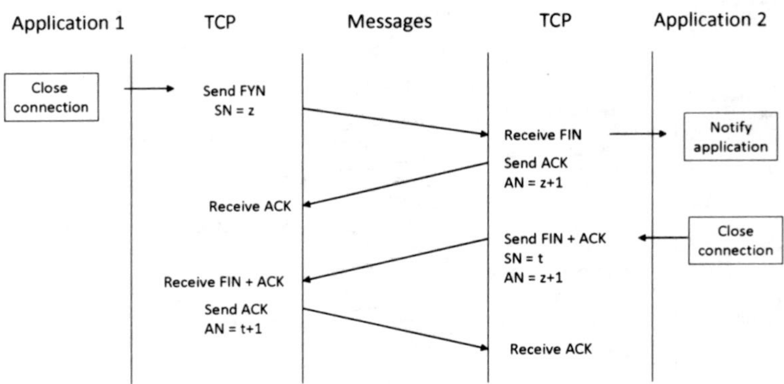

Figure 1.83 TCP connection termination phase.

of application 2 notifies the application about the reception of the connection termination message and then waits for the delivery of all data that is still in his buffer before notifying the end of the connection. Only after being notified by the application about the end of the connection, the TCP sends the appropriate segment that allows to completely terminate the connection on both directions. The Sequence Number and Acknowledge Number fields of these segments are used by the protocol stacks to mutually notify each other about the last byte that was sent.

1.11.4 Slow Start

For each connection, TCP maintains a congestion window, limiting the total number of unacknowledged packets that may be in transit end-to-end [95]. TCP uses a mechanism called slow start to increase the congestion window after a connection is initialized and after a timeout. It starts with a window of two times the Maximum Segment Size (MSS). Although the initial rate is low, the rate of increase is very rapid: for every packet acknowledged, the congestion window increases by 1 MSS, so that the congestion window effectively doubles for every Round Trip Time (RTT). When the congestion window exceeds a threshold (that we will designate as *ssthresh*), the algorithm enters a new state, called congestion avoidance. In some implementations, the initial *ssthresh* is large, and so the first slow start usually ends after a loss. However, *ssthresh* is updated at the end of each slow start and will often affect subsequent slow starts triggered by timeouts.

1.11.5 Congestion Avoidance

There may be a point during Slow Start where the network is forced to drop one or more packets due to overload or congestion. If this happens, Congestion Avoidance is used to slow the transmission rate [95]. In the Congestion Avoidance algorithm a retransmission timer expiring or the reception of duplicate acknowledgments can implicitly signal the sender that a network congestion situation is occurring. The sender immediately sets its transmission window to one half of the current window size (the minimum of the congestion window and the receiver's advertised window size), but to at least two segments. If congestion was indicated by a timeout, the congestion window is reset to one segment, which automatically puts the sender into Slow Start mode. If congestion was indicated by duplicate acknowledgments, the Fast Retransmit and Fast Recovery algorithms (explained below) are invoked.

As data is received during Congestion Avoidance, the congestion window is increased. However, Slow Start is only used up to the halfway point where congestion originally occurred. This halfway point was recorded earlier as the new transmission window. After this halfway point, the congestion window is increased by one segment for all segments in the transmission window that are acknowledged. This mechanism will force the sender to slowly grow its transmission rate, as it will approach the point where congestion had previously been detected.

1.11.6 Fast Retransmit

When a duplicate acknowledgment is received, the sender does not know if it is because a TCP segment was lost or simply that a segment was delayed and received out of order at the receiver. If the receiver can re-order segments, it should not be long before the receiver sends the latest expected acknowledgment. Typically no more than one or two duplicate acknowledgments should be received when simple out of order conditions exist. However, if more than two duplicate acknowledgments are received by the sender, it is a strong indication that at least one segment has been lost. The TCP sender will assume enough time has lapsed for all segments to be properly re-ordered by the fact that the receiver had enough time to send three duplicate acknowledgments. When three or more duplicate acknowledgments are received, the sender does not even wait for a retransmission timer to expire before retransmitting the segment. This process is called the Fast Retransmit algorithm [95].

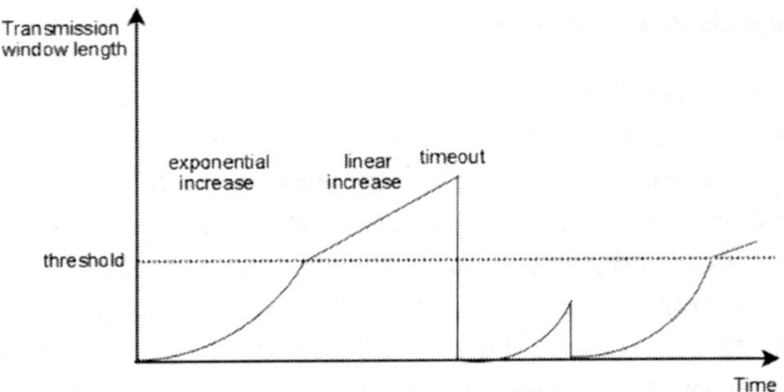

Figure 1.84 TCP congestion control mechanism.

1.11.7 Fast Recovery

Since the Fast Retransmit algorithm is used when duplicate acknowledgments are being received, the TCP sender has implicit knowledge that there is data still flowing to the receiver, because duplicate acknowledgments can only be generated when a segment is received. This is a strong indication that serious network congestion may not exist and that the lost segment was a rare event. So, instead of reducing the flow of data abruptly by going all the way into Slow Start, the sender only enters Congestion Avoidance mode [95]. Rather than start at a window of one segment as in Slow Start mode, the sender resumes transmission with a larger window, incrementing as if in Congestion Avoidance mode. This allows for higher throughput under the condition of only moderate congestion.

Figure 1.84 illustrates a typical data transfer phase using TCP congestion control, where periods of exponential window size increase, linear increase and drop-off are clearly visible.

1.11.8 Some of the Different TCP Versions

The basic TCP mechanisms have been implemented in different ways, leading to several TCP variants. The following are some of the most popular TCP variants:

- TCP Tahoe, the base TCP, is characterized by using the Slow Start and Congestion Avoidance mechanisms; after a timeout or three duplicate acknowledgments, the congestion window is set to 1 and the algorithm enters the slow start phase.

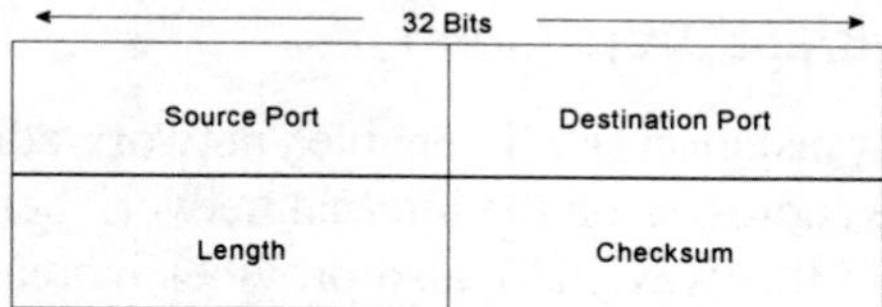

Figure 1.85 UDP packet format.

- TCP Reno, which uses the Slow Start, Congestion Avoidance, Fast Retransmit and Fast Recovery mechanisms; after a timeout, the congestion window is set to 1 and the algorithm enters the Slow Start phase; after three duplicate acknowledgments, the algorithm uses the Fast Recovery and Congestion Avoidance mechanisms.
- TCP New-Reno, which stays in Fast Recovery mode until all packet losses in a window are recovered; this variant is able to recover one packet loss per RTT without causing a timeout.
- Selective Acknowledgments (Sacks), which provides information about out-of-order packets received and is able to recover multiple packet losses per RTT.

1.12 The User Datagram Protocol

The User Datagram Protocol (UDP) is a connectionless transport-layer protocol [83]. UDP protocol ports distinguish multiple applications running on a single device. Unlike TCP, UDP adds no reliability, flow-control, or error-recovery functions to IP. Because of its simplicity, UDP headers contain fewer bytes and consume less network overhead than TCP. UDP is useful in situations where the reliability mechanisms of TCP are not necessary, such as in cases where a higher-layer protocol might provide error and flow control.

UDP is the transport protocol for several well-known application-layer protocols, including Network File System (NFS), Simple Network Management Protocol (SNMP), Domain Name System (DNS), and Trivial File Transfer Protocol (TFTP). The UDP packet format contains four fields, as shown in Figure 1.85, which include source and destination ports, length, and checksum fields.

Source and destination ports contain the 16-bit UDP protocol port numbers used to demultiplex datagrams for receiving application-layer processes. The Length field specifies the length of the UDP header and data, while the Checksum provides an (optional) integrity check on the UDP header and data.

1.13 Network Address Translation

Network Address Translation (NAT) enables network administrators to hide unregistered private addresses on the internal network behind registered public addresses [93]. In this way, NAT also provides increased security for the internal network. When users send traffic out to the Internet, they are not using their actual TCP/IP addresses. This makes malicious attacks over the internal network more difficult.

As already said, Internal or private addresses should not be routed onto the Internet, so border routers should implement NAT to translate these private addresses into registered public addresses. There are three main types of Network Address Translation: Static Address Translation, Dynamic Source Address Translation and Port Address Translation.

Static Address Translation is a one-to-one mapping from a private internal address to a registered public address. In Static Address Translation each addition, deletion, or change to the NAT must be done manually by the administrator. To configure a Static Address Translation, the following global configuration mode command should be used:

```
ip nat inside source static <local-ip><global-ip>
```

Dynamic Source Address Translation associates an internal host automatically with a public address from a pool of addresses. This would be implemented in an environment where there is a group of public addresses to be used for NAT and several users that may access the Internet at any time. This feature is dynamic, decreasing the administrative overhead. The following global configuration mode command is used to establish a NAT pool:

```
ip nat pool <name> <start-ip> <end-ip> {netmask <netmask> |
prefixlength <prefix length>}
```

Port Address Translation (PAT) gives the administrator the option to conserve public addresses in the address pool by enabling source ports in TCP or UDP connections to be translated. This provides the opportunity for numerous different private addresses to be associated with one public address by using port translation to distinguish between distinct communications. When more detailed translation is required, the new port number is assigned from the same pool as the original. Keyword *overload* enables UDP and TCP port translation. To configure Port Address Translation, the following global configuration mode command should be used:

```
ip nat inside source list <acl> pool <name> [overload]
```

To enable a NAT on the interface, we should use the following interface configuration mode command:

```
ip nat inside | outside
```

This command applies the NAT to the interface and designates that interface to be an "inside" interface. The *inside* keyword in the command means that this interface is using private addresses, which need to be translated to "outside" or public addresses.

1.13.1 Controlling NAT/PAT with Access Control Lists

In order to control which hosts apply to the established NAT/PAT, we have to implement Access Control Lists (ACLs). ACLs will be discussed in detail later in this book. To assign an ACL to a NAT configuration, first we have to establish the NAT and the ACL and then relate the ACL to the NAT. This can be done as in the following example:

```
access-list 1 permit 192.168.20.0 0.0.0.255
ip nat inside source list 1 pool <pool-name>
```

In this example, the IP NAT pool *pool-name* and ACL 1 are established. Then, the list variable is used within the *ip nat* command to connect the *pool-name* NAT to ACL 1. In this way, only hosts from network 192.168.20.0 can be translated from the NAT pool.

1.13.2 Controlling NAT/PAT with Route Maps

Route maps are intricate access lists that allow criteria to be tested against network traffic by using the *match* command. If the criterion is met then an action is taken against the traffic. This action is specified using the *set* command.

By matching traffic against ACLs, route maps allow a fine control of how routes are redistributed among route processes. The *route-map* command enables the router to create a route-map, defining a criterion as well as the action to be taken once the criteria match is successful. Route maps can be used for redistribution, routing, policies and traffic shaping.

When route maps are used for NAT, an entry is created for the inside and the outside interfaces and for both private and global addresses. Any TCP or UDP port information is included in the translation. Route maps allow network administrators to arrange any combination of access lists, output interface and next-hop TCP/IP address to determine which NAT pool to use.

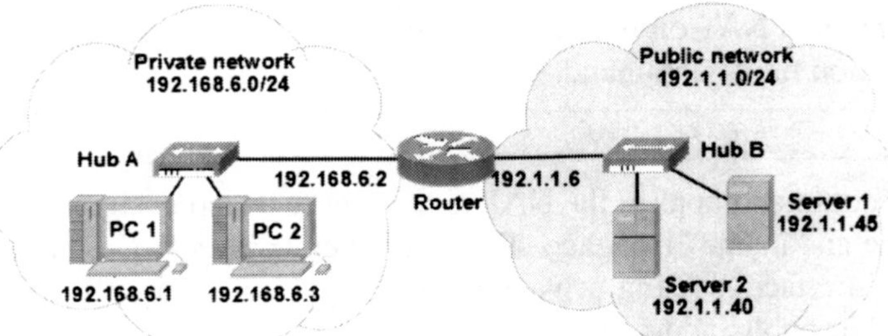

Figure 1.86 Network used to illustrate the functioning principles of NAT/PAT.

In this way, it is possible to create different combinations that provide the specific control we want.

Let us look at an example. First of all, a NAT pool should defined. Then, a route map should be associated to that NAT pool using the following global configuration mode command:

```
ip nat inside source route-map <route-map-name>
pool <pool-name>
```

Once the pool has been defined, the NAT should be assigned to an interface both on the inside and the outside locations. After the pool has been established and has been assigned to an interface, an access list must be configured to define which traffic will be used.

Finally, the route map criteria can be configured using the match command:

```
route-map <route-map-name> permit 10
match ip address <access-list-number>
```

Exercise - NAT/PAT Operation

Let us consider the network depicted in the Figure 1.86, representing the communications network of a small company. Suppose that the company decided to configure IP private addressing using the network 192.168.6.0/24 and the NAT mechanism (without PAT) to manage all Internet accesses. Consider that the company only has 2 public addresses, 192.1.1.6/24 and 192.1.1.16/24 (we will use this one for the NAT mechanism). The dynamic NAT configuration commands are shown in Box 1.9. Using this configuration, only one of the private PCs (the first one that attempts) will be able to access the public network.

By inserting the command:

```
# ip nat pool MYNATPOOL 192.1.1.16 192.1.1.16
netmask 255.255.255.0
# access-list 2 permit 192.168.6.0 0.0.0.255
# ip nat inside source list 2 pool MYNATPOOL
# interface FastEthernet 0/0
  # ip address 192.168.6.1 255.255.255.0
  # ip nat inside
  # no shutdown
  # exit
# interface FastEthernet 0/1
  # ip address 192.1.1.6 255.255.255.0
  # ip nat outside
  # no shutdown
  # end
```

Box 1.9 Dynamic NAT configuration.

```
Router#show ip nat translations verbose
Pro Inside global      Inside local       Outside local      Outside global
--- 192.1.1.16         192.168.6.1        ---                ---
    create 00:06:47, use 00:02:38 timeout:86400000, left 23:57:21, Map-Id(In): 1
,
    flags:
none, use_count: 0, entry-id: 1, lc_entries: 0
```

Figure 1.87 The result of the *show ip nat translations verbose* command.

```
Router#show ip nat statistics
Total active translations: 1 (0 static, 1 dynamic; 0 extended)
Outside interfaces:
  FastEthernet0/0
Inside interfaces:
  FastEthernet0/1
Hits: 20  Misses: 1
CEF Translated packets: 14, CEF Punted packets: 0
Expired translations: 2
Dynamic mappings:
-- Inside Source
[Id: 1] access-list 2 pool smile refcount 1
 pool smile: netmask 255.255.255.0
        start 192.1.1.16 end 192.1.1.16
        type generic, total addresses 1, allocated 1 (100%), misses 0
Queued Packets: 0
```

Figure 1.88 The result of the *show ip nat statistics* command.

```
show ip nat translations verbose
```

it is possible to see the active translation and the default NAT timeout (24 hours) (Figure 1.87). The other internal host will be able to access the public only after the timeout expiration.

The statistics of the translations can be consulted using the command:

```
show ip nat statistics
```

obtaining the output that is shown in Figure 1.88.

Now, let us activate port address translation by replacing the third command of Box 1.9 by this one:

```
Router#show ip nat translations verbose

Pro Inside global      Inside local        Outside local       Outside global
tcp 192.1.1.16:1035    192.168.6.1:1035    192.1.1.45:22       192.1.1.45:22
    create 00:02:04, use 00:00:32 timeout:86400000, left 23:59:27, Map-Id(In): 1
,
    flags:
extended, use_count: 0, entry-id: 44, lc_entries: 0
Router#show ip nat translations verbose
Pro Inside global      Inside local        Outside local       Outside global
tcp 192.1.1.16:1035    192.168.6.1:1035    192.1.1.45:22       192.1.1.45:22
    create 00:02:33, use 00:01:01 timeout:86400000, left 23:58:58, Map-Id(In): 1
,
    flags:
extended, use_count: 0, entry-id: 44, lc_entries: 0
tcp 192.1.1.16:1036    192.168.6.1:1036    192.1.1.45:80       192.1.1.45:80
    create 00:00:09, use 00:00:08 timeout:86400000, left 23:59:51, Map-Id(In): 1
,
    flags:
extended, use_count: 0, entry-id: 45, lc_entries: 0
tcp 192.1.1.16:1037    192.168.6.1:1037    192.1.1.40:80       192.1.1.40:80
    create 00:00:05, use 00:00:05 timeout:86400000, left 23:59:54, Map-Id(In): 1
,
    flags:
extended, use_count: 0, entry-id: 46, lc_entries: 0
```

Figure 1.89 Active translations using PAT.

```
Router#show ip nat statistic
Total active translations: 3 (0 static, 3 dynamic; 3 extended)
Outside interfaces:
  FastEthernet0/0
Inside interfaces:
  FastEthernet0/1
Hits: 287  Misses: 41
CEF Translated packets: 320, CEF Punted packets: 8
Expired translations: 42
Dynamic mappings:
-- Inside Source
[Id: 1] access-list 2 pool smile refcount 3
 pool smile: netmask 255.255.255.0
        start 192.1.1.16 end 192.1.1.16
        type generic, total addresses 1, allocated 1 (100%), misses 16
Queued Packets: 0
```

Figure 1.90 Statistics using PAT.

```
ip nat inside source list 2 pool MYNATPOOL OVERLOAD
```

Now, both private hosts are able to access the public network.

From PC 1, we will access the HTTP servers located on machines 192.1.1.40 and 192.1.1.45, as well as server 192.1.1.45 through SSH. Figures 1.89 and 1.90 show the new active NAT translations and NAT activity statistics, respectively. As can be seen, for each new communication that takes place a new port number is used by the PAT mechanism (using only one public IP address), making the (IP address, Port number) pair different for each new communication/translation.

Finally, suppose that PC 1 is located on the public network, as shown in Figure 1.91. If we want to access the web server of the private host (PC 2)

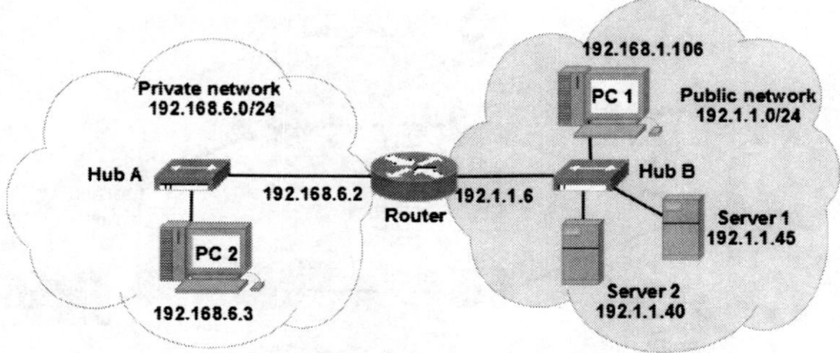

Figure 1.91 Network used to illustrate static NAT/PAT.

from PC 1 (supposing that the web server uses TCP port number 80), a static NAT/PAT mapping should be configured at the router using the following command:

```
ip nat inside source static tcp 192.168.6.3 80 192.1.1.16 80
```

This statement performs the static address translation for the Web server; so, users that try to reach 192.1.1.16, port 80 (www) are automatically redirected to 192.168.6.3, port 80 (www), which is the Web server.

1.13.3 NAT with Multiple ISPs and Asymmetric Routing

Figure 1.92 illustrates a private network that is connected to two different ISPs. Asymmetric routing is assumed, that is, routing from the private network to outside is done via ISP B while ISP A is used to route traffic in the opposite direction. Assuming that the NAT tables are not synchronized, a translation is only known by the router that routed the first packet. So, communication between a private host and an outside host will not be possible.

A possible solution to this problem is Stateful NAT (SNAT), illustrated in Figure 1.93. SNAT is a Cisco proprietary solution that provides data synchronization over TCP. In order to force data synchronization between the involved routers, it is necessary to define a primary server (the others are backup servers).

SNAT configuration at both routers should be similar to the configuration example shown in Box 1.10. We are supposing that the IP address of the private interface of the top router is 10.0.0.2, while the IP address of the private interface of the bottom router is 10.0.0.3; besides, address pools named

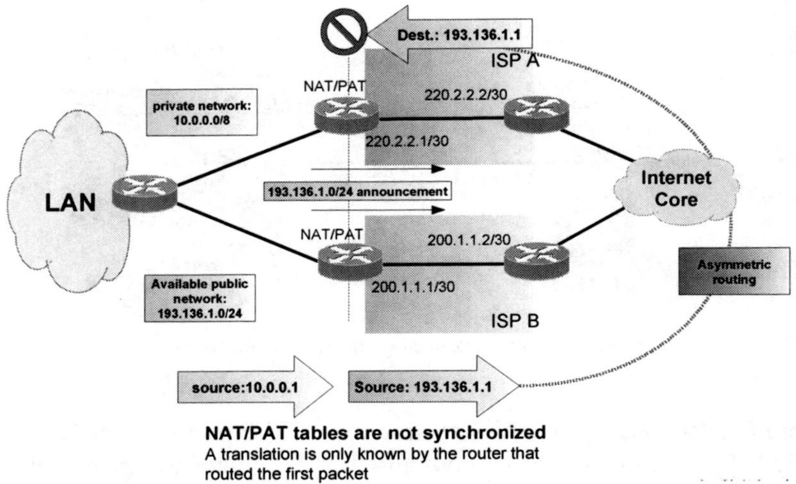

Figure 1.92 NAT with multiple ISPs and asymmetric routing.

POOLtop and POOLbottom were already configured at the top and bottom routers, respectively.

1.14 Use Case

1.14.1 Problem Description

In order to map the addressed concepts to a real network scenario, let us present a specific problem corresponding to the design of the communications infrastructure of a small to medium generic company.

CorporateX is a generic company whose physical facilities include two buildings, identified by numbers 0 and 1, each one having two floors (also identified as 0 and 1). The IP communications infrastructure of the company should be redundant and support data communication services, Voice over IP (VoIP) and a High-Definition (HD) video conference system. The company should also have a private data center (with servers and storage devices) and a public data center that will support the public services of the company.

The main objective of this exercise is to design the communications infrastructure of CorporateX, including the internal infrastructure of each building, the infrastructure that interconnects different floors of the same building and different buildings and the Internet access.

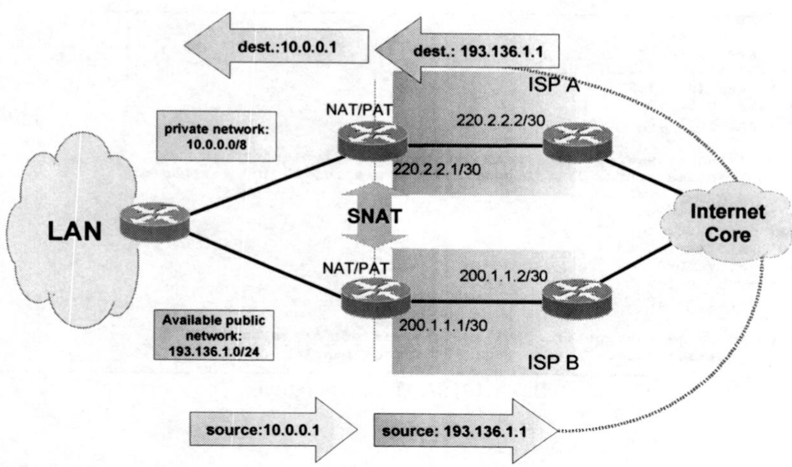

Figure 1.93 SNAT solution.

Figure 1.94 presents a high level picture of the CorporateX communication infrastructure. The Internet connection will be supported by two different Internet Service Providers.

A more detailed architecture of the network is illustrated in Figure 1.95. At each floor, the network should accommodate the necessary number of networked devices, which will require an appropriate number of layer 2 switches (for readability reasons, the figure only shows some layer 2 switches but the exact number will depend on the specific needs). The access, distribution, and core layers are all present in this architecture: layer 2 switching is used in the access layer; the distribution layer for each building is implemented using layer 3 switches, while routers are used for Internet access; layer 3 switching is used in the core layer of the infrastructure. This network uses a full mesh topology at the distribution and core layers, providing the maximum degree of redundancy.

The layer 3 switched backbone has several advantages: reduced router peering, flexible topology with no spanning-tree loops, multicast and broadcast control in the backbone and scalability to arbitrarily large size. Since the layer 3 switched backbone has dual links to the backbone from each distribution layer switch, each distribution layer switch maintains two equal-cost paths to every destination network, so recovery from any link failure is fast. This design also provides a double trunking capacity into the backbone.

```
#In the top router
#SNAT configuration
# ip nat Stateful id 1
  # backup 10.0.1.2
  # peer 10.0.2.2
  # mapping-id 10

#Association between public pool address and private networks
# ip nat inside source list 1 pool POOLtop mapping-id 10 overload

#In the bottom router
#SNAT configuration
# ip nat Stateful id 1
  # primary 10.0.2.2
  # peer 10.0.1.2
  # mapping-id 10

#Association between public pool address and private networks
# ip nat inside source list 1 pool POOLbottom mapping-id 10 overload
```

Box 1.10 SNAT configuration.

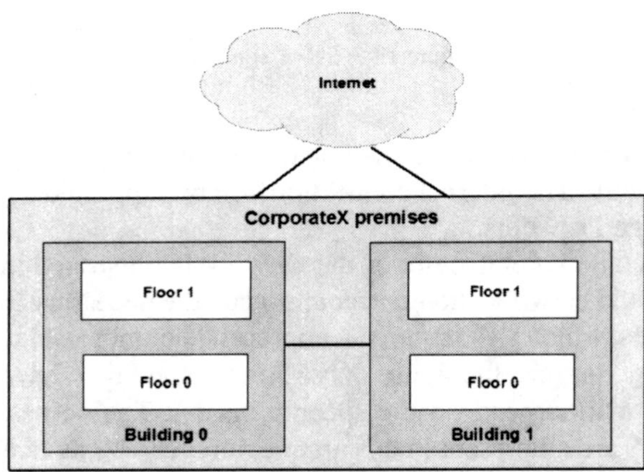

Figure 1.94 High level view of the CorporateX communications infrastructure.

Since Figure 1.95 is hard to read, Table 1.5 shows all the network links and the corresponding routers' interfaces.

1.14.2 Addressing Scheme

The best practice is to organize IP addresses in a such a way that it is possible to aggregate subnetworks based on the most appropriate logic (aggregation by service, building, etc). In this exercise, we will consider the generic IP addressing scheme that is illustrated in Figure 1.96. In IPv4, we will consider private IP addresses starting with 10 in the first byte, while IPv6 addresses

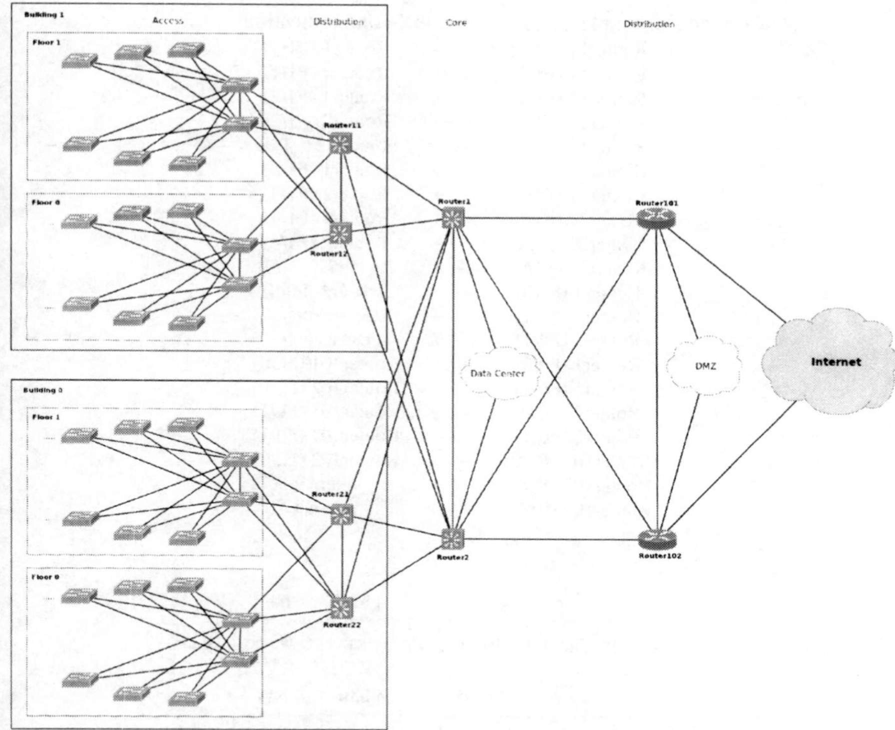

Figure 1.95 Architecture of the global CorporateX IP network.

will start by 2001:A:A. Four aggregation levels are considered, corresponding (from the most specific to the most general, that is, from right to left) to: (i) the building identification; (ii) a specific (or low level) sub-division of each high level service; (ii) each high level service that the company should support; (iv) each physical site of the company. Four bits are reserved for each one of these fields.

Several services should be supported by the company: Voice over IP (VoIP), Video Conference (VC), Administration Data, Engineering Data, Administrative Data (these data services correspond to three different types of users/employees), Core, DMZ (DeMilitarized Zone), Data Center and NAT (Network Address Translation). With the exception of NAT, all the other services should use private addresses. The complete addressing of the CorporateX network is represented in Figure 1.97.

Table 1.5 Network links of the CorporateX communications infrastructure.

Router11 (F1/1)	⟷	Router1 (F1/1)
Router11 (F1/2)	⟷	Router2 (F1/1)
Router12 (F1/1)	⟷	Router1 (F1/2)
Router12 (F1/2)	⟷	Router2 (F1/2)
Router11 (F1/0)	⟷	Router12 (F1/0)
Router21 (F1/1)	⟷	Router1 (F1/3)
Router21 (F1/2)	⟷	Router2 (F1/3)
Router22 (F1/1)	⟷	Router1 (F1/4)
Router22 (F1/2)	⟷	Router2 (F1/4)
Router21 (F1/0)	⟷	Router22 (F1/0)
Router1 (F2/0)	⟷	Router2 (F2/0)
Router1 (F2/1)	⟷	Datacenter
Router2 (F2/1)	⟷	Datacenter
Router1 (F0/0)	⟷	Router101 (F0/0)
Router1 (F0/1)	⟷	Router102 (F0/0)
Router2 (F0/0)	⟷	Router101 (F0/1)
Router2 (F0/1)	⟷	Router102 (F0/1)
Router101 (F2/0)	⟷	Router102 (F2/0)
Router101 (F1/0)	⟷	Internet
Router101 (F1/0)	⟷	Internet

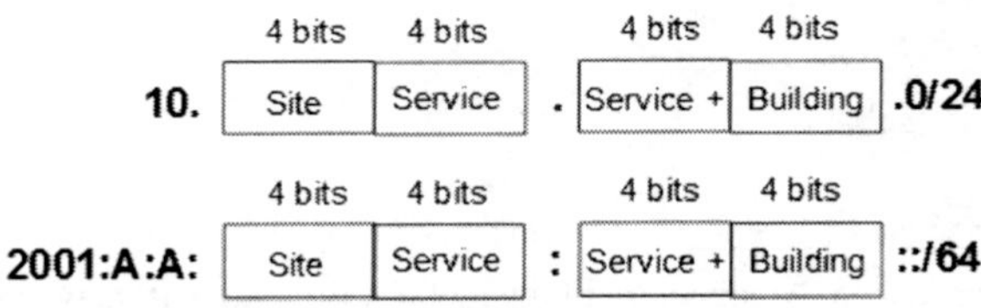

Figure 1.96 IP addressing scheme for the CorporateX network.

Traffic corresponding to each service is conveniently managed considering that each service corresponds to a Virtual LAN (VLAN), defined in such a way that allows the differentiation between services running on different buildings. So, several VLANs will be configured with the following rationale:

- Building 0 - VLAN 11 is the administrative VLAN; VLAN 12 is used for VoIP; VLAN 13 is used for VC; VLANs 14, 15 and 16 are used for Administration, Engineering and Administrative Data support, respectively;
- Building 1 - VLAN 21 is the administrative VLAN; VLAN 22 is used for VoIP; VLAN 23 is used for VC; VLANs 2, 25 and 26 are used for Administration, Engineering and Administrative Data support, respectively.

VLAN 1 is the native VLAN of all the infrastructure.

	VLAN	Private IPv4	Public IPv4	IPv6
Building 0	11	10.1.0.0/24	----------	2001:A:A:100::/64
	12	10.2.0.0/24	----------	2001:A:A:200::/64
	13	10.3.0.0/24	192.1.1.128/28	2001:A:A:300::/64
	14	10.4.0.0/24	----------	2001:A:A:400::/64
	15	10.5.0.0/24	----------	2001:A:A:500::/64
	16	10.6.0.0/24	----------	2001:A:A:600::/64
Building 1	21	10.1.1.0/24	----------	2001:A:A:101::/64
	22	10.2.1.0/24	----------	2001:A:A:201::/64
	23	10.3.1.0/24	192.1.1.136/28	2001:A:A:301::/64
	24	10.4.1.0/24	----------	2001:A:A:401::/64
	25	10.5.1.0/24	----------	2001:A:A:501::/64
	26	10.6.1.0/24	----------	2001:A:A:601::/64
Core		10.0.0.0/18	----------	2001:A:A:00::/58
DMZ		10.0.64.0/18	192.1.1.64/26	2001:A:A:40::/58
Datacenter		10.0.128.0/17	----------	2001:A:A:80::/57
NAT		----------	192.1.1.0/26	----------

Figure 1.97 IP addressing for the CorporateX network.

Since the *Site* field of the IP addressing scheme that was explained above will be zero (the company has only one site, at least for now), as well as the *Service* + field (since we will not sub-divide each considered service), the IP addresses of the IP networks corresponding to the different VLANs are the ones illustrated in Figure 1.97. The Core, DMZ and Data Center services are transverse to both buildings and will use private IP addresses. Public IP addresses are needed for the VC, DMZ and NAT services, since their networks or some of their hosts need to be accessible from the outside/Internet.

Besides the described VLANs, which are related to company services, interconnection VLANs are considered to differentiate the traffic that circulate between switches L3 1 and 2 and both corporate buildings: VLAN 101 is used to interconnect both buildings to switch L3 1, while VLAN 102 is used to interconnect both buildings to switch L3 2. All interconnection ports are access ports.

In the practical implementation of the use case in GNS3, we will only configure the access layer VLANs 11, 12 and 13. In fact, the configuration of the remaining VLANs is made exactly in the same way.

In order to define names for the different routers, a configuration similar to the one illustrated in Box 1.11 can be used.

```
Router> enable
Router# configure terminal
Router(config)# hostname Router11
Router11(config)# end
Router11# write
```

Box 1.11 Router name configuration at Router11.

2

Data Link Layer

2.1 Introduction

This chapter is dedicated to data link layer network design issues, covering topics like Virtual LANs (VLANs) and protocols that were specifically designed to avoid bridging loops: the Spanning Tree Protocol (STP), the Rapid STP and the Multiple STP. At the end of the chapter, a brief overview of the functioning principles of the IEEE 802.11 standard is also provided, reflecting the relevance that Wireless LANs have nowadays.

2.2 Virtual LANs

A Virtual LAN is an emulation of a standard LAN that allows data transfer to occur without the traditional physical constraints placed on a network. A network administrator can group users into a VLAN so they can communicate as if they were attached to the same physical segment, when in fact they are located on different LAN segments. Because VLANs are based on logical connections, they are very flexible.

The VLAN concept was created to address two main problems: the scalability of a flat network topology and the simplification of network management by making network reconfigurations (equipment moves and changes) easier. A VLAN consists of a single broadcast domain and solves the scalability problems of large flat networks by breaking a single broadcast domain into several smaller broadcast domains (or VLANs). VLANs offer easier moves and changes in a network design than traditional networks. LAN switches can be used to segment networks into logically defined virtual workgroups. This logical segmentation, commonly referred to as VLAN communication, offers a fundamental change in how LANs are designed, administered, and managed.

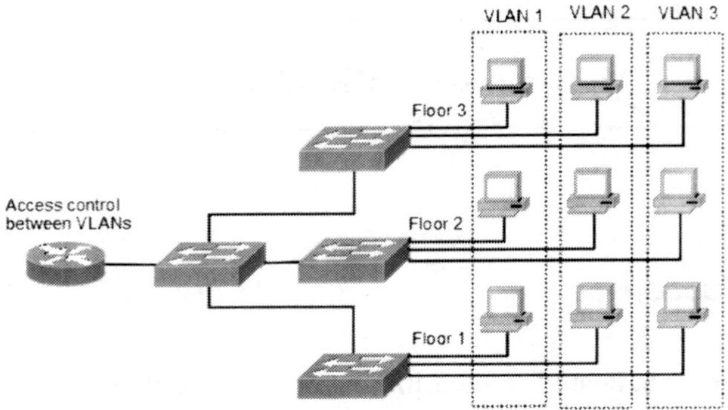

Figure 2.1 Typical network architecture to support VLANs.

Switches control individual VLANs, while routers provide inter-VLAN communication, as shown in Figure 2.1. Switches overcome the physical constraints imposed by shared-hub architectures because they are able to logically group users and ports across the network. Besides their traditional roles of firewalls, broadcast suppression, policy-based control, route processing and distribution, routers provide communication between VLANs and provide VLAN access to shared resources, such as servers and hosts.

VLAN assignment

VLAN assignment can be based on applications, protocols, performance requirements, security requirements, traffic loading characteristics, or other factors.

In practice, VLANs allow the sub-division of a large flat network into subnets, dividing broadcast domains: instead of flooding all broadcasts out every port, a VLAN-enabled switch can flood a broadcast out only the ports that are part of the same subnet as the sending station.

VLANs offer several advantages over traditional LANs:

- Increased performance - VLANs can reduce the need to send broadcast and multicast traffic to unnecessary destinations.
- Creation of Virtual Workgroups - Nowadays, it is common to find cross-functional product development teams with members from different departments such as marketing, sales, accounting, and research. These workgroups are usually formed for a short period of time and communication between members will be high. To contain broadcasts and multicasts within the workgroup, a VLAN can be set up for them.

- Simplified administration - Most of the network costs are a result of adds, moves and changes of users in the network due to re-cabling, new station addressing and reconfiguration of hubs and routers. Some of these tasks can be simplified with the use of VLANs.
- Reduced cost - VLANs can be used to create broadcast domains which eliminate the need for expensive routers.
- Security - Placing only users who can have access to sensitive data on a VLAN can reduce the chances of an outsider gaining access to that data. VLANs can also be used to control broadcast domains, set up firewalls, restrict access, and inform the network manager of an intrusion.

2.2.1 End-to-End and Local VLANs

An end-to-end VLAN is a single VLAN that is associated with switch ports widely dispersed throughout an enterprise network on multiple switches. Figure 2.2 shows an example of three end-to-end VLANs. Note that the layer 2 trunks connecting the switches of the different buildings carry the traffic of all VLANs. This kind of end-to-end VLAN model has the following main characteristics: each VLAN is dispersed geographically throughout the network; users are grouped into each VLAN regardless of their physical location; as a user moves throughout a campus, the VLAN membership of that user remains the same, regardless of the physical switch to which this user attaches; users are typically associated with a given VLAN for network management reasons; all devices on a given VLAN typically have addresses on the same IP subnet.

VLANs that are local to one access switch and trunk to the distribution switch are designated by local VLANs, being generally confined to a wiring closet as shown in Figure 2.3. If users move from one location to another in the campus, their connection changes to the new VLAN at the new physical location. In this VLAN model, layer 2 switching is implemented at the access level and routing is implemented at the distribution and core levels. Local VLANs should be created according to physical boundaries rather than to the job functions of the users on the end devices. Generally, local VLANs exist between the access and distribution levels. So, traffic from a local VLAN is routed at the distribution and core levels to reach destinations on other networks. A network that consists entirely of local VLANs can benefit from increased convergence times offered via routing protocols, instead of a spanning tree for layer 2 networks. It is usually recommended to have one to three VLANs per access layer switch.

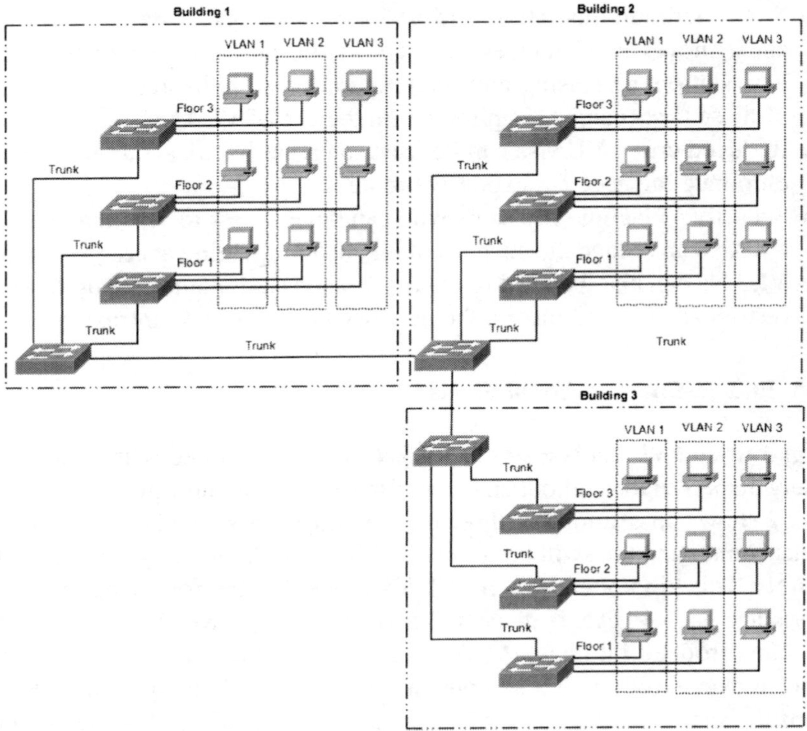

Figure 2.2 End-to-end VLANs.

We can identify several reasons to implement end-to-end VLANs:

- Users can be grouped on a common IP segment, even though they are geographically dispersed. With the recent trend towards virtualization, it is necessary to have end-to-end VLANs spread across segments of the campus.
- For security reasons, a VLAN can contain resources that should not be accessible to all users on the network, or there might be a reason to confine certain traffic to a particular VLAN.
- It is usually easier to apply quality of service (QoS) policies based on VLANs, giving a higher or lower access priority to certain network resources.
- If much of the VLAN user traffic is destined for devices on that same VLAN, and routing to those devices is not desirable, users can access re-

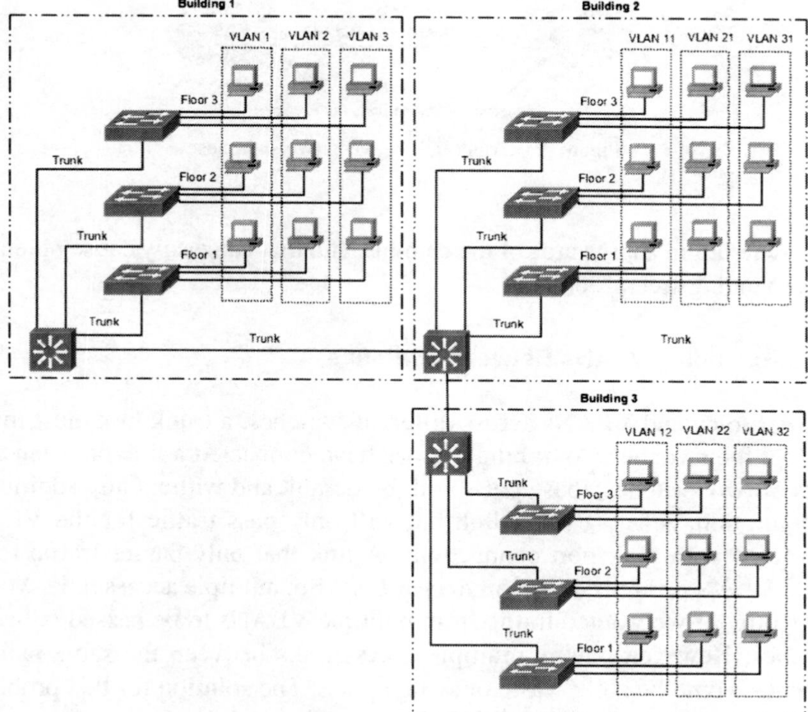

Figure 2.3 Local VLANs.

sources on their VLAN without their traffic being routed off the VLAN, even though the traffic might traverse multiple switches.

- Sometimes a special purpose VLAN is provisioned to carry a single type of traffic that must be dispersed throughout the campus (for example, multicast, voice, or visitor VLANs).

However, the network administrator should take several considerations into account when implementing end-to-end VLANs:

- Since users on an end-to-end VLAN can be anywhere in the network, all switches must be aware of that VLAN.
- Flooded traffic for the VLAN is, by default, passed to every switch even if it does not currently have any active ports in the particular end-to-end VLAN.
- Troubleshooting devices on a campus with end-to-end VLANs can be challenging because the traffic for a single VLAN can traverse multiple

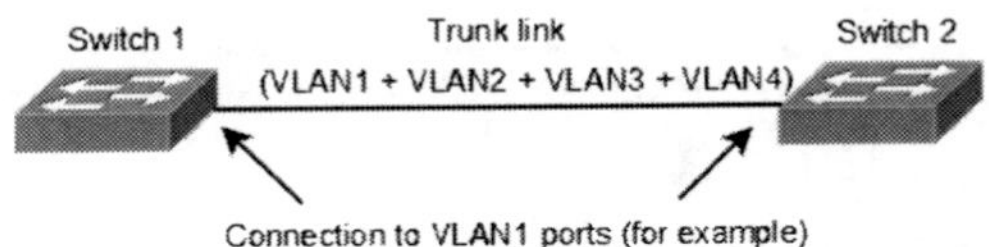

Figure 2.4 Trunk link between two switches.

switches in a large area of the campus, and that can easily cause potential spanning-tree problems.

2.2.2 Extending VLANs Between Switches

In order to extend VLANs across different switches, a trunk link must interconnect the switches. Assuming that we have connected a link between two ports of two switches, these ports will, by default and without any additional configuration, act as a trunk link but will only pass traffic for the VLAN associated with their port connections. A link that only passes traffic for a single VLAN is referred to as an *Access Link*. So, multiple access links would be required if we wanted traffic from multiple VLANs to be passed between switches. However, having multiple access links between the same pair of switches would be a big waste of switch ports. The solution for this problem comes through the use of VLAN tagging (Figure 2.4). The most popular tagging methods are the InterSwitch Link (ISL) [96] and IEEE 802.1q [33]. ISL is a Cisco proprietary VLAN tagging method, while IEEE 802.1q is an open standard.

The IEEE 802.1q tagging mechanism is based on the addition of a 4-byte overhead squeezed between the Source Address and Type/Length field of the Ethernet II frame (Figure 2.5). Obviously, the process of inserting the tag into an Ethernet II frame results in the original Frame Check Sequence (FCS) field to become invalid, since the frame was changed. Hence, it is essential to recalculate a new FCS based on the new frame that contains the IEEE 802.1q field. This process is automatically performed by the switch, right before it sends the frame down a trunk link.

The 802.1q header is only 4 bytes in length and contains all the necessary information required to successfully identify the frame's VLAN and ensure it arrived to the correct destination. The fields of this header are:

- Tag Protocol Identifier - This field is 16 bit long with a value of 0x8100. It is used to identify the frame as an IEEE 802.1q tagged frame.

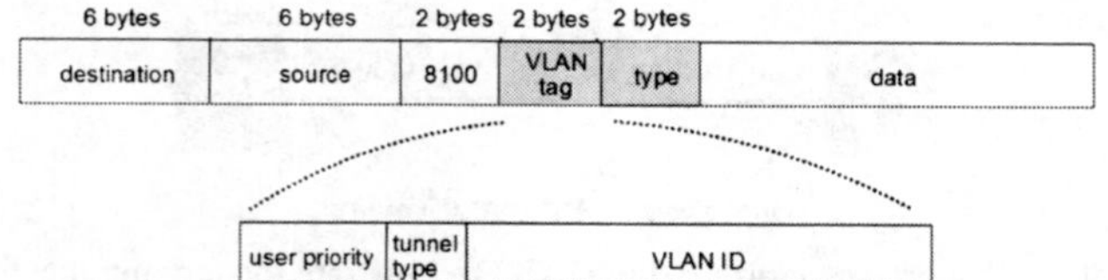

Figure 2.5 The IEEE 802.1q tagging method.

- Priority - This field is only 3 bits long and is used for prioritization of the data this frame is carrying. This field allows a total of 8 different priorities for each frame, that is, from level zero (0) to seven (7).
- Canonical Format Indicator (CFI) - This field is only 1 bit long. If set to "1", it means the MAC Address is in non-canonical format, if set to "0" means it is in a canonical format. For Ethernet switches, this field is always set to zero (0).
- Virtual Local Area Network Identifier (VLAN ID) - The VLAN ID field is perhaps the most important field out of all because it allows identifying which VLAN the frame belongs to, letting the receiving switch decide which ports the frame is allowed to exit depending on the switch configuration.

The Priority, CFI and VLAN ID fields are also known as the TCI (Tag Control Information) field.

Exercise - The IEEE 802.1q Tagging Mechanism

Let us consider the example of a multiple VLAN scenario, similar to the one illustrated in Figure 2.4. In GNS3, in order to have advanced switching features (that the simple Ethernet switch does not provide) we have to add a router with an EtherSwitch card or insert an EtherSwitch card in a slot of an already existent router. The card will function similarly to a switch. The EtherSwitch card that is supported is the NM-16ESW and it includes the following main features, among some others: layer 2 Ethernet interfaces, trunking (only according to the 802.1q standard) and the basic Spanning Tree Protocol. This scenario uses EtherSwitch cards that were inserted into router slots, as shown in Figure 2.6, resulting effectively in the use of layer three switches.

The configuration of the different VLANs on both layer 3 switches and the configuration of the IP addresses are shown in Box 2.1. Lines 1-5 define the different VLANs in both layer 3 switches. Lines 7-18 configure the

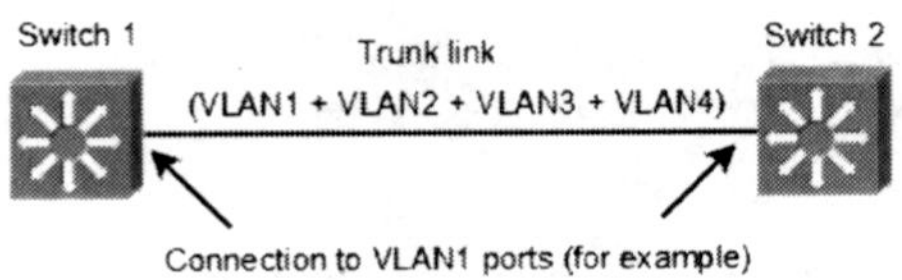

Figure 2.6 Network scenario used in GNS3 to illustrate the tagging mechanism.

No.	Time	Source	Destination	Protocol	Info
1	0.000000	10.1.1.1	10.1.1.2	ICMP	Echo (ping) request (id=0x0003, seq(be/le)=0/0, ttl=255)
2	0.002276	10.1.1.2	10.1.1.1	ICMP	Echo (ping) reply (id=0x0003, seq(be/le)=0/0, ttl=255)
3	1.296075	10.2.2.1	10.2.2.2	ICMP	Echo (ping) request (id=0x0004, seq(be/le)=0/0, ttl=255)
4	1.298261	10.2.2.2	10.2.2.1	ICMP	Echo (ping) reply (id=0x0004, seq(be/le)=0/0, ttl=255)
5	52.744554	10.3.3.1	10.3.3.2	ICMP	Echo (ping) request (id=0x0005, seq(be/le)=1/256, ttl=255)
6	52.746685	10.3.3.2	10.3.3.1	ICMP	Echo (ping) reply (id=0x0005, seq(be/le)=1/256, ttl=255)

```
▷ Frame 1: 114 bytes on wire (912 bits), 114 bytes captured (912 bits)
▷ Ethernet II, Src: c2:06:6a:68:00:00 (c2:06:6a:68:00:00), Dst: c2:07:6a:68:00:00 (c2:07:6a:68:00:00)
▷ Internet Protocol, Src: 10.1.1.1 (10.1.1.1), Dst: 10.1.1.2 (10.1.1.2)
▷ Internet Control Message Protocol
```

Figure 2.7 Packet belonging to the native VLAN.

different router interfaces as access or interswitch ports. Access ports are configured using command *switchport access <vlan vlan_number>*, while interswitch ports are configured using command *switchport mode trunk*. Lines 20-54 correspond to the configuration of the IP addresses. The *no autostate* command is used to force a port to be always up, even if it does not have any active connected terminals.

After completing the configuration, a *ping* command was executed from the IP address that was configured on each VLAN of Switch 1 to the corresponding IP address on Switch 2. All exchanged packets were captured using Wireshark. Figure 2.7 shows the Echo Request corresponding to VLAN 1, which is the native VLAN. A native VLAN is the untagged VLAN on an 802.1q trunked switchport. The native VLAN and the management VLAN can be the same, but it is always a good security practice that they are not.

Figures 2.8 and 2.9 illustrate the Echo Request and Reply packets that were exchanged in VLAN 2. As can be seen, both packets include the IEEE 802.1q tag, where a VLAN ID value of 2 was obviously set in the corresponding field.

Finally, Figure 2.10 shows the Echo Request corresponding to VLAN 3, where a VLAN ID value of 3 was set on the IEEE 802.1q tag.

2.3 Basic Transparent Switching

When a switch has no initial knowledge of the location of any end device, it must look at frames coming into each of its ports to figure out on which

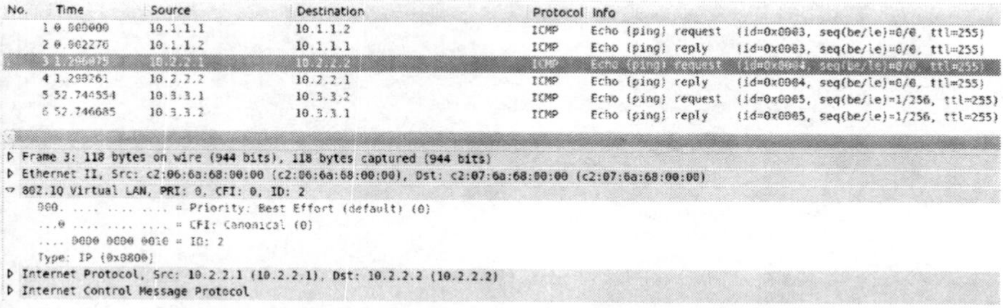

No.	Time	Source	Destination	Protocol	Info
1	0.000000	10.1.1.1	10.1.1.2	ICMP	Echo (ping) request (id=0x0003, seq(be/le)=0/0, ttl=255)
2	0.002276	10.1.1.2	10.1.1.1	ICMP	Echo (ping) reply (id=0x0003, seq(be/le)=0/0, ttl=255)
3	1.296075	10.2.2.1	10.2.2.2	ICMP	Echo (ping) request (id=0x0004, seq(be/le)=0/0, ttl=255)
4	1.298261	10.2.2.2	10.2.2.1	ICMP	Echo (ping) reply (id=0x0004, seq(be/le)=0/0, ttl=255)
5	52.744554	10.3.3.1	10.3.3.2	ICMP	Echo (ping) request (id=0x0005, seq(be/le)=1/256, ttl=255)
6	52.746685	10.3.3.2	10.3.3.1	ICMP	Echo (ping) reply (id=0x0005, seq(be/le)=1/256, ttl=255)

```
▷ Frame 3: 118 bytes on wire (944 bits), 118 bytes captured (944 bits)
▷ Ethernet II, Src: c2:06:6a:68:00:00 (c2:06:6a:68:00:00), Dst: c2:07:6a:68:00:00 (c2:07:6a:68:00:00)
▽ 802.1Q Virtual LAN, PRI: 0, CFI: 0, ID: 2
    000. .... .... .... = Priority: Best Effort (default) (0)
    ...0 .... .... .... = CFI: Canonical (0)
    .... 0000 0000 0010 = ID: 2
    Type: IP (0x0800)
▷ Internet Protocol, Src: 10.2.2.1 (10.2.2.1), Dst: 10.2.2.2 (10.2.2.2)
▷ Internet Control Message Protocol
```

Figure 2.8 Echo Request corresponding to VLAN 2.

No.	Time	Source	Destination	Protocol	Info
1	0.000000	10.1.1.1	10.1.1.2	ICMP	Echo (ping) request (id=0x0003, seq(be/le)=0/0, ttl=255)
2	0.002276	10.1.1.2	10.1.1.1	ICMP	Echo (ping) reply (id=0x0003, seq(be/le)=0/0, ttl=255)
3	1.296075	10.2.2.1	10.2.2.2	ICMP	Echo (ping) request (id=0x0004, seq(be/le)=0/0, ttl=255)
4	1.298261	10.2.2.2	10.2.2.1	ICMP	Echo (ping) reply (id=0x0004, seq(be/le)=0/0, ttl=255)
5	52.744554	10.3.3.1	10.3.3.2	ICMP	Echo (ping) request (id=0x0005, seq(be/le)=1/256, ttl=255)
6	52.746685	10.3.3.2	10.3.3.1	ICMP	Echo (ping) reply (id=0x0005, seq(be/le)=1/256, ttl=255)

```
▷ Frame 4: 118 bytes on wire (944 bits), 118 bytes captured (944 bits)
▷ Ethernet II, Src: c2:07:6a:68:00:00 (c2:07:6a:68:00:00), Dst: c2:06:6a:68:00:00 (c2:06:6a:68:00:00)
▽ 802.1Q Virtual LAN, PRI: 0, CFI: 0, ID: 2
    000. .... .... .... = Priority: Best Effort (default) (0)
    ...0 .... .... .... = CFI: Canonical (0)
    .... 0000 0000 0010 = ID: 2
    Type: IP (0x0800)
▷ Internet Protocol, Src: 10.2.2.2 (10.2.2.2), Dst: 10.2.2.1 (10.2.2.1)
▷ Internet Control Message Protocol
```

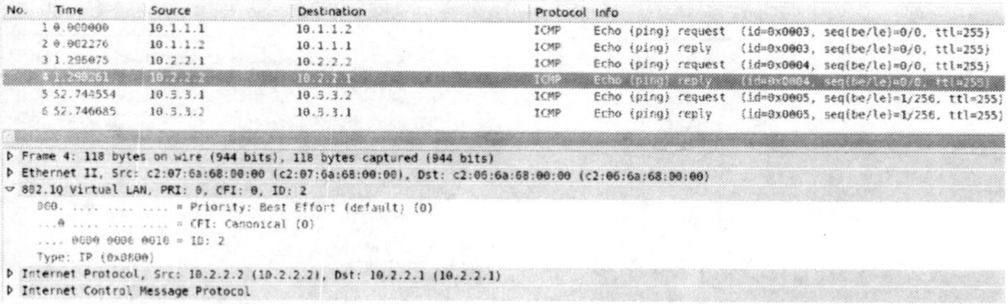

Figure 2.9 Echo Reply corresponding to VLAN 2.

network each device resides. The switch assumes that a device using the source MAC address is located behind the port that the frame arrives on. As the listening process continues, the switch builds a table that correlates source MAC addresses with the switch port numbers where they were detected. The switch can constantly update its bridging table upon detecting the presence of a new MAC address or upon detecting a MAC address that has changed location from one port to another. The switch can then forward frames by looking at the destination MAC address, looking up that address in the switch table, and sending the frame out the port where the destination device is known to be located. Box 2.2 shows an example of a MAC Address Table: for each destination MAC address that was learned, the table indicates the VLAN it belongs to, how it was learned (either dynamically or statically), the aging time of the corresponding entry and the port that should be used to forward packets to that destination.

If a frame arrives with the broadcast address as the destination address, the switch must forward, or flood, the frame out all available ports. However, the frame is not forwarded out the port that initially received the frame. In this

No.	Time	Source	Destination	Protocol	Info
1	0.000000	10.1.1.1	10.1.1.2	ICMP	Echo (ping) request (id=0x0003, seq(be/le)=0/0, ttl=255)
2	0.002276	10.1.1.2	10.1.1.1	ICMP	Echo (ping) reply (id=0x0003, seq(be/le)=0/0, ttl=255)
3	1.296075	10.2.2.1	10.2.2.2	ICMP	Echo (ping) request (id=0x0004, seq(be/le)=0/0, ttl=255)
4	1.298261	10.2.2.2	10.2.2.1	ICMP	Echo (ping) reply (id=0x0004, seq(be/le)=0/0, ttl=255)
5	52.744554	10.3.3.1	10.3.3.2	ICMP	Echo (ping) request (id=0x0005, seq(be/le)=1/256, ttl=255)
6	52.746685	10.3.3.2	10.3.3.1	ICMP	Echo (ping) reply (id=0x0005, seq(be/le)=1/256, ttl=255)

```
▷ Frame 5: 118 bytes on wire (944 bits), 118 bytes captured (944 bits)
▷ Ethernet II, Src: c2:06:6a:68:00:00 (c2:06:6a:68:00:00), Dst: c2:07:6a:68:00:00 (c2:07:6a:68:00:00)
▽ 802.1Q Virtual LAN, PRI: 0, CFI: 0, ID: 3
    000. .... .... .... = Priority: Best Effort (default) (0)
    ...0 .... .... .... = CFI: Canonical (0)
    .... 0000 0000 0011 = ID: 3
    Type: IP (0x0800)
▷ Internet Protocol, Src: 10.3.3.1 (10.3.3.1), Dst: 10.3.3.2 (10.3.3.2)
▷ Internet Control Message Protocol
```

Figure 2.10 Echo Request corresponding to VLAN 3.

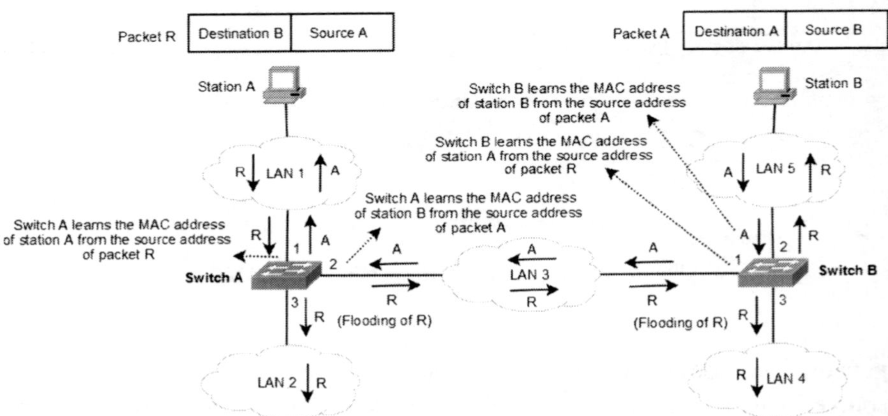

Figure 2.11 Bridge address learning process.

way, broadcasts can reach all available Layer 2 networks. A switch segments only collision domains but it does not segment broadcast domains.

If a frame arrives with a destination address that is not found in the switch table, the switch cannot determine which port to forward the frame to. In this case, the switch treats the frame as if it were a broadcast and floods it out all remaining ports. When a reply to that frame is received, the switch can learn the location of the unknown station and can add it to the switch table for future use. Frames forwarded across the switch cannot be modified by the switch itself. Therefore, the bridging process is transparent.

Figure 2.11 illustrates the basic transparent switching operations (flooding, forwarding and address learning).

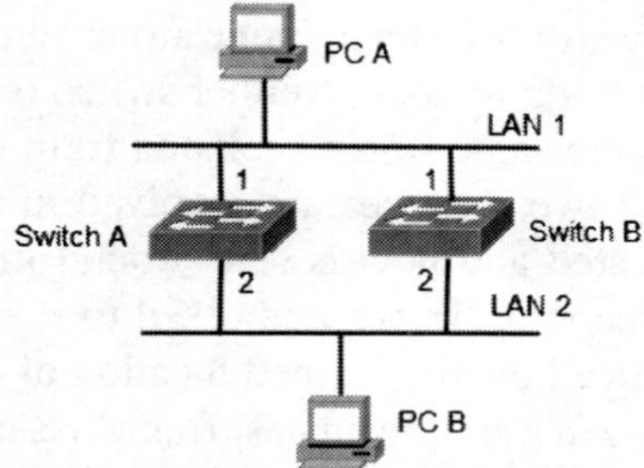

Figure 2.12 The existence of a physical bridging loop.

2.4 Bridging Loops

In the network of Figure 2.12, let us assume that neither switch A or switch B knows anything about the location of PCs A and B. The following sequence of events will take place:

- Both switches A and switch B receive the frame on their upper ports and record the PC A MAC address in its address table along with the receiving port number.
- Because the location of PC B is unknown, both switches correctly decide that they must flood the frame out all available ports.
- Each switch floods or copies the frame to its lower port on LAN 2. PC B, located on Segment B, receives the two frames destined for it. However, on LAN 2, switch A now hears the new frame forwarded by switch B, and switch B hears the new frame forwarded by switch A.
- Switch A sees that this "new" frame is from PC A to PC B. From the address table, the switch previously learned that PC A was on port 1, or LAN 1. However, the source address of PC A has just been heard on port 2, or LAN 2. By definition, the switch must relearn the location of PC A with the most recent information, which it now incorrectly assumes to be LAN 2. (Switch B follows the same procedure.)
- At this point, neither switch A nor switch B has learned the location of PC B because no frames have been received with PC B as the source address. Therefore, the new frame must be flooded out all available ports in an attempt to find PC B. This frame then is sent out switch A's 1 port and onto LAN 1, as well as switch B's 1 port and onto LAN 1.
- Now both switches relearn the location of PC A as LAN 1 and forward the "new" frames back onto LAN 2; then the entire process repeats.

This process of forwarding a single frame around and around between two switches is known as a bridging loop. Neither switch is aware of the other, so each happily forwards the same frame back and forth between its segments. Also note that because two switches are involved in the loop, the original frame has been duplicated and now is sent around in two counter-rotating loops. PC B begins receiving frames addressed to it as fast as the switches can forward them. Notice how the learned location of PC A keeps changing as frames get looped. Even a simple unicast frame has caused a bridging loop to form, and each switch's bridge table repeatedly is corrupted with incorrect data.

If, instead, PC A sent a broadcast frame, the bridging loop would be formed exactly as before. The broadcast frames continue to circulate forever. Now, however, every end-user device located on both LANs 1 and 2 receives and processes every broadcast frame. This type of broadcast storm can easily saturate the network segments and bring every host on the segments to a halt.

> **⚠ Ending bridging loops**
> The only way to end the bridging loop condition is to physically break the loop by disconnecting switch ports, shutting down a switch or configuring the Spanning Tree Protocol (STP).

2.5 The Spanning Tree Protocol

As seen in the previous section, if redundant paths or loops exist at the physical level they cannot be reflected at the logical level, otherwise the same packet will travel forever around the same path in a circle. It is always a good design practice to break up layer 2 networks with routers. However, implementing layer 2 redundancy is sometimes necessary and, in these cases, the Spanning Tree Protocol (STP) should be configured [91]. With this protocol, only one of the redundant paths is active at a time. When the single path goes down, another alternate path becomes active. STP is based on an algorithm invented by Radia Perlman while working for Digital Equipment Corporation [80, 81].

The STP enables switches to become aware of each other and negotiate a loop-free path through the network. Loops are discovered before they are made available for use and redundant links are effectively shut down to prevent the loops from forming. Besides, switches are aware that a link shut

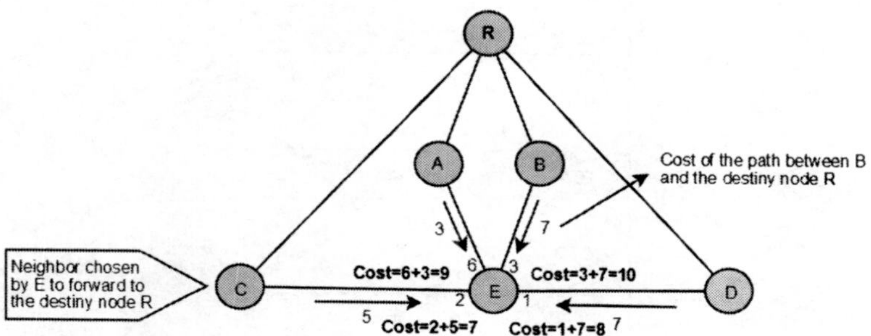

Figure 2.13 Distributed and asynchronous Bell-Ford algorithm.

down for loop prevention should be brought up quickly in case of a link failure.

Each switch executes the spanning tree algorithm based on information received from other neighboring switches. This algorithm is based on the asynchronous and distributed version of the Bellman-Ford algorithm. As can be seen in Figure 2.13, in the Bellman-Ford algorithm each node periodically transmits its estimate of the path cost between itself and the destination node. Then, each node recalculates the estimate of the path cost to the destination node by summing the estimates that were received from its neighbors to the cost of the port/link from which it has received the announcement from the neighbor. The smallest value is selected as the best neighbor to route messages to the destination node.

In a quite similar way, the Spanning Tree algorithm chooses a reference point in the network and calculates all the redundant paths to that reference point. When redundant paths are found, the Spanning Tree algorithm chooses one path by which to forward frames and disables, or blocks, forwarding on the other redundant paths. In other words, STP computes a tree structure that spans all switches in a subnet or network. Redundant paths are placed in a blocking or standby state to prevent frame forwarding. The switched network is then in a loop-free condition. However, if a forwarding port fails or becomes disconnected, the Spanning Tree algorithm recomputes the spanning tree topology so that the appropriate blocked links can be reactivated.

Data messages are exchanged in the form of Bridge Protocol Data Units (BPDU). A switch sends a BPDU frame out a port using the unique MAC address of the port itself as the source address and the well-known STP multicast address 01-80-c2-00-00-00 as the destination address. There are two types of BPDUs: configuration BPDU, used for spanning-tree computation

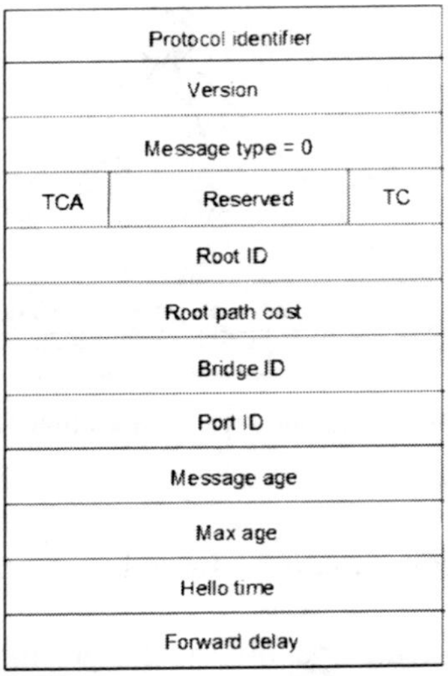

Protocol identifier		
Version		
Message type = 0		
TCA	Reserved	TC
Root ID		
Root path cost		
Bridge ID		
Port ID		
Message age		
Max age		
Hello time		
Forward delay		

Figure 2.14 Structure of the Configuration BPDU.

Protocol identifier
Version
Message type = 1

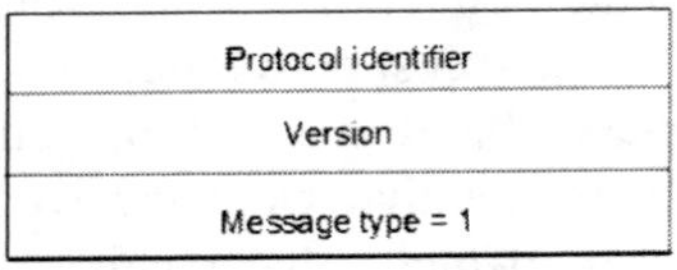

Figure 2.15 Structure of the Topology Change Notification BPDU.

and Topology Change Notification (TCN) BPDU, used to announce changes in the network topology. The Configuration BPDU message contains the fields shown in Figure 2.14, whereas the TCN BPDU is illustrated in Figure 2.15.

By default, BPDUs are sent out all switch ports every 2 seconds so that current topology information is exchanged and loops are identified quickly.

2.5.1 The Root Bridge

The Root Bridge is the common reference that allows all switches in a network to agree on a loop-free topology. The Root Bridge is chosen by an election process among all connected switches. Each switch has a unique Bridge ID that identifies it to other switches. The Bridge ID is an 8-byte value consisting of the following fields:

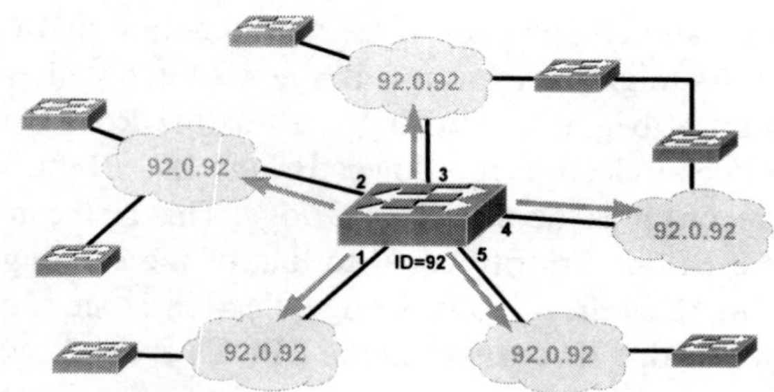

Figure 2.16 Messages sent by switch with ID equal to 92 at its start up.

- Bridge Priority (2 bytes)- is the priority or weight of a switch in relation to all other switches. The priority field can have a value of 0 to 65535 and usually defaults to 32768 (or 0x8000).
- MAC Address (6 bytes) - this address is hard-coded and unique, and the user cannot change it.

When a switch first powers up, it assumes that it is the Root Bridge itself. The election process then proceeds in the following way: every switch begins by sending out BPDUs with a Root Bridge ID equal to its own Bridge ID and a Sender Bridge ID that is its own Bridge ID and, of course, a Root Path Cost of zero. This situation is illustrated in Figure 2.16, where the notation *Root Bridge ID.Root Path Cost.Sender Bridge ID* is used in the generated BPDUs. The Sender Bridge ID simply tells other switches who is the actual sender of the BPDU message.

(8) Relevant fields of the Configuration BPDUs

Configuration BPDU messages include four relevant fields, "Root Bridge ID.Root Path Cost.Sender Bridge ID.Sender Port Number", that are used by switches to determine the highest priority message.

Received BPDU messages are analyzed to see if a better Root Bridge is being announced. A Root Bridge is considered better if the Root Bridge ID value is lower than another. Again, think of the Root Bridge ID as being broken into Bridge Priority and MAC address fields. If two Bridge Priority values are equal, the lower MAC address makes the Bridge ID better. When a switch hears of a better Root Bridge, it replaces its own Root Bridge ID with the Root Bridge ID announced in the BPDU. The switch then is required to recommend or advertise the new Root Bridge ID in its own BPDU messages, although it still identifies itself as the Sender Bridge ID. Sooner or later, the election converges and all switches agree on the notion that one of them is the Root Bridge. As might be expected, if a new switch with a lower Bridge Priority powers up, it begins advertising itself as the Root Bridge. Because the new switch does indeed have a lower Bridge ID, all the switches soon reconsider and record it as the new Root Bridge. This also can happen if the new switch has a Bridge Priority equal to that of the existing Root Bridge but has a lower MAC address. Root Bridge election is an ongoing process, triggered by Root Bridge ID changes in the BPDUs every 2 seconds.

2.5.2 The Root Path Cost

Now that a reference point has been elected for the entire switched network, each non-root switch must figure out where it is in relation to the Root Bridge. This action can be performed by selecting only one Root Port on each non-root switch. The Root Port always points towards the current Root Bridge. STP uses the concept of cost to determine many things. Selecting a Root Port involves evaluating the Root Path Cost. This value is the cumulative cost of all the links leading to the Root Bridge. A particular switch link also has a cost associated with it, called the Path Cost. To understand the difference between these values, remember that only the Root Path Cost is carried inside the BPDU. As the Root Path Cost travels along, other switches can modify its value to make it cumulative. The Path Cost, however, is not contained in the BPDU. It is known only to the local switch where the port resides.

Path Costs are defined as a 1-byte value. Generally, the higher the bandwidth of a link, the lower the cost of transporting data across it. The original IEEE 802.1D standard defined Path Cost as 1000 Mbps divided by the link bandwidth in megabits per second. Modern switched networks commonly use Gigabit Ethernet, and IEEE now uses a nonlinear scale for Path Cost, as shown in the Table 2.1.

The Root Path Cost value is determined in the following way:

- The Root Bridge sends out a BPDU with a Root Path Cost value of 0 because its ports belong to the Root Bridge.
- When the next-closest neighbor receives the BPDU, it adds the Path Cost of its own port where the BPDU arrived.
- The neighbor sends out BPDUs with this new cumulative value as the Root Path Cost.
- The Root Path Cost is incremented by the ingress port Path Cost as the BPDU is received at each switch down the line.

Table 2.1 Old and new STP Path Cost values.

Link Bandwidth	Old STP Cost	New STP Cost
4 Mbps	250	250
10 Mbps	100	100
16 Mbps	63	62
45 Mbps	22	39
100 Mbps	10	19
155 Mbps	6	14
622 Mbps	2	6
1 Gbps	1	4
10 Gbps	0	2

After incrementing the Root Path Cost, a switch also records the value in its memory. When a BPDU is received on another port and the new Root Path Cost is lower than the previously recorded value, this lower value becomes the new Root Path Cost. In addition, the lower cost tells the switch that the path to the Root Bridge must be better using this port than it was on other ports. The switch now has determined which of its ports has the best path to the Root: the Root Port.

A tree structure is beginning to emerge, but links have only been identified at this point. All links are still connected and could be active, causing bridging loops. To remove the possibility of bridging loops, STP makes a final computation to identify one Designated Port on each network segment. Suppose that two or more switches have ports connected to a single common network segment. If a frame appears on that segment, all the bridges attempt to forward it to its destination. Recall that this behavior was the basis of a bridging loop and should be avoided. Instead, only one of the links on a segment should forward traffic to and from that segment - the one that is selected as the Designated Port. Switches choose a Designated Port based on the lowest cumulative Root Path Cost to the Root Bridge. For example, a switch always has an idea of its own Root Path Cost, which it announces in its own BPDUs. If a neighboring switch on a shared LAN segment sends a BPDU announcing a lower Root Path Cost, the neighbor must have the Designated Port. If a switch learns only of higher Root Path Costs from other BPDUs received on a port, however, it then correctly assumes that its own receiving port is the Designated Port for the segment.

Figure 2.17 illustrates the best messages received so far by switch with ID equal to 92. Based on these messages, the different estimates of this switch

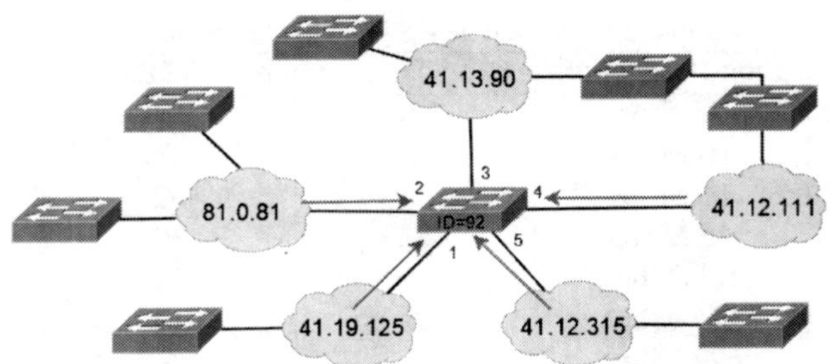

Figure 2.17 Best messages received so far by switch with ID equal to 92.

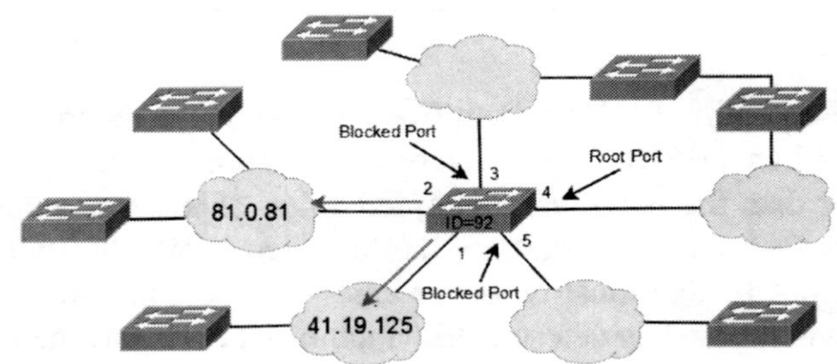

Figure 2.18 Messages sent by switch with ID equal to 92.

are: Root Bridge = 41; Root Port = 4; Root Path Cost = 12+1. So, ports 3 and 5 will be inactive, while ports 1 and 2 will be designated ports. This final port status situation is illustrated in Figure 2.18.

2.5.3 STP States

Each port of a switch must progress through several states. A port begins its life cycle in a Disabled state, moving through several passive states and, finally, into an active state if allowed to forward traffic. The STP port states are as follows:

- Disabled: ports that are administratively shut down by the network administrator, or by the system because of a fault condition, are in the Disabled state. This state is special and is not part of the normal STP progression for a port.
- Blocking: after a port initializes, it begins in the Blocking state so that no bridging loops can form. In the Blocking state, a port cannot receive or transmit data and cannot add MAC addresses to its address table. Instead, a port is allowed to receive only BPDUs so that the switch can

hear from other neighboring switches. In addition, ports that are put into standby mode to remove a bridging loop enter the Blocking state.

- Listening: a port is moved from Blocking to Listening if the switch thinks that the port can be selected as a Root Port or Designated Port. In other words, the port is on its way to begin forwarding traffic. In the Listening state, the port still cannot send or receive data frames. However, the port is allowed to receive and send BPDUs so that it actively can participate in the Spanning Tree topology process. Here, the port finally is allowed to become a Root Port or Designated Port because the switch can advertise the port by sending BPDUs to other switches. If the port loses its Root Port or Designated Port status, it returns to the Blocking state.
- Learning: after a period of time called the Forward Delay in the Listening state, the port is allowed to move into the Learning state. The port still sends and receives BPDUs as before. In addition, the switch now can learn new MAC addresses to add to its address table. This gives the port an extra period of silent participation and allows the switch to assemble at least some address table information. The port cannot yet send any data frames, however.
- Forwarding: after another Forward Delay period of time in the Learning state, the port is allowed to move into the Forwarding state. The port now can send and receive data frames, collect MAC addresses in its address table, and send and receive BPDUs. The port is now a fully functioning switch port within the spanning-tree topology.

Forwarding state condition

Remember that a switch port is allowed into the Forwarding state only if no redundant links (or loops) are detected and if the port has the best path to the Root Bridge as the Root Port or Designated Port.

Figure 2.19 represents the different STP states and the possible transitions between them.

2.5.4 STP Timers

BPDUs take a finite amount of time to travel from switch to switch. In addition, news of a topology change (such as a link or Root Bridge failure) can suffer from propagation delays as the announcement travels from one side of a network to the other. Because of the possibility of these delays, keeping the

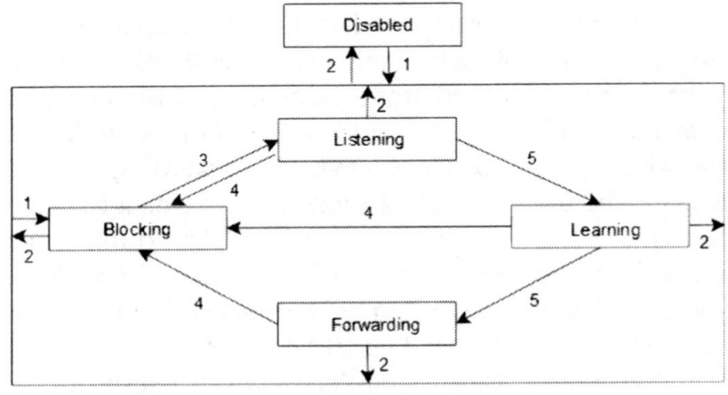
Disabled
Listening
Blocking
Learning
Forwarding
1. Port activated by management or initialization
2. Port deactivated by management or failure
3. Algorithm selects as designated or root port
4. Algorithm selects as non-designated and non-root port
5. Forwarding timer expires

Figure 2.19 Port states diagram.

spanning-tree topology from settling out or converging until all switches have had time to receive accurate information is important. STP uses three timers to make sure that a network converges properly before a bridging loop can form:

- Hello Time: the time interval between Configuration BPDUs sent by the Root Bridge. The Hello Time value configured in the Root Bridge switch determines the Hello Time for all nonroot switches because they just relay the Configuration BPDUs as they are received from the root. However, all switches have a locally configured Hello Time that is used to time TCN BPDUs when they are retransmitted. The IEEE 802.1D standard specifies a default Hello Time value of 2 seconds.
- Forward Delay: the time interval that a switch port spends in both the Listening and Learning states. The default value is 15 seconds.
- Max Age: the time interval that a switch stores a BPDU before discarding it. While executing the STP, each switch port keeps a copy of the best BPDU that it has heard. If the switch port loses contact with the BPDU source (no more BPDUs are received from it), the switch assumes that a topology change must have occurred after the Max Age time elapsed and so the BPDU is aged out. The default Max Age value is 20 seconds.

The timer values should be changed only on the Root Bridge switch. Since the timer values are advertised in fields within the BPDU, the Root Bridge ensures that they propagate to all other switches.

2.5.5 Topology Changes

To announce a change in the active network topology, switches send a TCN BPDU (Figure 2.15). A topology change occurs when a switch either moves a port into the Forwarding state or moves a port from the Forwarding or Learning states into the Blocking state. In other words, a port on an active switch comes up or goes down (action marked as 1).

Figure 2.20 illustrates the actions associated to a topology chance notification. Suppose that port 1 of switch 2 will change its state from forwarding to blocking. Switch 2 sends a TCN BPDU out its Root Port in order to notify the Root Bridge about the topology change (action 2). The switch continues sending TCN BPDUs every Hello Time interval until it gets an acknowledgment from its upstream neighbor (action 3). As the upstream neighbors receive the TCN BPDU, they propagate it on towards the Root Bridge (action 4) and send their own acknowledgments. When the Root Bridge receives the TCN BPDU, it also sends out an acknowledgment (action 5) to the bridge from where it has received the TCN BPDU. However, the Root Bridge sets the Topology Change flag in its Configuration BPDU, which is relayed to every other bridge in the network (action 6). This is done to signal the topology change and cause all other bridges to shorten their bridge table aging times from the default (300 seconds) to only the Forward Delay value (default 15 seconds). This condition causes the learned locations of MAC addresses to be flushed out much sooner than they normally would, easing the bridge table corruption that might occur because of the change in topology. However, any stations that actively are communicating during this time are kept in the bridge table. This condition lasts for the sum of the Forward Delay and the Max Age (default 15 + 20 seconds). Finally, bridge 1 also relays the Configuration BPDU with the Topology Change flag set to its downstream neighbor (action 7).

Next, we will present examples of different types of topology changes, along with the sequence of STP events. Each type has a different cause and a different effect.

2.5.5.1 Direct Topology Changes

A direct topology change is one that can be detected on a switch interface. For example, if a trunk link suddenly goes down, the switch on each end of the link immediately can detect a link failure. The absence of that link changes the bridging topology, so other switches should be notified. Figure 2.21 shows a network that has converged into a stable STP topology and the

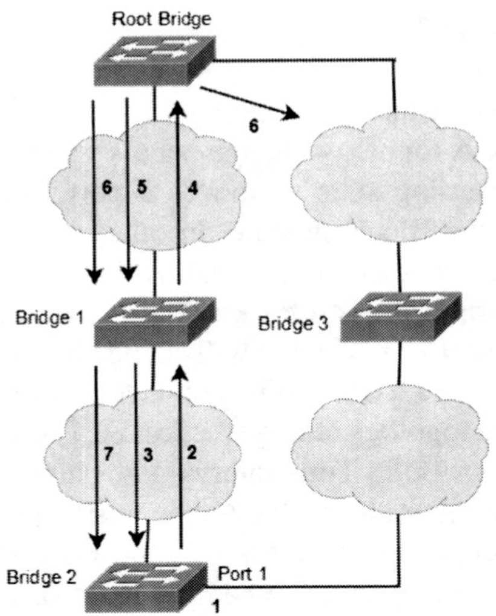

Figure 2.20 Topology change notification process.

VLAN is forwarding on all trunk links except port 2 on switch C, which is in the Blocking state.

This network has just suffered a link failure between switches A and C. The sequence of events is the following:

- Switch C detects a link down on its port 1; Switch A detects a link down on its port 2.
- Switch C removes the previous "best" BPDU it had received from the Root over port 1. Port 1 is now down so that BPDU is no longer valid. Normally, Switch C would try to send a TCN message out its Root Port, to reach the Root Bridge. However, the Root Port is broken, so that is not possible. Also, Switch A is aware of the link down condition on its own port 2.
- The Root Bridge sends a Configuration BPDU with the TCN bit set out its port 1. This is received and relayed by each switch along the way, informing each one of the topology change.
- Switches B and C receive the TCN message. The only reaction these switches take is to shorten their bridging table aging times to the Forward Delay time. At this point, they don't know how the topology has changed; they only know to force fairly recent bridging table entries to age out.

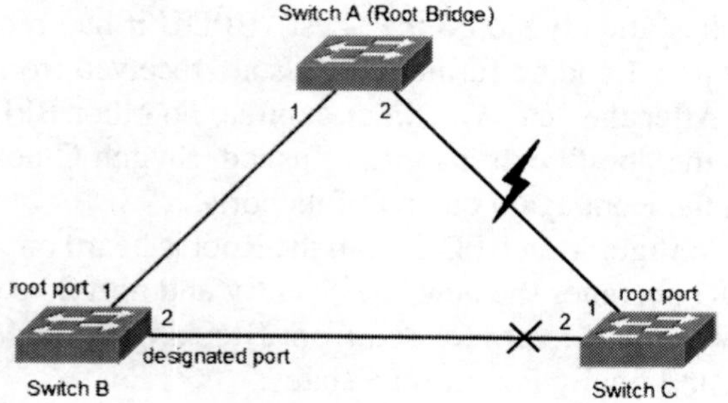

Figure 2.21 A direct topology change.

- Switch C basically just sits and waits to hear from the Root Bridge again. The BPDU TCN message is received on port 2, which was previously in the Blocking state. This BPDU becomes the "best" one received from the Root, so port 2 becomes the new Root Port. Switch C now can progress port 2 from Blocking through the Listening, Learning and Forwarding states.

As a result of a direct link failure, the topology has changed and STP has converged again. Notice that only Switch C has undergone any real effects from the failure. Switches A and B heard the news of the topology change but did not have to move any links through the STP states. In other words, the whole network did not go through a huge STP reconvergence. The total time that users on Switch C lost connectivity was roughly the time that port 2 spent in the Listening and Learning states. With the default STP timers, this amounts to about two times the Forward Delay period (15 seconds), 30 seconds total.

2.5.5.2 Indirect Topology Changes

Now, suppose that the link failure indirectly involves switches A and C. The link status at each switch stays up, but something between them has failed or is filtering traffic (for example, a firewall). STP can detect and recover from indirect failures thanks to timer mechanisms. In this case, the sequence of events is the following:

- Switches A and C both show a link up condition; data begins to be filtered elsewhere on the link.
- No link failure is detected, so no TCN messages are sent.

- Switch C has already stored the "best" BPDU it had received from the Root over port 1 and no further BPDUs are received from the Root over that port. After the MaxAge timer expires, no other BPDU is available to refresh the "best" entry, so it is flushed. Switch C now must wait to hear from the Root again on any of its ports.
- The next Configuration BPDU from the Root is heard on switch C port 2. This BPDU becomes the new "best" entry and port 2 becomes the Root Port. Now the port is progressed from Blocking through the Listening, Learning, and finally Forwarding states.

As a result of the indirect link failure, the topology does not change immediately. The absence of BPDUs from the Root causes switch C to take some action. Because this type of failure relies on STP timer activity, it generally takes longer to detect and mitigate. In this example, the total time that users on switch C lost connectivity was roughly the time until the MaxAge timer expired (20 seconds), plus the time until the next Configuration BPDU was received (2 seconds) on port 2, plus the time that port 2 spent in the Listening (15 seconds) and Learning (15 seconds) states. In other words, 52 seconds elapse if the default timer values are used.

2.5.5.3 Insignificant Topology Changes

Now, suppose that a user PC is connected to switch C, according to Figure 2.22. If the link status at the user's switch port, port number 12, goes up or down, the switch must view that as a topology change and inform the Root Bridge.

TCN messages are sent by the switch, just as if a trunk link between switches had changed state. To see what effect this has on the STP topology and the network, consider the following sequence of events:

- The PC on switch port 12 is turned off. The switch detects the link status going down.
- Switch C begins sending TCN BPDUs toward the Root, over its Root Port 1.
- The Root sends a TCN acknowledgment back to switch C and then sends a Configuration BPDU with the TC bit set to all downstream switches. This is done to inform every switch of a topology change somewhere in the network.
- The TC flag is received from the Root, and both switches B and C shorten their bridge table aging times. This causes recently idle entries to be flushed, leaving only the actively transmitting stations in the table.

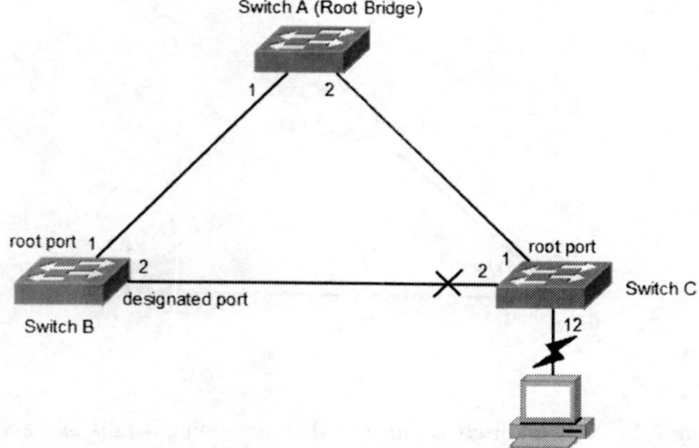

Figure 2.22 Insignificant topology change.

The aging time stays short for the duration of the Forward Delay and Max Age timers.

Notice that no actual topology change occurred because none of the switches had to change port states to reach the Root Bridge. Instead, powering off the PC caused all the switches to age out entries from their ARP tables much sooner than normal. At first, this does not seem like a major problem because the PC link state affects only the "newness" of the table contents. If the table entries are flushed as a result, they will be probably learned again. This becomes a problem when every user PC is considered. Now, every time any PC in the network powers up or down, every switch in the network must age out its table entries. Besides, remember that when a switch does not have a ARP table entry for a destination, the packet must be flooded out all its ports. Flushed tables mean more unknown unicasts, which mean more broadcasts or flooded packets throughout the network. Fortunately, some recent switches have a feature that can designate a port as a special case: the STP PortFast feature can be enabled on ports with attached PCs and, as a result, TCNs are not sent when the port changes state, and the port is brought right into the Forwarding state when the link comes up.

Exercise - Illustration of the STP

The network represented in Figure 2.23 will be used to illustrate the functioning details of the Spanning Tree Protocol. Once again, the network scenario was implemented in GNS3 using EtherSwitch cards inserted into router slots.

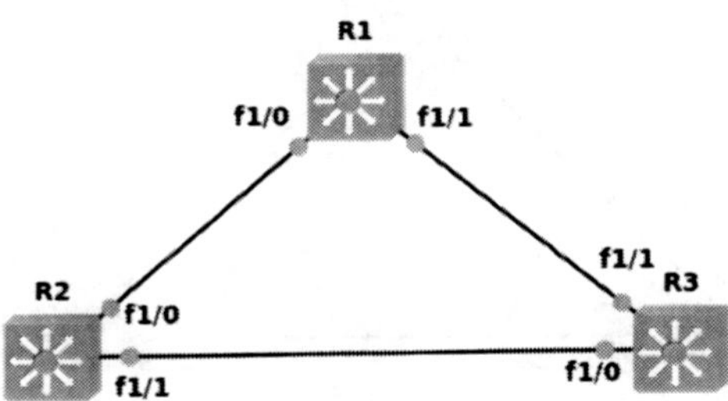

Figure 2.23 Network used to illustrate the functioning details of the STP.

No.	Time	Source	Destination	Protocol	Info
1	0.000000	c2:00:4a:02:f1:00	Spanning-tree-(for-bridges)_00	STP	Conf. Root = 32768/c2:11:11:11:00:00 Cost = 0 Port = 0x8029
2	1.975669	c2:00:4a:02:f1:00	Spanning-tree-(for-bridges)_00	STP	Conf. Root = 32768/c2:11:11:11:00:00 Cost = 0 Port = 0x8029
3	4.000687	c2:00:4a:02:f1:00	Spanning-tree-(for-bridges)_00	STP	Conf. Root = 32768/c2:11:11:11:00:00 Cost = 0 Port = 0x8029
4	5.979330	c2:00:4a:02:f1:00	Spanning-tree-(for-bridges)_00	STP	Conf. Root = 32768/c2:11:11:11:00:00 Cost = 0 Port = 0x8029
5	7.984004	c2:00:4a:02:f1:00	Spanning-tree-(for-bridges)_00	STP	Conf. Root = 32768/c2:11:11:11:00:00 Cost = 0 Port = 0x8029
6	9.978541	c2:00:4a:02:f1:00	Spanning-tree-(for-bridges)_00	STP	Conf. Root = 32768/c2:11:11:11:00:00 Cost = 0 Port = 0x8029

```
▷ Frame 1: 60 bytes on wire (480 bits), 60 bytes captured (480 bits)
▷ IEEE 802.3 Ethernet
▷ Logical-Link Control
▽ Spanning Tree Protocol
    Protocol Identifier: Spanning Tree Protocol (0x0000)
    Protocol Version Identifier: Spanning Tree (0)
    BPDU Type: Configuration (0x00)
  ▷ BPDU flags: 0x00
  ▷ Root Identifier: 32768 / c2:11:11:11:00:00
    Root Path Cost: 0
  ▷ Bridge Identifier: 32768 / c2:11:11:11:00:00
    Port identifier: 0x8029
    Message Age: 0
    Max Age: 20
    Hello Time: 2
    Forward Delay: 15
```

Figure 2.24 Configuration BPDUs captured on a segment that is connected to the Root Bridge.

After activating the STP at all switches (actually, it is active by default, so no action is necessary), Configuration BPPUs were captured using Wireshark: on a segment connected to the Root Bridge, Configuration BPDUs are sent by the Root itself (Figure 2.24); on a segment that is not connected to the Root Bridge, BPDUs are sent by the Designated Bridge, instead of the Root Bridge (Figure 2.25). From these captures, we can see that the source address of the different packets is different from the MAC address of the sending bridge. This is a consequence of using switching modules incorporated into router slots: although all bridge IDs have been configured, the source address of the different packets is the MAC address of the physical router interface.

The STP configuration on the different switches can be consulted using the *show spanning-tree* command. The result is shown in Boxes 2.3 and 2.4:

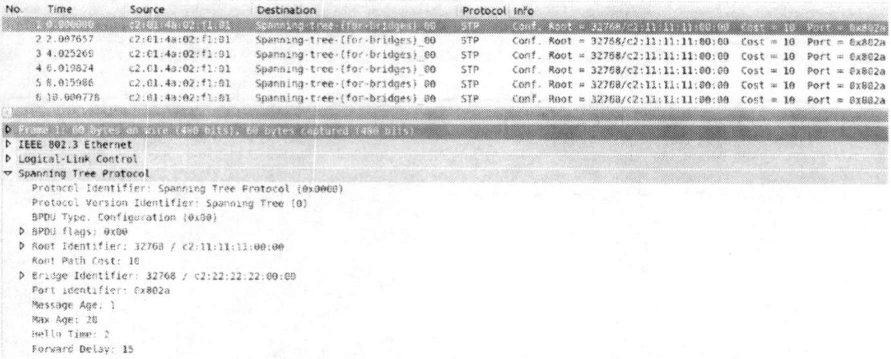

No.	Time	Source	Destination	Protocol	Info
1	0.000000	c2:01:4a:02:f1:01	Spanning-tree-(for-bridges)_00	STP	Conf. Root = 32768/c2:11:11:11:00:00 Cost = 10 Port = 0x802a
2	2.007657	c2:01:4a:02:f1:01	Spanning-tree-(for-bridges)_00	STP	Conf. Root = 32768/c2:11:11:11:00:00 Cost = 10 Port = 0x802a
3	4.025269	c2:01:4a:02:f1:01	Spanning-tree-(for-bridges)_00	STP	Conf. Root = 32768/c2:11:11:11:00:00 Cost = 10 Port = 0x802a
4	6.019824	c2:01:4a:02:f1:01	Spanning-tree-(for-bridges)_00	STP	Conf. Root = 32768/c2:11:11:11:00:00 Cost = 10 Port = 0x802a
5	8.015986	c2:01:4a:02:f1:01	Spanning-tree-(for-bridges)_00	STP	Conf. Root = 32768/c2:11:11:11:00:00 Cost = 10 Port = 0x802a
6	10.000778	c2:01:4a:02:f1:01	Spanning-tree-(for-bridges)_00	STP	Conf. Root = 32768/c2:11:11:11:00:00 Cost = 10 Port = 0x802a

```
▷ Frame 1: 60 bytes on wire (480 bits), 60 bytes captured (480 bits)
▷ IEEE 802.3 Ethernet
▷ Logical-Link Control
▽ Spanning Tree Protocol
    Protocol Identifier: Spanning Tree Protocol (0x0000)
    Protocol Version Identifier: Spanning Tree (0)
    BPDU Type: Configuration (0x00)
  ▷ BPDU flags: 0x00
  ▷ Root Identifier: 32768 / c2:11:11:11:00:00
    Root Path Cost: 10
  ▷ Bridge Identifier: 32768 / c2:22:22:22:00:00
    Port identifier: 0x802a
    Message Age: 1
    Max Age: 20
    Hello Time: 2
    Forward Delay: 15
```

Figure 2.25 Configuration BPDUs captured on a segment that is not connected to the Root Bridge.

as can be seen, switch 1 is the Root Bridge, its two ports are in the forwarding state and are Designated ports; the Root Port of switch 2 is port 41, with a Root Path Cost of 20, while port 42 is in the blocking state; the Root Port of switch 3 is port 42, with a Root Path Cost of 10, while port 41 is a Designated port and has also a Root Path Cost of 10. We can also see the Port Identifiers (Port IDs) and the different timers of the STP (in this case, the default values).

The second experiment intends to illustrate the consequences of changing the Root Bridge. A new Root Bridge can be elected by simply decreasing its Priority value. Initially, the Root Bridge has a MAC address equal to c2:11:11:11:00:00 and the Configuration BPDUs have both the TC and TCA flags unset (Figure 2.26).

Once a new Root Bridge is detected, the Root Identifier and Bridge Identifier fields are updated to c2:22:22:22:00:00 and the TC flag is set to one (Figure 2.27). This flag will remain set in the forthcoming BPDUs, during a time interval approximately equal to Max Age + Forward Delay (35 seconds, by default). After this time interval, the TC flag is again set to zero (Figure 2.28).

In the third experiment, we will change the Designated Bridge of the segment that is not connected to the Root Bridge and capture all BPDUs that circulate on a segment that is connected to the Root Bridge. Remember that the change of the Designated Bridge on the segment that is not connected to the Root Bridge was obtained by simply establishing a lower Root Path Cost in the chosen minimum cost path.

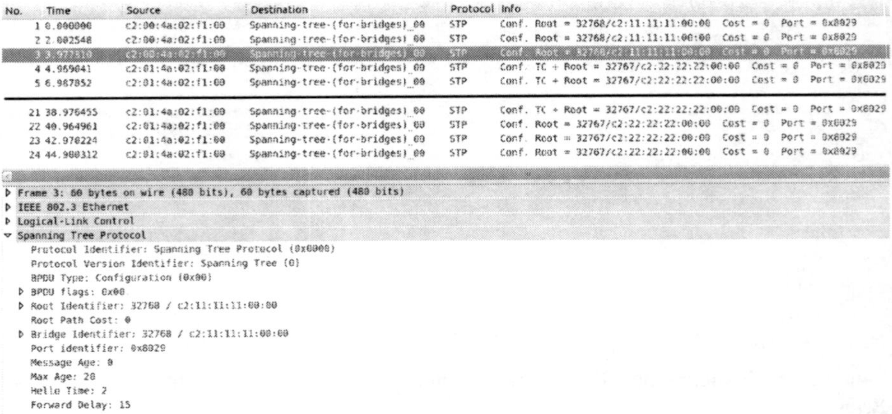

No.	Time	Source	Destination	Protocol	Info
1	0.000000	c2:00:4a:02:f1:00	Spanning-tree-(for-bridges)_00	STP	Conf. Root = 32768/c2:11:11:11:00:00 Cost = 0 Port = 0x8029
2	2.002548	c2:00:4a:02:f1:00	Spanning-tree-(for-bridges)_00	STP	Conf. Root = 32768/c2:11:11:11:00:00 Cost = 0 Port = 0x8029
3	3.977310	c2:00:4a:02:f1:00	Spanning-tree-(for-bridges)_00	STP	Conf. Root = 32768/c2:11:11:11:00:00 Cost = 0 Port = 0x8029
4	4.969041	c2:01:4a:02:f1:00	Spanning-tree-(for-bridges)_00	STP	Conf. TC + Root = 32767/c2:22:22:22:00:00 Cost = 0 Port = 0x8029
5	6.987852	c2:01:4a:02:f1:00	Spanning-tree-(for-bridges)_00	STP	Conf. TC + Root = 32767/c2:22:22:22:00:00 Cost = 0 Port = 0x8029
21	38.976455	c2:01:4a:02:f1:00	Spanning-tree-(for-bridges)_00	STP	Conf. TC + Root = 32767/c2:22:22:22:00:00 Cost = 0 Port = 0x8029
22	40.964961	c2:01:4a:02:f1:00	Spanning-tree-(for-bridges)_00	STP	Conf. Root = 32767/c2:22:22:22:00:00 Cost = 0 Port = 0x8029
23	42.978224	c2:01:4a:02:f1:00	Spanning-tree-(for-bridges)_00	STP	Conf. Root = 32767/c2:22:22:22:00:00 Cost = 0 Port = 0x8029
24	44.980312	c2:01:4a:02:f1:00	Spanning-tree-(for-bridges)_00	STP	Conf. Root = 32767/c2:22:22:22:00:00 Cost = 0 Port = 0x8029

```
▷ Frame 3: 60 bytes on wire (480 bits), 60 bytes captured (480 bits)
▷ IEEE 802.3 Ethernet
▷ Logical-Link Control
▽ Spanning Tree Protocol
    Protocol Identifier: Spanning Tree Protocol (0x0000)
    Protocol Version Identifier: Spanning Tree (0)
    BPDU Type: Configuration (0x00)
  ▷ BPDU flags: 0x00
  ▷ Root Identifier: 32768 / c2:11:11:11:00:00
    Root Path Cost: 0
  ▷ Bridge Identifier: 32768 / c2:11:11:11:00:00
    Port identifier: 0x8029
    Message Age: 0
    Max Age: 20
    Hello Time: 2
    Forward Delay: 15
```

Figure 2.26 Configuration BPDUs immediately before the change of the Root Bridge.

The first switch that changes the status of one of its ports sends a TCN BPDU towards the Root over its root port, as shown in Figure 2.29. Then, the Root Bridge sends a Configuration BPDU back to that switch, acknowledging the reception of the TCN BPDU and "telling" the switch to stop sending these messages. So, the acknowledgment Configuration BPDU has both TC and TCA flags set to one, as illustrated in Figure 2.30. Finally, the Root Bridge starts sending Configuration BPDUs with the TC bit set to all downstream switches, in order to inform them that a topology change occurred somewhere in the network. Once these packets are received from the Root, they are forwarded by the designated bridges of the different LANs (Figure 2.31), making all switches to short their bridge table aging times. This situation lasts for the duration of the Forward Delay and Max Age timers.

2.6 The Rapid Spanning Tree Protocol

The Rapid Spanning Tree Protocol (RSTP) is based on the IEEE 802.1w standard [35, 98]. There are several differences between RSTP and STP. RSTP requires full-duplex point-to-point connection between adjacent switches. STP and RSTP also have port designation differences: RSTP has the Alternate and Backup port designations, while ports not participating in the spanning tree are known as edge ports. An edge port becomes a non-edge port immediately if a BPDU is heard on the port. Non-edge ports participate in the spanning-tree algorithm; hence, only non-edge ports generate Topology

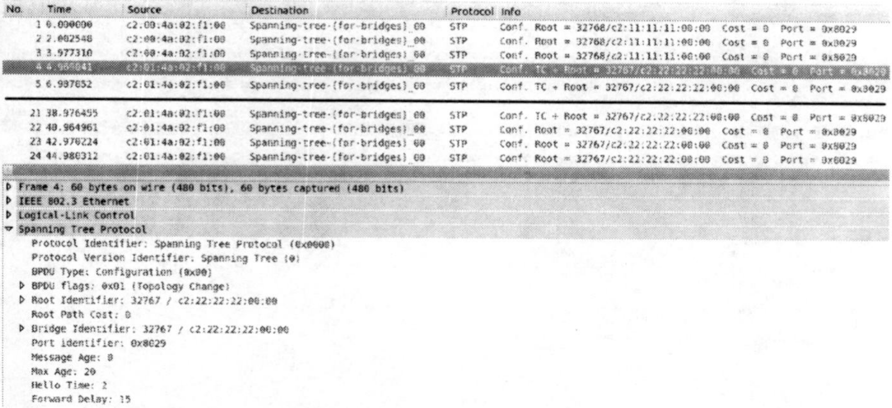

No.	Time	Source	Destination	Protocol	Info
1	0.000000	c2:00:4a:02:f1:00	Spanning-tree-(for-bridges)_00	STP	Conf. Root = 32768/c2:11:11:11:00:00 Cost = 0 Port = 0x8029
2	2.002548	c2:00:4a:02:f1:00	Spanning-tree-(for-bridges)_00	STP	Conf. Root = 32768/c2:11:11:11:00:00 Cost = 0 Port = 0x8029
3	3.977310	c2:00:4a:02:f1:00	Spanning-tree-(for-bridges)_00	STP	Conf. Root = 32768/c2:11:11:11:00:00 Cost = 0 Port = 0x8029
4	4.988841	c2:01:4a:02:f1:00	Spanning-tree-(for-bridges)_00	STP	Conf. TC + Root = 32767/c2:22:22:22:00:00 Cost = 0 Port = 0x8029
5	6.987852	c2:01:4a:02:f1:00	Spanning-tree-(for-bridges)_00	STP	Conf. TC + Root = 32767/c2:22:22:22:00:00 Cost = 0 Port = 0x8029
21	38.976455	c2:01:4a:02:f1:00	Spanning-tree-(for-bridges)_00	STP	Conf. TC + Root = 32767/c2:22:22:22:00:00 Cost = 0 Port = 0x8029
22	40.964961	c2:01:4a:02:f1:00	Spanning-tree-(for-bridges)_00	STP	Conf. Root = 32767/c2:22:22:22:00:00 Cost = 0 Port = 0x8029
23	42.970224	c2:01:4a:02:f1:00	Spanning-tree-(for-bridges)_00	STP	Conf. Root = 32767/c2:22:22:22:00:00 Cost = 0 Port = 0x8029
24	44.980312	c2:01:4a:02:f1:00	Spanning-tree-(for-bridges)_00	STP	Conf. Root = 32767/c2:22:22:22:00:00 Cost = 0 Port = 0x8029

```
▷ Frame 4: 60 bytes on wire (480 bits), 60 bytes captured (480 bits)
▷ IEEE 802.3 Ethernet
▷ Logical-Link Control
▽ Spanning Tree Protocol
    Protocol Identifier: Spanning Tree Protocol (0x0000)
    Protocol Version Identifier: Spanning Tree (0)
    BPDU Type: Configuration (0x00)
  ▷ BPDU flags: 0x01 (Topology Change)
  ▷ Root Identifier: 32767 / c2:22:22:22:00:00
    Root Path Cost: 0
  ▷ Bridge Identifier: 32767 / c2:22:22:22:00:00
    Port identifier: 0x8029
    Message Age: 0
    Max Age: 20
    Hello Time: 2
    Forward Delay: 15
```

Figure 2.27 Configuration BPDUs after the change of the Root Bridge - the TC flag is set to one.

Changes (TCs) on the network when transitioning to forwarding state only. TCs are not generated for any other RSTP states.

RSTP port designations are the following:

- Root Port (RP) - Defined as the port that is closer (in the metric point of view) to the Root. This designation is similar to the STP one.
- Alternate Port - Alternate path to get to the Root. Alternate ports do not forward traffic, so they are equivalent to a backup of the RP.
- Designated Port (DP) - Port used to forward the best BPDU on each segment.
- Backup Port - This port is a backup to the DP on the segment. It does not forward traffic.

These port designations are illustrated in Figure 2.32. The Root bridge will forward its BPDUs to both switches 2 and 3. Switch 2 and Switch 3 RPs are directly connected to the Root, so they will be receiving configuration BPDUs from the Root and will be in forwarding state. Switches 2 and 3 will be competing as to which switch will forward BPDUs on segment C. The decision process is the same for both RSTP and STP and involves: the lowest path cost to the Root, the lowest sender Bridge ID (BID) and the lowest Port ID. In this example, the lowest BID will determine which switch will be the DP for segment C because the cost to the Root is the same for both switches. Since switch 2 has a lower MAC address than switch 3, it will be forwarding on segment C. The DP is associated with switch 2 and the Backup Port on

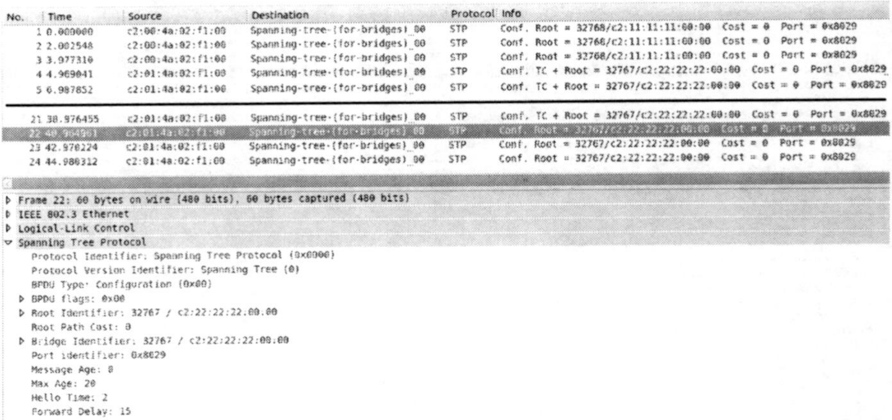

```
No.  Time        Source             Destination                    Protocol Info
 1 0.000000    c2:00:4a:02:f1:00  Spanning-tree-(for-bridges)_00  STP      Conf. Root = 32768/c2:11:11:11:00:00  Cost = 0  Port = 0x8029
 2 2.002548    c2:00:4a:02:f1:00  Spanning-tree-(for-bridges)_00  STP      Conf. Root = 32768/c2:11:11:11:00:00  Cost = 0  Port = 0x8029
 3 3.977310    c2:00:4a:02:f1:00  Spanning-tree-(for-bridges)_00  STP      Conf. Root = 32768/c2:11:11:11:00:00  Cost = 0  Port = 0x8029
 4 4.969041    c2:01:4a:02:f1:00  Spanning-tree-(for-bridges)_00  STP      Conf. TC + Root = 32767/c2:22:22:22:00:00  Cost = 0  Port = 0x8029
 5 6.987852    c2:01:4a:02:f1:00  Spanning-tree-(for-bridges)_00  STP      Conf. TC + Root = 32767/c2:22:22:22:00:00  Cost = 0  Port = 0x8029
------------------------------------------------------------------------------------------------------------------------------------------
21 38.976455   c2:01:4a:02:f1:00  Spanning-tree-(for-bridges)_00  STP      Conf. TC + Root = 32767/c2:22:22:22:00:00  Cost = 0  Port = 0x8029
22 40.984981   c2:01:4a:02:f1:00  Spanning-tree-(for-bridges)_00  STP      Conf. Root = 32767/c2:22:22:22:00:00  Cost = 0  Port = 0x8029
23 42.970224   c2:01:4a:02:f1:00  Spanning-tree-(for-bridges)_00  STP      Conf. Root = 32767/c2:22:22:22:00:00  Cost = 0  Port = 0x8029
24 44.980312   c2:01:4a:02:f1:00  Spanning-tree-(for-bridges)_00  STP      Conf. Root = 32767/c2:22:22:22:00:00  Cost = 0  Port = 0x8029

▷ Frame 22: 60 bytes on wire (480 bits), 60 bytes captured (480 bits)
▷ IEEE 802.3 Ethernet
▷ Logical-Link Control
▽ Spanning Tree Protocol
    Protocol Identifier: Spanning Tree Protocol (0x0000)
    Protocol Version Identifier: Spanning Tree (0)
    BPDU Type: Configuration (0x00)
  ▷ BPDU flags: 0x00
  ▷ Root Identifier: 32767 / c2:22:22:22:00:00
    Root Path Cost: 0
  ▷ Bridge Identifier: 32767 / c2:22:22:22:00:00
    Port identifier: 0x8029
    Message Age: 0
    Max Age: 20
    Hello Time: 2
    Forward Delay: 15
```

Figure 2.28 Configuration BPDUs after the change of the Root Bridge - the TC flag is again set to zero.

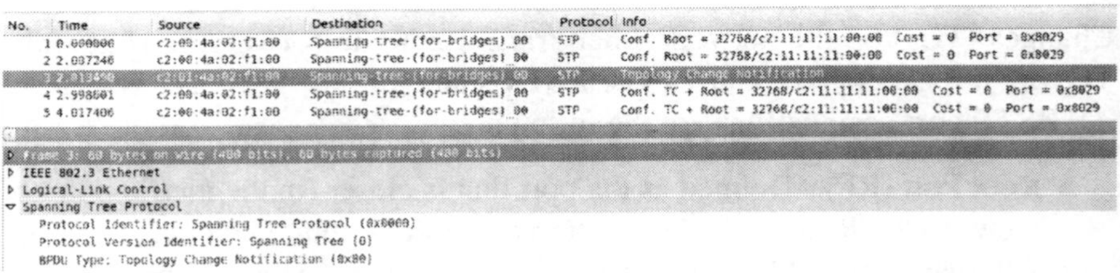

```
No.  Time       Source             Destination                    Protocol Info
 1 0.000000   c2:00:4a:02:f1:00  Spanning-tree-(for-bridges)_00  STP      Conf. Root = 32768/c2:11:11:11:00:00  Cost = 0  Port = 0x8029
 2 2.007246   c2:00:4a:02:f1:00  Spanning-tree-(for-bridges)_00  STP      Conf. Root = 32768/c2:11:11:11:00:00  Cost = 0  Port = 0x8029
 3 2.813450   c2:01:4a:02:f1:00  Spanning-tree-(for-bridges)_00  STP      Topology Change Notification
 4 2.998601   c2:00:4a:02:f1:00  Spanning-tree-(for-bridges)_00  STP      Conf. TC + Root = 32768/c2:11:11:11:00:00  Cost = 0  Port = 0x8029
 5 4.017406   c2:00:4a:02:f1:00  Spanning-tree-(for-bridges)_00  STP      Conf. TC + Root = 32768/c2:11:11:11:00:00  Cost = 0  Port = 0x8029

▷ Frame 3: 60 bytes on wire (480 bits), 60 bytes captured (480 bits)
▷ IEEE 802.3 Ethernet
▷ Logical-Link Control
▽ Spanning Tree Protocol
    Protocol Identifier: Spanning Tree Protocol (0x0000)
    Protocol Version Identifier: Spanning Tree (0)
    BPDU Type: Topology Change Notification (0x80)
```

Figure 2.29 TCN BPDU sent to the Root Bridge after a port status change in another switch.

switch 2 will be discarding. It will be backing up DP should it go down. The Alternate Port is in discarding state and will be backing up the RP on switch 3. The only forwarding port on segment C will be the DP on switch 2.

Table 2.2 shows the different port states for the RSTP and STP protocols. The three port states in RSTP are the following: Discarding, Learning, Forwarding.

The BPDU packet has also changed with RSTP. The version field in legacy STP was set at 1, but in RSTP, the version is set at 2. The Flag field in the STP BPDU packet contained TCN and TCA. In RSTP, the Flag field, 1 byte long, has been modified to accommodate port designations and proposal/agreement between adjacent switches. The different bits have the following meaning (from left to right):

- 0: Topology change
- 1: Proposal

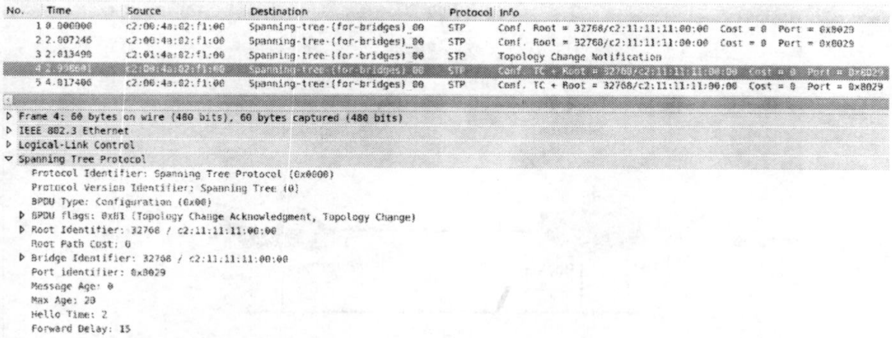

```
No. Time       Source             Destination                      Protocol Info
  1 0.000000   c2:00:4a:02:f1:00  Spanning-tree-(for-bridges)_00   STP      Conf. Root = 32768/c2:11:11:11:00:00  Cost = 0  Port = 0x8029
  2 2.007246   c2:00:4a:02:f1:00  Spanning-tree-(for-bridges)_00   STP      Conf. Root = 32768/c2:11:11:11:00:00  Cost = 0  Port = 0x8029
  3 2.013490   c2:01:4a:02:f1:00  Spanning-tree-(for-bridges)_00   STP      Topology Change Notification
  4 2.998601   c2:00:4a:02:f1:00  Spanning-tree-(for-bridges)_00   STP      Conf. TC + Root = 32768/c2:11:11:11:00:00  Cost = 0  Port = 0x8029
  5 4.017406   c2:00:4a:02:f1:00  Spanning-tree-(for-bridges)_00   STP      Conf. TC + Root = 32768/c2:11:11:11:00:00  Cost = 0  Port = 0x8029

▷ Frame 4: 60 bytes on wire (480 bits), 60 bytes captured (480 bits)
▷ IEEE 802.3 Ethernet
▷ Logical-Link Control
▽ Spanning Tree Protocol
    Protocol Identifier: Spanning Tree Protocol (0x0000)
    Protocol Version Identifier: Spanning Tree (0)
    BPDU Type: Configuration (0x00)
  ▷ BPDU flags: 0x81 (Topology Change Acknowledgment, Topology Change)
  ▷ Root Identifier: 32768 / c2:11:11:11:00:00
    Root Path Cost: 0
  ▷ Bridge Identifier: 32768 / c2:11:11:11:00:00
    Port identifier: 0x8029
    Message Age: 0
    Max Age: 20
    Hello Time: 2
    Forward Delay: 15
```

Figure 2.30 Acknowledgment sent by the Root Bridge - both TC and TCA flags are set to one.

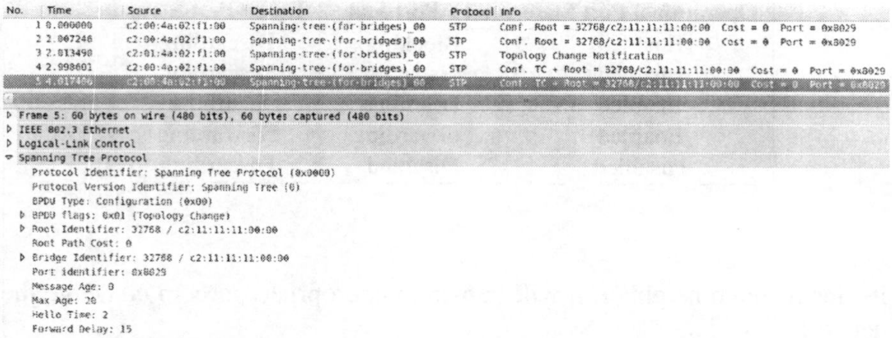

```
No. Time       Source             Destination                      Protocol Info
  1 0.000000   c2:00:4a:02:f1:00  Spanning-tree-(for-bridges)_00   STP      Conf. Root = 32768/c2:11:11:11:00:00  Cost = 0  Port = 0x8029
  2 2.007246   c2:00:4a:02:f1:00  Spanning-tree-(for-bridges)_00   STP      Conf. Root = 32768/c2:11:11:11:00:00  Cost = 0  Port = 0x8029
  3 2.013490   c2:01:4a:02:f1:00  Spanning-tree-(for-bridges)_00   STP      Topology Change Notification
  4 2.998601   c2:00:4a:02:f1:00  Spanning-tree-(for-bridges)_00   STP      Conf. TC + Root = 32768/c2:11:11:11:00:00  Cost = 0  Port = 0x8029
  5 4.017406   c2:00:4a:02:f1:00  Spanning-tree-(for-bridges)_00   STP      Conf. TC + Root = 32768/c2:11:11:11:00:00  Cost = 0  Port = 0x8029

▷ Frame 5: 60 bytes on wire (480 bits), 60 bytes captured (480 bits)
▷ IEEE 802.3 Ethernet
▷ Logical-Link Control
▽ Spanning Tree Protocol
    Protocol Identifier: Spanning Tree Protocol (0x0000)
    Protocol Version Identifier: Spanning Tree (0)
    BPDU Type: Configuration (0x00)
  ▷ BPDU flags: 0x01 (Topology Change)
  ▷ Root Identifier: 32768 / c2:11:11:11:00:00
    Root Path Cost: 0
  ▷ Bridge Identifier: 32768 / c2:11:11:11:00:00
    Port identifier: 0x8029
    Message Age: 0
    Max Age: 20
    Hello Time: 2
    Forward Delay: 15
```

Figure 2.31 Configuration BPDUs sent after the acknowledgment - the TC flag is set to one.

- 2,3: Portrole (00-Unknown; 01-Alternate/Backup; 10-Root; 11-Designated)
- 4: Learning
- 5: Forwarding
- 6: Agreement
- 7: Topology Change Acknowledgment

BPDUs are sent every 2 seconds. Unlike legacy STP, in RSTP each switch generates its own BPDUs regardless if it hears BPDUs from the Root. In RSTP mode, the switch needs only to worry about its immediate neighbors. Hence, BPDUs also serve as keepalive mechanisms between adjacent switches. If the switch does not hear three consecutive BPDUs from

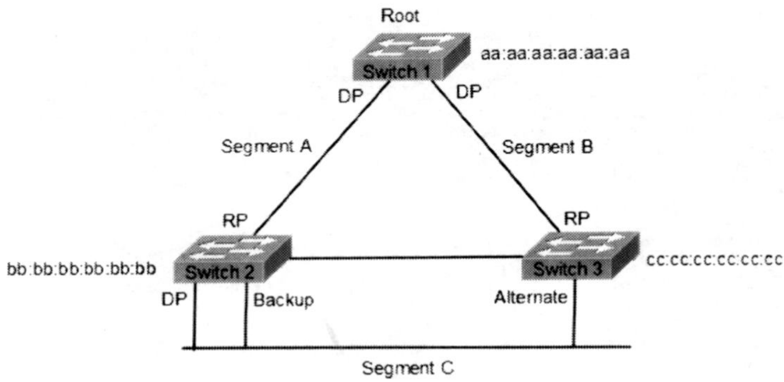

Figure 2.32 RSTP port designations.

Table 2.2 Comparing Port States between STP and RSTP.

Operational Port State	STP Port State	RSTP Port State
Enabled	Blocking	Discarding
Enabled	Listening	Discarding
Enabled	Learning	Learning
Enabled	Forwarding	Forwarding
Disabled	Disabled	Discarding

its downstream neighbor, it will transition appropriate ports to converge the network.

RSTP switches require BPDUs from their connected neighbors to keep the link up. The mechanism involved in proposal/agreement between adjacent switches is very fast. It takes less than few seconds to transition a port to the appropriate state, whereas in STP it took a minimum of 30 seconds. In Figure 2.33, BPDU exchange between switches 1 and 2 has not yet taken place. Proposal BPDUs will only be sent in the discarding and learning states. Assume that the ports connecting the two switches are in the learning state: switch 1, with lower Bridge ID, sends a proposal BPDU to switch 2. Switch 2, having received the proposal, sees that switch 1 has better BPDU; it will accept switch 1 as the Root for the VLAN. Switch 2 will send an agreement BPDU back to switch 1. In a situation where switch 1 does not receive an agreement BPDU, it will fall back to legacy STP mode.

Now, suppose that in the network of Figure 2.34 a new connection has been set up between switch 1 and switch 3. When this connection comes up, switch 3 will receive a better BPDU from switch 1. It must, therefore,

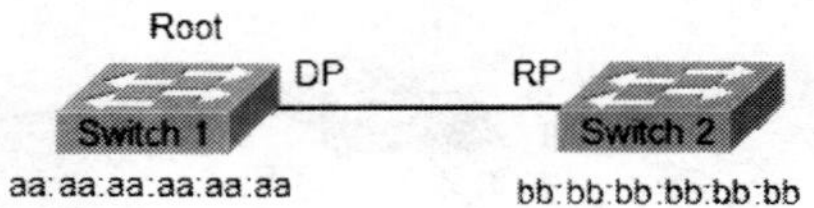

Figure 2.33 RSTP proposal/agreement.

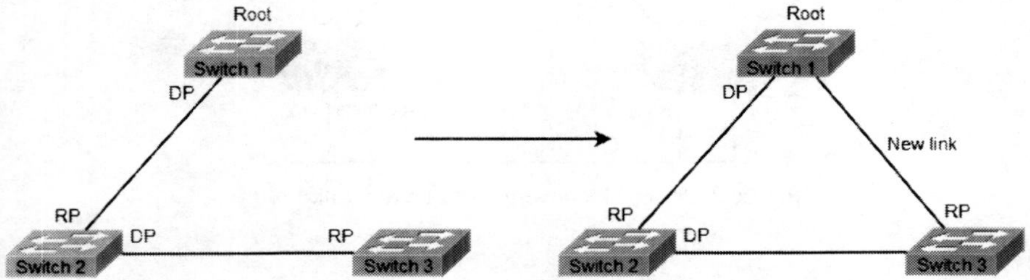

Figure 2.34 A New Connection Between Switches.

transition its current RP and designate a new RP. The following steps outline how switch 3 chooses a new RP:

1. A new connection between switch 1 and switch 3 has been set up.
2. Switch 3 receives a better BPDU from switch 1. It keeps the new port in blocking state.
3. Switch 3 transitions the current RP to Alternate port (discarding state).
4. Switch 3 sends an agreement BPDU to switch 1.
5. Switch 3 transitions the new RP to forwarding state.
6. Switch 1 receives the agreement BPDU and transitions its port to forwarding state as well.

RSTP is backward compatible with legacy STP. Typically, RSTP generates BPDUs every 2 seconds on the wire informing the segment of which bridge is the Root. Based on this information, the downstream bridges will appropriately designate their spanning-tree ports. The problem is that legacy STP devices do not understand these BPDU messages that they are receiving from RSTP-enabled devices. As a result, legacy STP devices will continue to generate their own BPDUs and ignore the RSTP BPDUs. Legacy STP BPDUs may have wrong information as to which bridge is the Root in the network. It becomes crucial for RSTP implementation to be backward compatible to prevent such a scenario.

RSTP handles the legacy STP device problem through a timer. The RSTP device will wait 4 seconds (2*Hello Interval) to see if legacy STP BPDUs will cease. If they do not, RSTP will fall back to legacy STP mode. Now, all

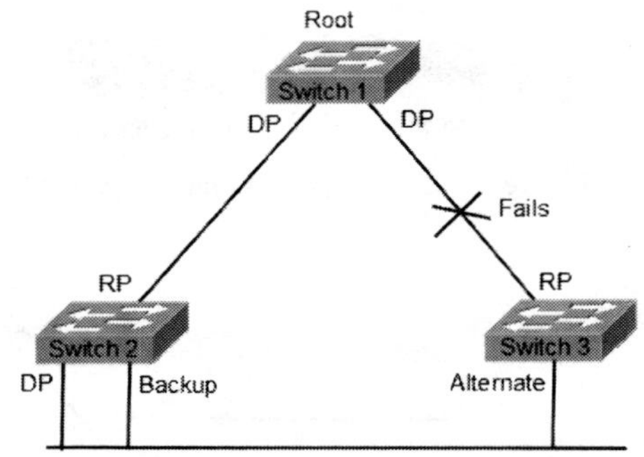

Figure 2.35 RSTP convergence due to a link failure.

the bridges will communicate via legacy STP specifications. Hence, the fast convergence and relative stability gained through RSTP is lost.

2.6.1 RSTP Direct and Indirect Failures

In Figure 2.35, when switch 3 loses its RP, it immediately transitions the Alternate port into forwarding mode. In legacy STP mode, a direct failure of this type would have taken 30 seconds. The TC BPDUs generated by RSTP to the upstream switch clear the appropriate ARP entries associated with the broken link.

When a TC bit is set, the switch starts a TC While timer equal to 4 seconds (2*Hello Interval) for all its non-edge ports. It flushes the MAC addresses that were associated with that port. The upstream switch that received the TC BPDU will flush its MAC addresses from all ports except the port that received the BPDU. This process accelerates and simplifies the convergence process. In legacy STP, the TCNs first needed to be propagated to the Root, which afterwards generated configuration BPDUs that were propagated back to the spanning-tree domain. The amount of time it took to converge the network was dependent on how big the spanning-tree domain was. In RSTP, the TCs are flooded quickly to non-edge ports and RPs, and the upstream switches flush their Content Address Memory entries, resulting in faster convergence time. The downside to this process is that some flooding does take place in the network.

When an indirect type of failure occurs (suppose that the link between switches 1 and 2 of Figure 2.21 goes down), RSTP takes an active role in bringing the network into convergence rather than passively waiting for timers to expire before transitioning a non-edge port. The following steps outline how RSTP handles indirect link failures:

1. Switch B loses its RP connection to the Root. As a result, switch B starts to generate BPDUs, informing other switches that it is the Root.
2. Switch C receives the inferior BPDUs from Switch B. Switch C knows through periodic BPDUs that it still has a connection to the Root.
3. Switch C sends BPDUs informing switch B that switch A is still the Root.
4. Upon receiving the superior BPDUs, switch B stops sending BPDUs. It transitions its DP to RP.

2.7 The Multiple Spanning Tree Protocol

The Multiple Spanning Tree Protocol (MSTP) is defined in the IEEE 802.1s standard [34, 97], is compatible with previous standards and was developed to address the limitations of the STP and RSTP protocols when dealing with multiple VLANs with topologies employing alternative physical links. Instead of running an STP instance for every VLAN, MSTP runs a number of VLAN-independent STP instances (representing logical topologies). The network administrator should map each VLAN to the most appropriate logical topology (STP instance). The number of STP instances is kept to minimum, but the network capacity is utilized in a more optimal way by using all possible paths for VLAN traffic.

The VLAN traffic forwarding logic is slightly different from the previously presented spanning tree protocols. In order for a frame to be forwarded out of a port, the VLAN must be active on this port and the STP instance the VLAN maps to must be in non-discarding state for this port. Note that due to multiple logical topologies active on a port, the port could be blocking for one instance and forwarding for another.

2.7.1 MSTP Regions and Instances

A MSTP region is a collection of switches that share the same view of the physical topology partitioning into a set of logical topologies. For two switches to become members of the same region, they must share the following attributes: configuration name, configuration revision number (a 16 bit value) and the table of 4096 elements that maps VLANs to STP instance numbers.

The IEEE 802.1s implementation does not send BDPUs for every active STP instance separately, nor does it encapsulate VLAN numbers list configuration messages. Instead, a special STP instance number 0, called Internal Spanning Tree (IST), is designated to carry all STP related information. The BPDUs for the IST contain all standard RSTP-style information for the IST itself, as well as additional information fields, like the configuration name, revision number and a hash value calculated over VLANs to MST instances mapping table contents. Using just this condensed information switches may detect mis-configuration in VLAN mappings by comparing the hash value received from the peer with the local value.

By default, all VLANs are mapped to the IST. This represents the case of classic IEEE RSTP with all VLANs sharing the same spanning tree. Other MSTP Instances (MSTIs) can be enabled, assigning their own priorities to the switches and using their own link costs to come up with a private logical topology, separate from the IST. Since MSTP does not send MSTI information in separate BPDUs, this information is piggybacked into the IST BPDUs using special M-Record fields (one for every active MSTI).

2.7.2 MSTI Tree Construction

Every switch sends its own configuration BPDUs, with a periodicity equal to Hello seconds. BDPUs have full information about the IST and carry M-Records for every active MSTI. Using the RSTP convergence mechanics, separate STP instances are constructed for the IST and every MSTI. Fundamental STP timers, such as Hello, ForwardTime and MaxAge could only be configured for the IST, with all other instances (MSTIs) inheriting the timers from it.

MSTP has a special mechanism to age out old information out of the domain. The IST BDPU has a special field called Remaining Hops. The IST root sends BPDUs with hop count equal to MaxHops (configurable value) and every downstream switch decrements the hop count field on reception of the IST BPDU. As soon as the hop count becomes zero, the information in the BPDU is ignored, and the switch may start declaring itself as a new IST root. The well known MaxAge and ForwardDelay timers are still used when MSTP interacts with RSTP or STP bridges.

As an example of a MSTI construction, consider the physical topology of three switches represented in Figure 2.36. Suppose that VLANs 1, 10, 20, 30, 40, 50 and 60 are supported and we have the following design objectives: VLANs 10, 20 and 30 should follow the link from switch 3 to switch 1;

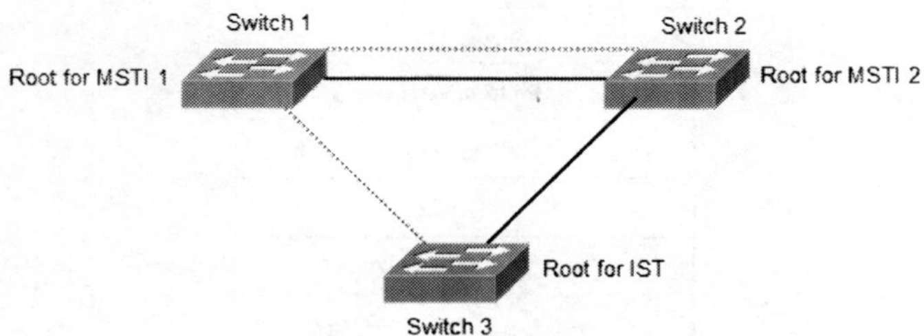

Figure 2.36 Example of a MSTP topology.

VLANs 40, 50 and 60 should follow the link from switch 3 to switch 2 and if one of these links fail, the affected VLANs should fall-back to the other link. To accomplish these goals, two MSTIs (number 1 and 2) are created: switch 1 will be the root for instance 1 and switch 2 will be the root for instance 2. Switch 3 will be the root switch for the IST (MSTI0). VLAN 1 will remain mapped to the IST, VLANs 10, 20 and 30 would map to MSTI1, while VLANs 40, 50 and 60 would map to MSTI2.

2.7.3 Common and Internal Spanning Tree (CIST)

As already explained, every MSTP region runs a special instance of spanning-tree known as Internal Spanning Tree (IST) or MSTI0. This instance mainly serves the purpose of disseminating STP topology information for MSTIs. IST has a root bridge, elected based on the lowest Bridge ID. The situation changes with multiple MSTP regions in the network. When a switch detects BPDU messages sourced from another region, it marks the corresponding port as a MSTP boundary. A switch that has boundary ports is known as a boundary switch.

When multiple regions connect together, every region needs to construct its own IST and all regions should build one common CIST spanning across the regions. Figure 2.37 shows the structure of the MSTP BPDU, which contains two important information blocks. One, the upper highlighted region, is related to CIST Root and CIST Regional Root election. The CIST Root is elected among all regions and CIST Regional Root is elected in every region. The lower highlighted block outlines the information about the CIST Regional Root (which becomes the IST Root in the presence of multiple regions). The CIST Internal Root path cost is the intra-region cost to reach the CIST Regional Root.

Protocol ID
Protocol Version
BPDU Type
CIST Flags
CIST Root ID
CIST External Root Path Cost
CIST Regional Root ID
CIST Port ID
Message Age
Max Age
Hello Time
Forward Delay
MIST Configuration ID
CIST Internal Root Path Cost
CIST Bridge ID
CIST Remaining Hops
MIST Configuration Messages aka M- Records

Figure 2.37 The MSTP BPDU format.

Note that the IST Root is the same as the CIST Regional Root in case where multiple regions interoperate. The CIST Root is the bridge that has the lowest Bridge ID among all regions: this could be a bridge inside a region or a boundary switch in a region. The CIST Regional Root is a boundary switch elected for every region based on the shortest external path cost to reach the CIST Root. Path cost is calculated based on costs of the links connecting the regions, excluding the internal regional paths. CIST Regional Root becomes the root of the IST for the given region as well.

2.7.4 CIST Root Bridges Election Process

When a switch boots up, it declares itself as CIST Root and CIST Regional Root and announces this fact in outgoing BPDUs. The switch will adjust its decision upon reception of better information and continue advertising the best known CIST Root and CIST Regional Root on all internal ports. On the boundary ports, the switch advertises only the CIST Root Bridge ID and CIST External Root Path Cost, thus hiding the details of the region's internal topology. The CIST External Root Path Cost is the cost to reach the CIST Root across the links connecting the boundary ports, that is, the inter-region links. When a BPDU is received on an internal port, this cost is not changed. When a BPDU is received on a boundary port, this cost is adjusted based on the receiving boundary port cost. In result, the CIST External Root Path Cost is propagated unmodified inside any region. Only a boundary switch could be elected as the CIST Regional Root, and this is the switch with the lowest cost to reach the CIST Root. If a boundary switch hears better CIST External Root Path cost received on its internal link, it will abandon its role of CIST Regional Root and start announcing the new metric out of its boundary ports. Every boundary switch needs to properly block its boundary ports. If the switch is a CIST Regional Root, it elects one of the boundary ports as the "CIST Root port" and blocks all other boundary ports. If a boundary switch is not the CIST Regional Root, it will mark the boundary ports as CIST Designated or Alternate. The boundary port on a non regional-root bridge becomes designated only if it has superior information for the CIST Root: a better External Root Path cost or, if costs are equal, a better CIST Regional Root Bridge ID. This follows the normal rules of the STP process.

As a result of CIST construction, every region will have one switch having a single port unblocked in the direction of the CIST Root. This switch is the CIST Regional Root. All boundary switches will advertise the region's CIST Regional Root Bridge ID out of their non-blocking boundary ports. From the outside perspective, the whole region will look like a single virtual bridge with the Bridge ID equal to the CIST Regional Root ID and single root port elected on the CIST Regional Root switch. The region that contains the CIST Root will have all boundary ports unblocked and marked as CIST designated ports. Effectively, the region would look like a virtual root bridge with the Bridge ID equal to CIST Root and all ports being designated. Notice that the region with CIST Root has CIST Regional Root equal to CIST Root as they share the same lowest bridge priority value across all regions.

Figure 2.38 illustrates this election process. First, SW1-1 is elected as the CIST Root as it has the lowest Bridge ID among all bridges in all regions. This automatically makes region 1 a virtual bridge with all boundary ports unblocked. Next, SW2-1 and SW3-1 are elected as the CIST Regional Roots in their respective regions. Notice that SW3-2 and SW2-3 have equal External Costs to reach the CIST Root but SW3-2 wins the CIST Regional Root role due to lower priority.

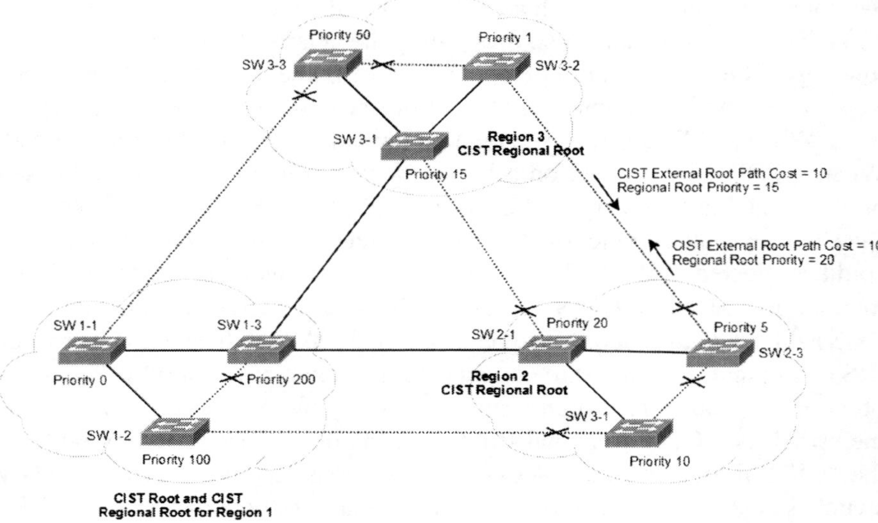

Figure 2.38 CIST topology calculation.

> **IST root election**
>
> Keep in mind that in the topology with multiple MSTP regions, every region that does not contain the CIST Root has to change the IST Root election process and make IST Root equal to CIST Regional Root.

CIST has essentially a two-level hierarchy organization. The first level treats all regions as "virtual bridges" and operates with the External Root Path Cost. The first-level spanning tree roots in the CIST Root Bridge and encompasses the virtual bridges. This spanning-tree is known as Common Spanning Tree (CST). The CST connects all boundary ports and perceives every region as a single virtual bridge with the Bridge ID equal to CIST Regional Root Bridge ID. CST is also the construct where MSTP inter-operates with the IEEE STP/RSTP regions as well. The legacy switch regions join their STP instance with the CST and perceive MSTP regions as "transparent" virtual bridges, staying unaware of their internal topology.

The second level of the CIST hierarchy consists of the various MSTP regional ISTs. Every MSTP region builds an IST instance using the internal path costs and following the optimal "internal" topology, using the CIST Regional Root as the IST Root. The changes to CST may affect the ISTs in every region, as those changes may result in the re-election of new CIST

Regional Roots. Changes to the regions internal topologies normally do not affect the CST, unless those changes partition the region.

2.7.5 Mapping MSTIs to CIST

MSTIs are constructed independently in every region, but they have to be mapped to the CIST at the boundary ports. This means inability to load-balance VLAN traffic on the boundary links by mapping VLANs to different instances. All VLANs use the same non-blocking boundary ports, which are either upstream or downstream with respect to the CIST Root. This is only valid with respect to the CST paths connecting the regional virtual bridges. Inside any region, VLANs follow the internal topology paths, based on the respective MSTI configurations.

MSTIs have no idea of the CIST Root: they only use internal paths and internal MSTI root to build the spanning trees. However, all MSTP instances see the root port (towards the CIST Root) of the CIST Regional Bridge as a special Master Port connecting them to the CIST Root bridge. This port serves the purpose of the "gateway" linking MSTIs to other regions. Recall that switches do not send M-Records (MSTI information) out of boundary ports, only CIST information. Thus, the CIST and MSTIs may converge independently and in parallel. The Master Port will only begin forwarding when all respective MSTI ports are in sync and forwarding to avoid temporary bridging loops.

2.7.6 MSTP Multi-Region Design Considerations

There are tree main problems with Ethernet that affect MSTP designs:

- Unknown unicast flooding results in traffic peaks under topology changes. Those are either result of asymmetric routing or persistent topology changes. Every topology change causes massive invalidation of MAC address tables and unicast traffic flooding. This process is the result of Ethernet topology unawareness, that is, the bridges don't know MAC addresses location.
- Broadcast and Multicast flooding. Since many core protocols rely on multicasting or broadcasting, a network under intense load can be congested at every point.
- Spanning-Tree Convergence. MSTP uses the RSTP procedure for STP re-negotiation. Since it is based on distance-vector behavior, it is prone to some convergence issues, such as counting to infinity. This is espe-

cially noticeable in larger topologies and under special conditions, such as the failure of the root bridge.

The concept of MSTP region allows for bounding STP re-computations. Since MSTIs in every region are independent, any change affecting MSTI in one region will not affect MSTIs in other regions. This is a direct result of the fact that M-Record information is not exchanged between regions. However, the CIST recalculations affect every region and might be slow converging. This is why it is a good idea not to map any VLAN to CIST and avoid connecting MSTP regions to STP domains.

Topology changes in MSTP are treated the same way as in RSTP. That is, only non-edge links going to forwarding state will cause a topology change and the switch detecting the change will flood this information through the domain. However, a single physical link may be forwarding for one MSTI and blocking for another. Thus, a single physical change may have different effect on MSTIs and the CIST. Topology changes in MSTIs are bounded to a single region, while topology changes to the CIST propagate through all regions. Every region treats the TC notification from another region as "external" and applies them to CIST-associated ports only.

A topology change to CST (the tree connecting the virtual bridges) will affect all MSTIs in all regions and the CIST. This is due to the fact that any new link that becomes forwarding between the virtual bridges may change all paths in the topology and thus require massive MAC address re-learning. Thus, from the standpoint of topology change, something happening to the CST will have most massive impact of flooding in the set of interconnected MSTP regions.

A good design rule for MSTP networks is to separate "meshy" topologies in their own regions and interconnect regions using "sparse" mesh, keeping in mind a balance between redundancy and topology changes effect. Besides, exposing a lot of links to CST will reduce load-balancing choices because CST supports only one STP instance.

Even though region partitioning offers better fault isolation, it still does not eliminate unicast and broadcast flooding. For example, unicast flooding could be caused by unidirectional traffic and broadcast flooding may be a result of transient bridging loops when a root bridge fails.

2.7.7 Wireless LANs

IEEE released the 802.11 Wireless Local Area Network (WLAN) standard in 1997 [52]. This standard defines MAC and physical layers specifications for wireless LANs. Three different physical layer specifications were defined: Frequency Hopping Spread Spectrum (FHSS), Direct Sequence Spread Spectrum (DSSS) and Infrared (IR), with the maximum data transmission rate of up to 2 Mbps. During the years, while the original MAC remained intact, the technology continued evolving with new physical layer specifications: 802.11b [54] and 802.11a [53] with data transmission rates of up to 11 and 54 Mbps, respectively, 802.11g [55] that extended the 802.11b physical layer to support data transmission rates of up to 54 Mbps, among others.

IEEE 802.11 gained a huge popularity due to its cost effectiveness and easy deployment. Today, IEEE 802.11 hotspots (the area where IEEE 802.11 access is available is known as an hotspot) are available at offices, campuses, airports, hotels, public transport stations and residential places, making it one of the most widely deployed wireless network technologies in the world.

2.7.7.1 The IEEE 802.11 Architecture

IEEE 802.11 defines two different architectures, BSS (Basic Service Set) and IBSS (Independent Basic Service Set). In a Basic Service Set, wireless stations (STA) are associated to an AP (Access Point). All communications take place through the AP. In an Independent Basic Service Set, wireless stations can communicate directly to each other, providing that they are within each other's transmission range. This form of architecture is facilitated to form a wireless ad-hoc network in absence of any network infrastructure. Several BSS can be connected together via a Distribution System (DS) to form an extended network, called Extended Service Set (ESS). So, an ESS is a set of two or more wireless APs connected to the same wired network that defines a single logical network segment bounded by a router (also known as a subnet). Figure 2.39 illustrates the architecture of IEEE 802.11 network.

In both infrastructure and ad-hoc operating modes, a Service Set Identifier (SSID), also known as the wireless network name, identifies the wireless network. The SSID is a name configured on the wireless AP (for infrastructure mode) or an initial wireless client (for ad hoc mode) that identifies the wireless network. The SSID is periodically advertised by the wireless AP or the initial wireless client using a special 802.11 MAC management frame known as a beacon frame.

So, in ad hoc mode wireless clients communicate directly with each other without the use of a wireless AP. This mode is also called peer-to-peer mode, where wireless clients form an IBSS. One of the wireless clients, the first one in the IBSS, takes over some of the responsibilities of the wireless AP. These responsibilities include the periodic beaconing process and the authentication

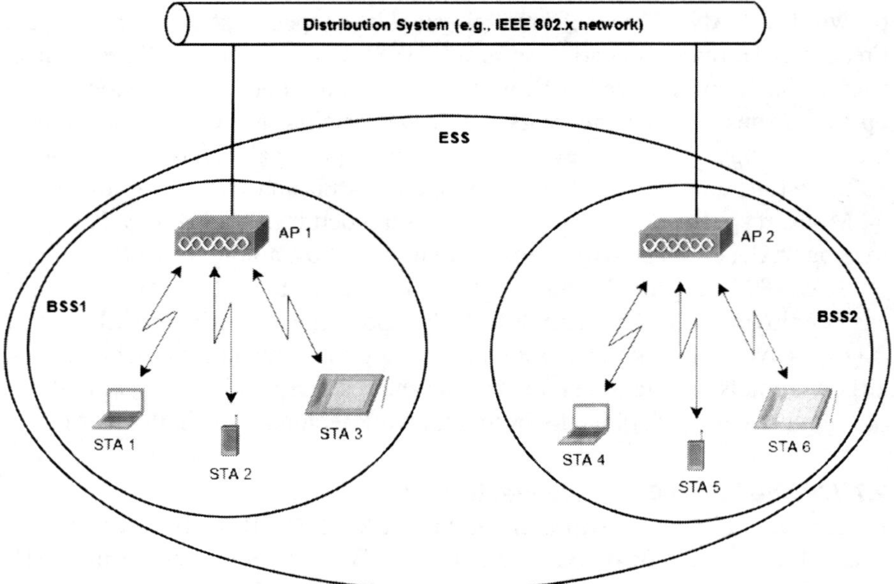

Figure 2.39 Architecture of the IEEE 802.11 network.

of new members. This wireless client does not act as a bridge to relay information between wireless clients. Ad hoc mode is used to connect wireless clients together when there is no wireless AP present. The wireless clients must be explicitly configured to use ad hoc mode. There can be a maximum of nine members in an ad hoc 802.11 wireless network.

In infrastructure mode, there is at least one wireless AP and one wireless client. The wireless client uses the wireless AP to access the resources of a traditional wired network. The wired network can be an organization intranet or the Internet, depending on the placement of the wireless AP.

A station willing to join a BSS must get in contact with the AP. This can happen through:

- Passive scanning, where the station scans the channels for a beacon frame that is sent periodically from an AP to announce its presence and provide the SSID and other parameters for Wireless NICs (WNICs) within range;
- Active scanning (the station tries to find an AP), where the station sends a Probe Request frame. This frame is sent from a station when it requires information from another station. All APs within reach reply with a

Probe Response frame containing capability information, supported data rates, among other relevant information.

Once an AP is found/selected, a station goes through an authentication process, which can be:

- Open system authentication (default, 2-step process) - Station sends authentication frame with its identity and AP sends frame as an Ack/Nack;
- Shared key authentication - Stations receive a shared secret key through a secure channel that is independent of 802.11. After the WNIC sends its initial authentication request, it will receive an authentication frame from the AP containing a challenge text. The WNIC sends an authentication frame containing the encrypted version of the challenge text to the AP. The AP ensures the text was encrypted with the correct key by decrypting it with its own key. The result of this process determines the WNIC's authentication status.

Once a station is authenticated, it starts the association process, i.e., information exchange about the AP/station capabilities and roaming:

- STA to AP - Associate Request frame, which enables the AP to allocate resources and synchronize. The frame carries information about the WNIC, including supported data rates and the SSID of the network the station wishes to associate with.
- AP to STA - Association Response frame, corresponding to the acceptance or rejection to an association request. If it is an acceptance, the frame will contain information such as as association ID and the supported data rates.
- New AP informs old AP, if it is a handover.

Only after association is complete, a station can transmit and receive data frames.

Hidden terminal and exposed station are two common problems in wireless networks, which can be summarily explained in the following way:

- Hidden terminal - Wireless stations have transmission ranges and not all stations are within radio range of each other, so simple CSMA will not work because if C transmits to B and A "senses" the channel, it will not hear C's transmission and will falsely conclude that A can begin a transmission to B.
- Exposed station - B wants to send to C and listens to the channel; when B hears A's transmission, it falsely assumes that it cannot send to C.

2.7.7.2 The IEEE 802.11 MAC Protocol

Whenever a wireless station is associated with an AP, it can start sending and receiving frames to and from the AP. When multiple stations want to transmit at the same time, and since they will use a shared channel, a multiple access protocol is needed to coordinate the transmissions. IEEE 802.11 designers chose a random access protocol, CSMA/CA (Carrier Sense Multiple Access with Collision Avoidance), inspired on the one used by Ethernet.

The IEEE 802.11 Media Access Control (MAC) layer must:

- Overcome the problem of hidden and exposed nodes;
- Support different physical layers;
- Allow overlapping of different networks in the same area;
- Support real-time services;
- Support roaming.

In order to accomplish these goals, the IEEE 802.11 MAC standard defines two different access mechanisms, the mandatory Distributed Coordination Function (DCF), which provides distributed channel access based on CSMA/CA (Carrier Sense Multiple Access with Collision Avoidance), and the optional Point Coordination Function (PCF), which provides centrally controlled channel access through polling.

In DCF, all stations contend for the access to the medium, in a distributive manner, based on the CSMA/CA protocol. For this reason, the access mechanisms is also referred to as contention-based channel access. It is appropriate for asynchronous data transfer of delay-tolerant traffic (like file transfer).

The collision detection mechanisms used in wired LANs cannot be used in wireless environments due to the following reasons: a collision detection mechanism requires a full duplex radio implementation, which increases the price significantly; on a wireless environment we cannot assume that all stations hear each other and the fact that a station willing to transmit and senses the medium free does not necessarily mean that the medium is free around the receiver area.

In order to overcome these problems, IEEE 802.11 uses a Collision Avoidance mechanism together with a positive acknowledgment scheme. A station willing to transmit senses the medium. If the medium is busy, then the station chooses a random backoff value and counts down this value when the channel is sensed idle. While the sender is sensed busy, the counter value remains frozen. When the counter reaches zero, the station transmits the entire frame and then waits for acknowledgment. If the medium is free for a specified short time (known as DIFS - Distributed Inter Frame Space), then

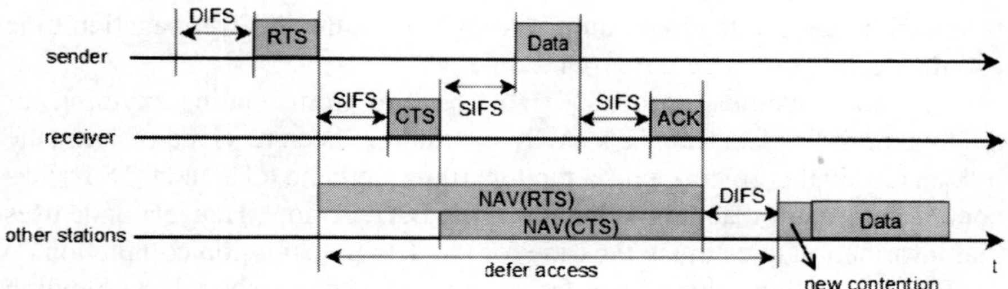

Figure 2.40 Contention in 802.11.

the station is allowed to transmit. After successful reception, the receiving station will send an acknowledgment (ACK) packet, after waiting for a short period of time known as the Short Inter-Frame Spacing (SIFS). Receipt of the ACK will notify the transmitter that no collision occurred. If the sender doe not receive the ACK within a given amount of time, it will retransmit the frame until it gets acknowledged or until it completes a fixed number of retransmissions.

A Virtual carrier sensing mechanism is used to reduce the probability of two stations colliding, because they do not hear each other. A station willing to transmit will first transmit a short control packet called Request to Send (RTS), which will include the source, destination and duration of the following transaction. If the medium is free, the destination station replies with a Clear To Send frame (CTS), including the same duration information. All stations receiving the RTS and/or CTS will set the Virtual Carrie Sense indicator (Network Allocation Vector - NAV) for the given duration, using it together with the Physical Carrie Sense in order to sense the medium. So, the CTS reserves the channel for the sender, notifying possibly hidden stations (Figure 2.40).

So, assuming that a node A has data to transfer to a node B, node A initiates the process by sending a RTS to node B. The destination node replies with a CTS and, after receiving the CTS, node A sends data. After successful reception, node B replies with an Acknowledgment frame (ACK). If node A has to send more than one data fragment, it has to wait a random time after each successful data transfer and compete with adjacent nodes for the medium using the RTS/CTS mechanism.

Any node overhearing an RTS frame refrains from sending anything until a CTS is received, or after waiting a certain time. If the captured RTS is not followed by a CTS, the maximum waiting time is the RTS propagation time and the destination node turnaround time.

Any node overhearing a CTS frame refrains from sending anything for the time until the data frame and ACK should have been received (solving the hidden terminal problem), plus a random time. Both the RTS and CTS frames contain information about the length of the DATA frame. Hence a node uses that information to estimate the time for the data transmission completion.

Before sending a long Data frame, node A sends a short Data-Sending frame (DS), which provides information about the length of the DATA frame. Every station that overhears this frame knows that the RTS/CTS exchange was successful. An overhearing station, which might have received RTS and DS but not CTS, defers its transmissions until after the ACK frame should have been received plus a random time.

In PCF, a Point Coordinator (PC), which is most often collocated in AP, controls the medium access based on the polling scheme, such that the PC polls individual stations to grant access to the medium based on their requirements. As in PCF stations do not contend for the medium and instead the medium access is controlled centrally, the access mechanism is sometimes referred to as contention-free channel access. It is appropriate for synchronous data transfer of real-time traffic (like audio and video).

The IEEE 802.11 standard defines four IFSs, which are used to provide different priorities (Figure 2.41):

- SIFS (Short IFS) - Is use to separate transmissions belonging to a single dialog. Gives the highest priority and is used for ACK, CTS and polling response;
- PIFS (PCF IFS) - Is used by the AP to gain access to the medium before any other station. Gives medium priority and is used for time-bounded service using PCF;
- DIFS (DCF IFS) - Used for a station willing to start a new transmission. Gives the lowest priority and is used for asynchronous data service;
- EIFS (Extended IFS) - Lowest priority interval used to report bad or unknown frame.

2.7.7.3 802.11 MAC Frame Format

The 802.11 MAC frame, shown in Figure 2.42, shares many similarities with an Ethernet frame and consists of a MAC header, the frame body, and a frame check sequence (FCS). The main fields are the following:

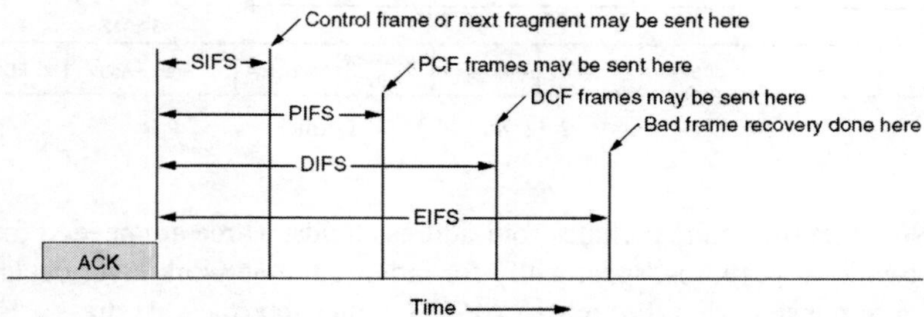

Figure 2.41 Interframe Spacing in 802.11 [99].

- Frame Control Field - Contains control information used for defining the type of 802.11 MAC frame and providing information necessary for the following fields to understand how to process the MAC frame.
- Duration/ID Field - This field is used for all control type frames, except with the subtype of Power Save (PS) Poll, to indicate the remaining duration needed to receive the next frame transmission. When the subtype is PS Poll, the field contains the association identity (AID) of the transmitting STA.
- Address Fields - Depending upon the frame type, the four address fields will contain a combination of the following address types: BSS Identifier, Destination Address (the MAC address of the final destination to receive the frame), Source Address (the MAC address of the original source that initially created and transmitted the frame), Receiver Address (the MAC address of the next immediate STA on the wireless medium to receive the frame), Transmitter Address (the MAC address of the STA that transmitted the frame onto the wireless medium).
- Sequence Control - Contains two subfields, the Fragment Number field (which indicates the number of each frame sent of a fragmented frame) and the Sequence Number field (which indicates the sequence number of each frame).
- Frame Body - Contains the data or information included in either management type or data type frames.
- Frame Check Sequence - The transmitting STA uses a cyclic redundancy check (CRC) over all the fields of the MAC header and the frame body field to generate the FCS value. The receiving STA then uses the same CRC calculation to determine its own value of the FCS field to verify whether or not any errors occurred in the frame during the transmission.

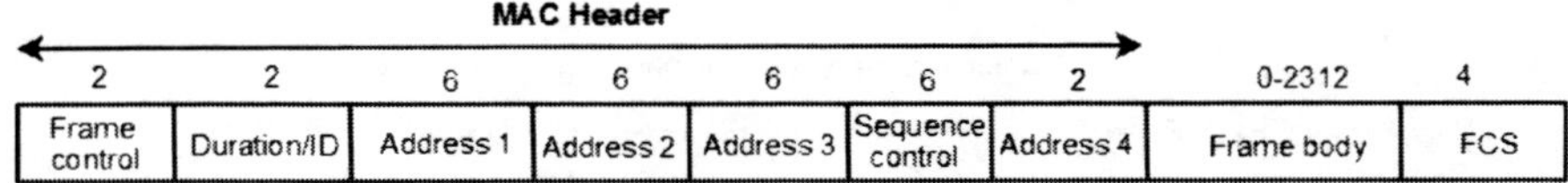

Figure 2.42 802.11 MAC Frame.

Note that the frame includes four address fields. Three are needed for internetworking purposes, specifically for moving the network layer datagram from a wireless station through an AP to a router interface. Address 2 is the MAC address of the station that transmits the frame. Address 1 is the MAC address of the wireless station that is to receive the frame. Address 3 contains the MAC address of the router interface that interconnects this subnet (recall that the BSS is part of a subnet) to other subnets. The forth address field is used when APs forward frames to each other in ad hoc mode.

2.7.7.4 Mobility in the Same IP Subnet

In order to increase the physical range of a wireless LAN, companies usually deploy several BSSs within the same IP subnet. So, how wireless stations seamlessly move from one BSS to another while maintaining ongoing transport layer sessions? Let us first consider the situation where all stations in the two BSSs, including the APs, belong to the same IP subnet, as illustrated in Figure 2.43.

In this case, as a station moves away from an AP, the signal from the AP weakens and the station starts to scan for a stronger signal from another AP. The station receives a beacon from the new AP (which usually has the same SSID as the previous AP), disassociates with the first AP and associates with the second one, while keeping its IP address and maintaining ongoing transport layer sessions.

When stations move between subnets, more sophisticated mobile management protocols are needed, such as mobile IP [79].

2.8 Use Case

2.8.1 VLAN and Port Mode Configuration

Let us first define the different VLANs in all routers. For Router 11, this definition/configuration can be made as illustrated in Box 2.5. A similar procedure should be used for all other routers, so there is no need to replicate this information.

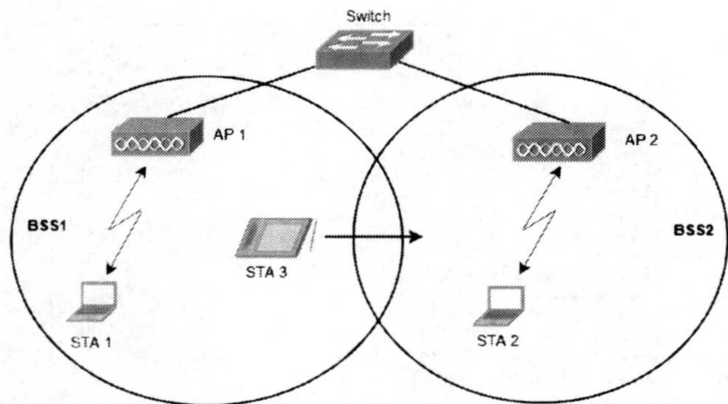

Figure 2.43 Mobility in the same subnet.

Now, the different router interfaces should be associated to the created VLANs, specifying if they correspond to access or interswitch ports. At the core of the network all interfaces should be configured as access ports (using the command `switchport access <vlan vlan_number>` (line 2, for example), while ports that connect to the access switches should be configured as interswitch ports (using command `switchport mode trunk` to define the port mode (lines 7 and 11) and command `switchport trunk allowed vlan except <vlan_range>` (lines 8 and 12) to establish which VLANs use this trunk connection). The configurations made at Router 11 are illustrated in Box 2.6, and similar configurations are made for the other routers.

The spanning tree protocol is enabled by default in all layer 3 switches. If this is not the case, the following command should be used:

```
# spanning-tree vlan <vlan-number>
```

2.8.2 IP Addressing Configuration

Finally, IPv4 and IPv6 addresses should be configured for the different VLANs according to the addressing methodology that was established in chapter 1. Box 2.7 shows the configuration commands at Router 11. Similar configurations should be made in the other routers.

```
vlan database
vlan 2
vlan 3
vlan 4
exit

#configure terminal
#interface fastethernet 1/0
  #switchport mode trunk
  #interface fastethernet 1/1
  #switchport access vlan 1
  #interface range fastethernet 1/2 - 5
    #switchport access vlan 2
    #interface range fastethernet 1/6 - 8
    #switchport access vlan 3
    #interface range fastethernet 1/9 - 15
    #switchport access vlan 4
    #end

!!Switch 1

#interface Vlan1
  #ip address 10.1.1.1 255.255.255.0
  #no autostate

#interface Vlan2
  #ip address 10.2.2.1 255.255.255.0
  #no autostate

#interface Vlan3
  #ip address 10.3.3.1 255.255.255.0
  #no autostate

#interface Vlan4
  #ip address 10.4.4.1 255.255.255.0
  #no autostate

!!Switch 2

#interface Vlan1
  #ip address 10.1.1.2 255.255.255.0
  #no autostate

#interface Vlan2
  #ip address 10.2.2.2 255.255.255.0
  #no autostate

#interface Vlan3
  #ip address 10.3.3.2 255.255.255.0
  #no autostate

#interface Vlan4
  #ip address 10.4.4.2 255.255.255.0
  #no autostate
```

Box 2.1 Configuration of the different VLANs on both layer 3 switches.

```
# show mac-address-table

VLAN   MAC Address      Type     Age    Port
1      0018.b967.3cd0   dynamic  10     Eth1/3
1      001c.b05a.5380   dynamic  200    Eth1/3
```

Box 2.2 Example of a switch MAC Address Table.

```
SW1#sh spanning-tree

 VLAN1 is executing the ieee compatible Spanning Tree protocol
  Bridge Identifier has priority 32768, address c211.1111.0000
  Configured hello time 2, max age 20, forward delay 15
  We are the root of the spanning tree
  Topology change flag set, detected flag set
  Number of topology changes 2 last change occurred 00:00:05 ago
  Times:  hold 1, topology change 35, notification 2
          hello 2, max age 20, forward delay 15
  Timers: hello 0, topology change 30, notification 0, aging 300

 Port 41 (FastEthernet1/0) of VLAN1 is forwarding
   Port path cost 10, Port priority 128, Port Identifier 128.41.
   Designated root has priority 32768, address c211.1111.0000
   Designated bridge has priority 32768, address c211.1111.0000
   Designated port id is 128.41, designated path cost 0
   Timers: message age 0, forward delay 0, hold 0
   Number of transitions to forwarding state: 1
   BPDU: sent 2009, received 561

 Port 42 (FastEthernet1/1) of VLAN1 is forwarding
   Port path cost 10, Port priority 128, Port Identifier 128.42.
   Designated root has priority 32768, address c211.1111.0000
   Designated bridge has priority 32768, address c211.1111.0000
   Designated port id is 128.42, designated path cost 0
   Timers: message age 0, forward delay 0, hold 0
   Number of transitions to forwarding state: 1
   BPDU: sent 2560, received 5

SW2#sh spanning-tree

 VLAN1 is executing the ieee compatible Spanning Tree protocol
  Bridge Identifier has priority 32768, address c222.2222.0000
  Configured hello time 2, max age 20, forward delay 15
  Current root has priority 32768, address c211.1111.0000
  Root port is 41 (FastEthernet1/0), cost of root path is 20
  Topology change flag not set, detected flag not set
  Number of topology changes 3 last change occurred 00:14:41 ago
          from FastEthernet1/1
  Times:  hold 1, topology change 35, notification 2
          hello 2, max age 20, forward delay 15
  Timers: hello 0, topology change 0, notification 0, aging 300

 Port 41 (FastEthernet1/0) of VLAN1 is forwarding
   Port path cost 20, Port priority 128, Port Identifier 128.41.
   Designated root has priority 32768, address c211.1111.0000
   Designated bridge has priority 32768, address c211.1111.0000
   Designated port id is 128.41, designated path cost 0
   Timers: message age 2, forward delay 0, hold 0
   Number of transitions to forwarding state: 1
   BPDU: sent 563, received 2504

 Port 42 (FastEthernet1/1) of VLAN1 is blocking
   Port path cost 10, Port priority 128, Port Identifier 128.42.
   Designated root has priority 32768, address c211.1111.0000
   Designated bridge has priority 32768, address c233.3333.0000
   Designated port id is 128.41, designated path cost 10
   Timers: message age 3, forward delay 0, hold 0
   Number of transitions to forwarding state: 1
   BPDU: sent 2626, received 448
```

Box 2.3 Details of the spanning tree configuration - Switches 1 and 2.

```
SW3#sh spanning-tree

 VLAN1 is executing the ieee compatible Spanning Tree protocol
  Bridge Identifier has priority 32768, address c233.3333.0000
  Configured hello time 2, max age 20, forward delay 15
  Current root has priority 32768, address c211.1111.0000
  Root port is 42 (FastEthernet1/1), cost of root path is 10
  Topology change flag not set, detected flag not set
  Number of topology changes 3 last change occurred 00:14:37 ago
          from FastEthernet1/0
  Times:  hold 1, topology change 35, notification 2
          hello 2, max age 20, forward delay 15
  Timers: hello 0, topology change 0, notification 0, aging 300

 Port 41 (FastEthernet1/0) of VLAN1 is forwarding
   Port path cost 10, Port priority 128, Port Identifier 128.41.
   Designated root has priority 32768, address c211.1111.0000
   Designated bridge has priority 32768, address c233.3333.0000
   Designated port id is 128.41, designated path cost 10
   Timers: message age 0, forward delay 0, hold 0
   Number of transitions to forwarding state: 2
   BPDU: sent 460, received 2620

 Port 42 (FastEthernet1/1) of VLAN1 is forwarding
   Port path cost 10, Port priority 128, Port Identifier 128.42.
   Designated root has priority 32768, address c211.1111.0000
   Designated bridge has priority 32768, address c211.1111.0000
   Designated port id is 128.42, designated path cost 0
   Timers: message age 2, forward delay 0, hold 0
   Number of transitions to forwarding state: 2
   BPDU: sent 7, received 3061
```

Box 2.4 Details of the spanning tree configuration - Switch 3.

```
vlan databse
  vlan 101
  vlan 102
  vlan 11
  vlan 12
  vlan 13
  apply
  exit
```

Box 2.5 VLAN configuration at Router 11.

```
interface FastEthernet1/1
  switchport access vlan 101
interface FastEthernet1/2
  switchport access vlan 102
!
interface range FastEthernet 1/3 - 15
  switchport mode trunk
  switchport trunk allowed vlan except 101-102
!
interface FastEthernet 1/0
  switchport mode trunk
  switchport trunk allowed vlan except 101-102
```

Box 2.6 Access and trunk ports configuration at Router 11.

```
interface Vlan11
  no shutdown
  ip address 10.1.0.254 255.255.255.0
  ipv6 enable
  ipv6 address 2001:A:A:100::FFFF/64
!
interface Vlan102
  no shutdown
  ip address 10.0.2.11 255.255.255.0
  ipv6 enable
  ipv6 address 2001:A:A:2::11/64
```

Box 2.7 Sample IPv4 and IPv6 addresses configuration at Router 11.

3

Network Link Layer

3.1 Routing and Routing Table

Routing is the process of learning all the paths through the Internet, known as routes, and use them to forward data from one network to another. Routes can be defined statically or calculated based on a dynamic algorithm.

The Internet is divided into Autonomous Systems (ASs), regions that are administered by a single entity. Routing is done differently within an AS (intradomain routing) and between ASs (interdomain routing). Intradomain routing ignores the Internet outside the AS. Protocols for intradomain routing are called Interior Gateway Protocols (IGPs). The most popular IGPs are the Routing Information Protocol (RIP) and the Open Shortest Path First (OSPF), that will be addressed in detail later in this chapter.

Interdomain routing assumes that the Internet consists of a collection of interconnected ASs. Normally, there is one dedicated router in each AS that handles interdomain traffic. Protocols for interdomain routing are also called Exterior Gateway Protocols (EGPs). The most popular EGP is the Border Gateway Protocol (BGP). EGPs are outside the scope of this book.

A routing table is a data file in RAM that is used to store route information about directly connected and remote networks. Since hop-by-hop routing is the fundamental characteristic of the IP layer, each routing table lists, for all reachable destinations, the address of the next device (the next hop) along the path to that destination. Assuming that the routing tables are consistent, the simple algorithm of relaying packets to their destination's next hop is sufficient to deliver data anywhere in a network.

The routing table consists of at least three information fields: the destination network identification, the cost or metric of the path through which the packet is to be sent and the next hop, that is, the address of the next equipment to which the packet is to be sent on the way to its final destination (Figure 3.1). Depending on the application and implementation, it can also contain

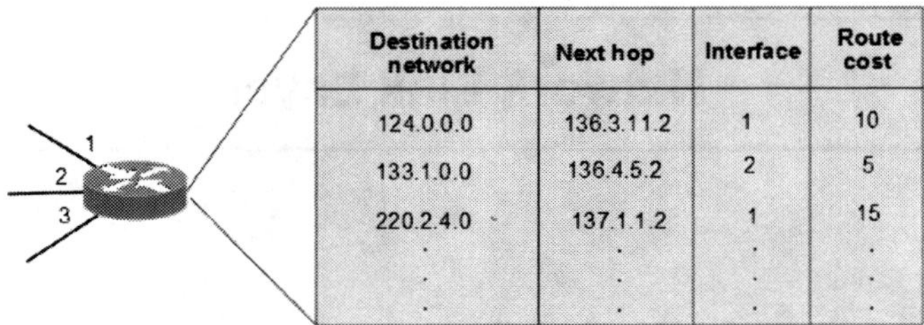

Destination network	Next hop	Interface	Route cost
124.0.0.0	136.3.11.2	1	10
133.1.0.0	136.4.5.2	2	5
220.2.4.0	137.1.1.2	1	15
.	.	.	.
.	.	.	.
.	.	.	.

Figure 3.1 Routing table example.

additional values that refine path selection: quality of service associated with the route, links to filtering criteria/access lists associated with the route, the interface, among others.

3.2 Static Routing

Static routing is a manual method to set up routing between IP networks. The network administrator configures static routes in a router by entering routes directly into its routing table. Static routing is supported on all routing devices, is simple to configure and easy to predict and understand in small networks. However, it does not dynamically adapt to network topology changes or equipment failures and does not scale well in large networks.

When there are two or more routes to the same destination, routers use an administrative distance to decide which routing protocol or static route to trust more. The lower the number, the more trustworthy the type of route is. The following values are used: 0 for a static route to a connected interface, 1 for a static route to an IP address, 110 for an OSPF route and 120 for a RIP route, just to mention the most popular ones. So, according to these values, static routes will always be used over dynamic routes.

The syntax of the Cisco global configuration command used to enter a static route is the following:

```
# ip route destination-prefix destination-prefix-mask interface
  OR forwarding-router's-IP-address
```

There are static routes that point to an interface on the router and others that point to an IP address on the network. Besides, there is a special kind of static route called *default route*, where the network and subnet mask specified as destination are all zeros. The default route has the following meaning: “any

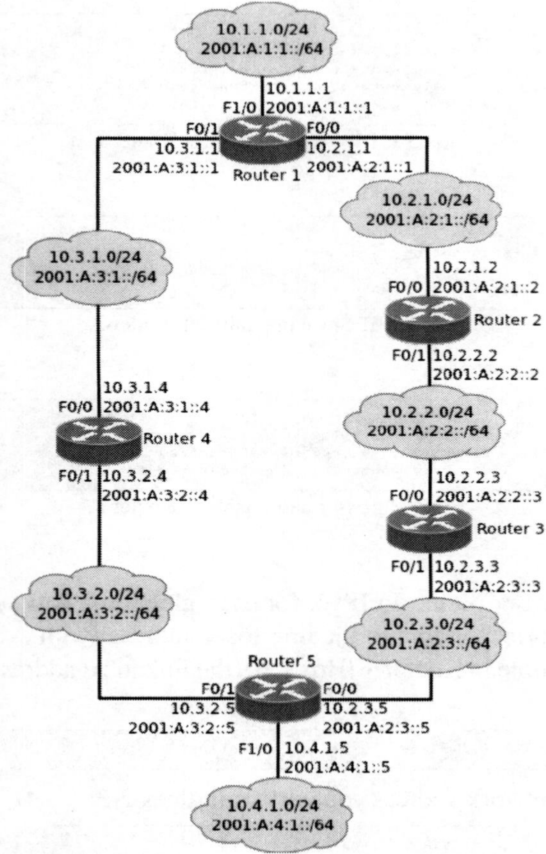

Figure 3.2 Network scenario used to illustrate routing concepts.

traffic that does not match a specific route in the routing table should be sent to this destination".

Exercise - Static Routing

Let us consider the network illustrated in Figure 3.2. After configuring the IPv4 and IPv6 addresses at all interfaces, the IPv4 routing tables of the different routers are the ones shown in Boxes 3.1, 3.2, 3.3, 3.4 and 3.5, while the IPv6 routing tables are shown in Boxes 3.6, 3.7, 3.8, 3.9 and 3.10. The IPv4 routing tables are quite simple to understand: there is an entry for each directly connected network, showing also the interface whose IP address in-

```
#show ip route
     10.0.0.0/24 is subnetted, 3 subnets
C       10.3.1.0 is directly connected, FastEthernet0/1
C       10.2.1.0 is directly connected, FastEthernet0/0
C       10.1.1.0 is directly connected, FastEthernet1/0
```

Box 3.1 IPv4 routing table at Router 1.

```
#show ip route
     10.0.0.0/24 is subnetted, 2 subnets
C       10.2.1.0 is directly connected, FastEthernet0/0
C       10.2.2.0 is directly connected, FastEthernet0/1
```

Box 3.2 IPv4 routing table at Router 2.

```
#show ip route
     10.0.0.0/24 is subnetted, 2 subnets
C       10.2.2.0 is directly connected, FastEthernet0/0
C       10.2.3.0 is directly connected, FastEthernet0/1
```

Box 3.3 IPv4 routing table at Router 3.

cludes that network prefix. In IPv6, for each global network address there is also an entry corresponding to the link local address (with a 128 bits subnet mask): for example in Router 1 (Box 3.6), the link local address entry in lines 4-5:

```
L 2001:A:1:1::1/128 [0/0] via ::, FastEthernet1/0
```

has its global network address counterpart in lines 2-3:

```
C 2001:A:1:1::/64 [0/0] via ::, FastEthernet1/0
```

Besides, the following entries (see Box 3.6 lines 14-17) are automatically inserted in each routing table:

```
L FE80::/10 [0/0] via ::, Null0
```

corresponding to the link-local prefix, and

```
L FF00::/8 [0/0] via ::, Null0
```

corresponding to a multicat prefix.

```
#show ip route
     10.0.0.0/24 is subnetted, 2 subnets
C       10.3.1.0 is directly connected, FastEthernet0/0
C       10.3.2.0 is directly connected, FastEthernet0/1
```

Box 3.4 IPv4 routing table at Router 4.

```
#show ip route
     10.0.0.0/24 is subnetted, 3 subnets
C       10.3.2.0 is directly connected, FastEthernet0/1
C       10.2.3.0 is directly connected, FastEthernet0/0
C       10.4.1.0 is directly connected, FastEthernet1/0
```

Box 3.5 IPv4 routing table at Router 5.

```
#show ipv6 route
C   2001:A:1:1::/64 [0/0]
     via ::, FastEthernet1/0
L   2001:A:1:1::1/128 [0/0]
     via ::, FastEthernet1/0
C   2001:A:2:1::/64 [0/0]
     via ::, FastEthernet0/0
L   2001:A:2:1::1/128 [0/0]
     via ::, FastEthernet0/0
C   2001:A:3:1::/64 [0/0]
     via ::, FastEthernet0/1
L   2001:A:3:1::1/128 [0/0]
     via ::, FastEthernet0/1
L   FE80::/10 [0/0]
     via ::, Null0
L   FF00::/8 [0/0]
     via ::, Null0
```

Box 3.6 IPv6 routing table at Router 1.

```
#show ipv6 route
C   2001:A:2:1::/64 [0/0]
     via ::, FastEthernet0/0
L   2001:A:2:1::2/128 [0/0]
     via ::, FastEthernet0/0
C   2001:A:2:2::/64 [0/0]
     via ::, FastEthernet0/1
L   2001:A:2:2::2/128 [0/0]
     via ::, FastEthernet0/1
L   FE80::/10 [0/0]
     via ::, Null0
L   FF00::/8 [0/0]
     via ::, Null0
```

Box 3.7 IPv6 routing table at Router 2.

```
#show ipv6 route
C    2001:A:2:2::/64 [0/0]
      via ::, FastEthernet0/0
L    2001:A:2:2::3/128 [0/0]
      via ::, FastEthernet0/0
C    2001:A:2:3::/64 [0/0]
      via ::, FastEthernet0/1
L    2001:A:2:3::3/128 [0/0]
      via ::, FastEthernet0/1
L    FE80::/10 [0/0]
      via ::, Null0
L    FF00::/8 [0/0]
      via ::, Null0
```

Box 3.8 IPv6 routing table at Router 3.

```
#show ipv6 route
C    2001:A:3:1::/64 [0/0]
      via ::, FastEthernet0/0
L    2001:A:3:1::4/128 [0/0]
      via ::, FastEthernet0/0
C    2001:A:3:2::/64 [0/0]
      via ::, FastEthernet0/1
L    2001:A:3:2::4/128 [0/0]
      via ::, FastEthernet0/1
L    FE80::/10 [0/0]
      via ::, Null0
L    FF00::/8 [0/0]
      via ::, Null0
```

Box 3.9 IPv6 routing table at Router 4.

```
#show ipv6 route
C    2001:A:2:3::/64 [0/0]
      via ::, FastEthernet0/0
L    2001:A:2:3::5/128 [0/0]
      via ::, FastEthernet0/0
C    2001:A:3:2::/64 [0/0]
      via ::, FastEthernet0/1
L    2001:A:3:2::5/128 [0/0]
      via ::, FastEthernet0/1
C    2001:A:4:1::/64 [0/0]
      via ::, FastEthernet1/0
L    2001:A:4:1::5/128 [0/0]
      via ::, FastEthernet1/0
L    FE80::/10 [0/0]
      via ::, Null0
L    FF00::/8 [0/0]
      via ::, Null0
```

Box 3.10 IPv6 routing table at Router 5.

```
1 # ip route 10.4.1.0 255.255.255.0 10.3.1.4
2 # ipv6 unicast-routing
3 # ipv6 route 2001:A:4:1::/64 2001:A:3:1::4
```

Box 3.11 Static Route creation at Router 1.

```
1 # ip route 10.4.1.0 255.255.255.0 10.3.2.5
2 # ipv6 unicast-routing
3 # ipv6 route 2001:A:4:1::/64 2001:A:3:2::5
```

Box 3.12 Static Route creation at Router 4.

```
1 # ip route 10.3.1.0 255.255.255.0 10.3.2.4
2 # ipv6 unicast-routing
3 # ipv6 route 2001:A:3:1::/64 2001:A:3:2::4
```

Box 3.13 Static Route creation at Router 5.

Now, let us configure several static routes at Routers 1, 4 and 5. The configuration details are shown in Boxes 3.11, 3.12 and 3.13. By default, IPv6 unicast routing is disabled; in order to activate it the command

```
# ipv6 unicast-routing
```

must be issued (see line 2 in Boxes 3.11, 3.12 and 3.13). This command enables forwarding of IPv6 packets. At Router 1, we have configured a static route to IPv4 network 10.4.1.0 pointing to next-hop 10.3.1.4 (line 1) and a static route to IPv6 network 2001:A:4:1::/64 pointing to next-hop 2001:A:3:1::4 (line 3); at Router 4, we have configured a static route to IPv4 network 10.4.1.0 pointing to next-hop 10.3.2.5 (line 1) and a static route to IPv6 network 2001:A:4:1::/64 pointing to next-hop 2001:A:3:2::5 (line 3); at Router 5, we have configured a static route to IPv4 network 10.3.1.0 pointing to next-hop 10.3.2.4 (line 1) and a static route to IPv6 network 2001:A:3:1::/64 pointing to next-hop 2001:A:3:2::4 (line 3).

The configuration of these static routes leads to changes on the routing tables of all routers. These changes are illustrated in Boxes 3.14, 3.15, 3.16, 3.17, 3.18 and 3.19: basically, there is a new entry at each routing table corresponding to the static route that was configured (the entry is marked with the S letter). This entry has a route cost of 1 and an administrative distance of 0.

```
#show ip route
     10.0.0.0/24 is subnetted, 4 subnets
...
S        10.4.1.0 [1/0] via 10.3.1.4
```

Box 3.14 Additional IPv4 routing table entries at Router 1 after static routes creation.

```
#show ip route
     10.0.0.0/24 is subnetted, 3 subnets
...
S        10.4.1.0 [1/0] via 10.3.2.5
```

Box 3.15 Additional IPv4 routing table entries at Router 4 after static routes creation.

```
#show ip route
     10.0.0.0/24 is subnetted, 4 subnets
...
S        10.3.1.0 [1/0] via 10.3.2.4
```

Box 3.16 Additional IPv4 routing table entries at Router 5 after static routes creation.

```
#show ipv6 route
...
S    2001:A:4:1::/64 [1/0]
     via 2001:A:3:1::4
```

Box 3.17 Additional IPv6 routing table entries at Router 1 after static routes creation.

```
#show ipv6 route
...
S    2001:A:4:1::/64 [1/0]
     via 2001:A:3:2::5
```

Box 3.18 Additional IPv6 routing table entries at Router 4 after static routes creation.

```
#show ipv6 route
...
S    2001:A:3:1::/64 [1/0]
     via 2001:A:3:2::4
```

Box 3.19 Additional IPv6 routing table entries at Router 5 after static routes creation.

No.	Time	Source	Destination	Protocol	Info
1	0.000000	10.3.1.1	10.4.1.5	ICMP	Echo (ping) request (id=0x0001, seq(be/le)=1/256, ttl=255)
2	0.006809	10.4.1.5	10.3.1.1	ICMP	Echo (ping) reply (id=0x0001, seq(be/le)=1/256, ttl=254)
3	0.008934	10.3.1.1	10.4.1.5	ICMP	Echo (ping) request (id=0x0001, seq(be/le)=2/512, ttl=255)
4	0.015112	10.4.1.5	10.3.1.1	ICMP	Echo (ping) reply (id=0x0001, seq(be/le)=2/512, ttl=254)
5	0.017229	10.3.1.1	10.4.1.5	ICMP	Echo (ping) request (id=0x0001, seq(be/le)=3/768, ttl=255)
6	0.023423	10.4.1.5	10.3.1.1	ICMP	Echo (ping) reply (id=0x0001, seq(be/le)=3/768, ttl=254)
7	0.025531	10.3.1.1	10.4.1.5	ICMP	Echo (ping) request (id=0x0001, seq(be/le)=4/1024, ttl=255)
8	0.034673	10.4.1.5	10.3.1.1	ICMP	Echo (ping) reply (id=0x0001, seq(be/le)=4/1024, ttl=254)

Figure 3.3 Result of the ping command.

No.	Time	Source	Destination	Protocol	Info
1	0.000000	2001:a:3:1::1	2001:a:4:1::5	ICMPv6	Echo (ping) request id=0x0bb7, seq=0
2	0.014545	2001:a:4:1::5	2001:a:3:1::1	ICMPv6	Echo (ping) reply id=0x0bb7, seq=0
3	0.018727	2001:a:3:1::1	2001:a:4:1::5	ICMPv6	Echo (ping) request id=0x0bb7, seq=1
4	0.024902	2001:a:4:1::5	2001:a:3:1::1	ICMPv6	Echo (ping) reply id=0x0bb7, seq=1
5	0.027007	2001:a:3:1::1	2001:a:4:1::5	ICMPv6	Echo (ping) request id=0x0bb7, seq=2
6	0.031131	2001:a:4:1::5	2001:a:3:1::1	ICMPv6	Echo (ping) reply id=0x0bb7, seq=2
7	0.033732	2001:a:3:1::1	2001:a:4:1::5	ICMPv6	Echo (ping) request id=0x0bb7, seq=3
8	0.037852	2001:a:4:1::5	2001:a:3:1::1	ICMPv6	Echo (ping) reply id=0x0bb7, seq=3

Figure 3.4 Result of the ping6 command.

No.	Time	Source	Destination	Protocol	Info
1	0.000000	10.3.1.1	10.4.1.5	ICMP	Echo (ping) request (id=0x0003, seq(be/le)=0/0, ttl=255)
2	0.001993	10.3.1.4	10.3.1.1	ICMP	Destination unreachable
3	0.004106	10.3.1.1	10.4.1.5	ICMP	Echo (ping) request (id=0x0003, seq(be/le)=1/256, ttl=255)
4	2.001635	10.3.1.1	10.4.1.5	ICMP	Echo (ping) request (id=0x0003, seq(be/le)=2/512, ttl=255)
5	2.003631	10.3.1.4	10.3.1.1	ICMP	Destination unreachable
6	2.005719	10.3.1.1	10.4.1.5	ICMP	Echo (ping) request (id=0x0003, seq(be/le)=3/768, ttl=255)
7	4.004823	10.3.1.1	10.4.1.5	ICMP	Echo (ping) request (id=0x0003, seq(be/le)=4/1024, ttl=255)
8	4.006851	10.3.1.4	10.3.1.1	ICMP	Destination unreachable

Figure 3.5 Ping result after link breakdown.

With the static routes that were configured, there is connectivity between Router 1 and networks 10.4.1.0/24 and 2001:A:4:1::/64. So, the *ping 10.4.1.5* and *ping6 2001:A:4:1::5* commands, executed at Router 1, are successful, as shown in Figures 3.3 and 3.4, respectively.

Finally, we shut down interface F0/1 of Router 4, thus breaking the physical path that supports the configured static routes. Obviously, Router 4 is aware of this occurrence, removing networks 10.3.1.0/24 and 2001:A:3:2::/64 from its routing tables, as can be shown in Boxes 3.20 and 3.22. However, Router 1 keeps its routing tables unchanged (Boxes 3.21 and 3.23), which illustrates one the main disadvantages of static routing: static routes are not announced, so routers are not able to react to topology changes.

If we repeat the *ping 10.4.1.5* and *ping6 2001:A:4:1::5* commands at Router 1, obviously there is no connectivity and we get a reply from Router 4: in IPv4 the reply is an ICMP Destination Unreachable message (Figure

No.	Time	Source	Destination	Protocol	Info
1	0.000000	2001:a:3:1::1	2001:a:4:1::5	ICMPv6	Echo (ping) request id=0x0270, seq=0
2	0.004113	2001:a:3:1::4	2001:a:3:1::1	ICMPv6	Unreachable (Route unreachable)
3	0.006278	2001:a:3:1::1	2001:a:4:1::5	ICMPv6	Echo (ping) request id=0x0270, seq=1
4	0.008327	2001:a:3:1::4	2001:a:3:1::1	ICMPv6	Unreachable (Route unreachable)
5	0.010481	2001:a:3:1::1	2001:a:4:1::5	ICMPv6	Echo (ping) request id=0x0270, seq=2
6	0.012530	2001:a:3:1::4	2001:a:3:1::1	ICMPv6	Unreachable (Route unreachable)
7	0.015789	2001:a:3:1::1	2001:a:4:1::5	ICMPv6	Echo (ping) request id=0x0270, seq=3
8	0.016708	2001:a:3:1::4	2001:a:3:1::1	ICMPv6	Unreachable (Route unreachable)

Figure 3.6 Ping6 result after link breakdown.

```
#show ip route
     10.0.0.0/24 is subnetted, 2 subnets
C       10.3.1.0 is directly connected, FastEthernet0/0
```

Box 3.20 IPv4 routing table at Router 4 after link down.

```
#show ip route
     10.0.0.0/24 is subnetted, 3 subnets
C       10.3.1.0 is directly connected, FastEthernet0/1
C       10.2.1.0 is directly connected, FastEthernet0/0
C       10.1.1.0 is directly connected, FastEthernet1/0
S       10.4.1.0 [1/0] via 10.3.1.4
```

Box 3.21 IPv4 routing table at Router 1 after link down.

3.5), while in IPv6 is an ICMP Unreachable (with a Route Unreachable code) message (Figure 3.6).

Now, at Router 1 let us configure static IPv4 and IPv6 default routes passing through Router 4, as shown in Box 3.24. Obviously, this configuration implies the appearance of an additional entry on both the IPv4 and IPv6 routing tables, as shown in Boxes 3.25 and 3.26.

```
#show ipv6 route
C    2001:A:3:1::/64 [0/0]
      via ::, FastEthernet0/0
L    2001:A:3:1::4/128 [0/0]
      via ::, FastEthernet0/0
L    FE80::/10 [0/0]
      via ::, Null0
L    FF00::/8 [0/0]
      via ::, Null0
```

Box 3.22 IPv6 routing table at Router 4 after link down.

```
#show ipv6 route
C    2001:A:1:1::/64 [0/0]
      via ::, FastEthernet1/0
L    2001:A:1:1::1/128 [0/0]
      via ::, FastEthernet1/0
C    2001:A:2:1::/64 [0/0]
      via ::, FastEthernet0/0
L    2001:A:2:1::1/128 [0/0]
      via ::, FastEthernet0/0
C    2001:A:3:1::/64 [0/0]
      via ::, FastEthernet0/1
L    2001:A:3:1::1/128 [0/0]
      via ::, FastEthernet0/1
S    2001:A:4:1::/64 [1/0]
      via 2001:A:3:1::4
L    FE80::/10 [0/0]
      via ::, Null0
L    FF00::/8 [0/0]
      via ::, Null0
```

Box 3.23 IPv6 routing table at Router 1 after link down.

```
# ip route 0.0.0.0 10.3.1.4
# ipv6 route ::/0 2001:A:3:1::4
```

Box 3.24 Static Route creation at Router 1.

```
#show ip route
...
S*    0.0.0.0/0 [1/0] via 10.3.1.4
```

Box 3.25 Aditional IPv4 routing table entry at Router 1 after static default route creation.

```
#show ipv6 route
...
S    ::/0 [1/0]
      via 2001:A:3:1::4
```

Box 3.26 Aditional IPv6 routing table entry at Router 1 after static default route creation.

```
# route-map route-map-name permit 10
  # match criteria-1
  # set perform-action-1
# route-map route-map-name permit 20
  # match criteria-2
  # set perform-action-2
```

Box 3.27 Example of a Route-map

```
# route-map TrafficToOutside 10
  # match ip 25
  # set next hop 10.15.15.1
```

Box 3.28 Route-map creation

```
# interface FastEthernet 0/0
  # ip policy TrafficToOutside
```

Box 3.29 Applying the route-map

3.3 Route-Maps

The Route-map mechanism allows changing the way traffic is routed, applying policies rather than traditional routing methodologies. Typically, routers just look at the destination address of a packet, compare it to their routing table and send the packet on their way. To change this basic routing procedure, thus creating a policy routing methodology, a route-map should be configured.

Route maps are mainly used for redistribution and policy routing [19, 20]. Redistribution is the use of a routing protocol to advertise routes that are learned by some other means, such as by another routing protocol, static routes, or directly connected routes. Multi-protocol routing is common for several reasons, such as company mergers, multiple departments managed by multiple network administrators, and multi-vendor environments. Having a multiple protocol environment makes redistribution a necessity. Policy routes are sophisticated static routes. Whereas static routes forward a packet to a specified next hop based on the destination address of the packet, policy routes can forward a packet to a specified next hop based on the source of the packet or other fields in the packet header. Policy routes can also be linked to extended IP access lists (that will be covered in detail in chapter 5) so that routing might be based on things such as protocol types and port numbers. Like a static route, a policy route influences routing only at the router on which it is configured.

Route-maps are similar to a scripting language for routers. They define traffic and then process it according to a defined list of statements, almost like a computer program. Route-maps inherit their structure from the *if-then* statements of programming languages. First, it creates a step (10 and 20 in the following example), matches a criteria in each step and then performs an action, using the syntax that is shown in Box 3.27.

Let us consider a simple example of a route-map for routing traffic by source. First, we have to define the traffic that is going to be processed, using a standard access-list:

```
access-list 25 10.10.25.0 0.0.0.255
```

Access lists will be explained in Chapter 5 but for now let us only say that a standard access list simply allows to control traffic based on its source address (in this case, 10.10.25.0 with mask 255.255.255.0, which is exactly the binary inverse of 0.0.0.255). The route-map should be now created, including its association to the previously defined access list and an action (in this case, the definition of the next hop to use), as illustrated in Box 3.28. Finally, the route-map should be applied to the router interface that the traffic enters into. This is done using the `ip policy <Route-Map-Name>` command, as shown in Box 3.29.

In this simple example traffic from network 10.10.25.0 will be forced to the router interface 10.15.15.1 rather than routed according to the decision taken by looking at the router routing table.

3.4 Dynamic Routing

Dynamic routing protocols dynamically learn network destinations and how to reach them, advertising those destinations to other routers. This advertisement function will in turn allow all routers to learn about existing destination networks and the best paths to reach them.

A router using dynamic routing will, first of all, "learn" the routes to all networks that are directly connected to it. Next, it will learn routes from other routers that run the same routing protocol. Each router will then sort its list of routes and select one or more "best" routes for each network destination the router knows or has learned. Dynamic routing protocols will then distribute this "best route" information to other routers running the same routing protocol, thereby extending the information on what networks exist and can be reached. This gives dynamic routing protocols the ability to adapt to logical network topology changes, equipment failures or network outages "on the fly".

3.5 Distance Vector versus Link State Protocols

There are two main types of IP routing algorithms: Distance Vector Routing and Link State Routing. Distance Vector protocols determine the best path based on the distance to the destination, while Link State protocols are able to use more complex methods that take into account link variables, such as bandwidth, delay, reliability and load.

Distance Vector routing protocols pass periodic copies of the routing table to neighbor routers and accumulate distance vectors. Distance can represent hops or a combination of metrics calculated to represent a distance value. In this type of protocols, routers discover the best path to destination from each neighbor. Routing updates are sent, step by step, from one router to another. The Routing Information Protocol (RIP), which will be studied in detail in this chapter, and the Interior Gateway Routing Protocol (IGRP) are important examples of Distance Vector routing protocols.

In Link State routing protocols, routers announce their closest neighbors to all other routers on the network. The entire routing table is not distributed from any router, only the part of the table containing its neighbors. Each router should maintain, at least, a partial map of the network. When a network link changes state, a notification called Link State Advertisement (LSA) is flooded throughout the network and all routers note the change and re-compute their routes accordingly. Some of the most relevant link-state routing protocols are Open Shortest Path First (OSPF), Intermediate System to Intermediate System (IS-IS) and Enhanced Interior Gateway Routing Protocol (EIGRP).

There are two major differences between Distance Vector and Link State routing protocols. First of all, Distance Vector exchanges the routing updates periodically, whether the topology is changed or not. This will maximize the convergence time, which also increases the chance of routing loops. Link State routing protocols send triggered change-based updates when there is a topology change. After initial flood, routers pass small event-based triggered link state updates to all other routers. This will minimize the convergence time, avoiding routing loops. Secondly, Distance Vector routing protocols rely on the information from their directly connected neighbors in order to calculate and accumulate route information. Distance Vector routing protocols require very little overhead (in terms of memory and processor power) as compared to Link State routing protocols, while Link State routing protocols do not rely solely on the information from the neighbors or adjacent router in order to calculate route information. Instead, they have a system of databases they use in order to calculate the best route to destinations in the network.

Besides these, there are other differences between both types of routing protocols:

- Distance Vector routing protocols are based on the Bellman-Ford algorithm.
- Distance Vector routing protocols are less scalable: for example, RIP only supports 16 hops while IGRP has a maximum of 100 hops.
- Distance Vector routing protocols uses hop count and composite metric.
- Link State routing protocols are based on the Dijkstra algorithm.
- Link State routing protocols are very much scalable, supporting an infinite number of hops.
- Link State routing protocols use cost as its metric.

3.6 Routing Information Protocol

The Routing Information Protocol (RIP) standard was formally defined in RFC 1058 [36], while its second version (RIPv2) was proposed in RFC 1723 [37] and is an extension of the RIP capabilities: RIPv2 enables RIP messages to carry more information, permitting the use of a simple authentication mechanism to secure table updates, and supports subnet masks, a critical feature that is unavailable in RIP.

RIP sends routing update messages at regular intervals and when the network topology changes. When a router receives a routing update that includes changes to an entry, it updates its routing table to reflect the new route. The metric value for the path is increased by 1, and the sender is indicated as the next hop. RIP routers maintain only the best route (the route with the lowest metric value) to a destination. After updating its routing table, the router immediately begins transmitting routing updates (triggered updates) to inform other network routers of the change. These updates are sent independently of the regularly scheduled updates that RIP routers send.

RIP uses a single routing metric (hop count) to measure the distance between the source and a destination network. Each hop in a path from source to destination is assigned a hop count value (typically 1). When a router receives a routing update that contains a new or changed destination network entry, the router adds 1 to the metric value indicated in the update and enters the

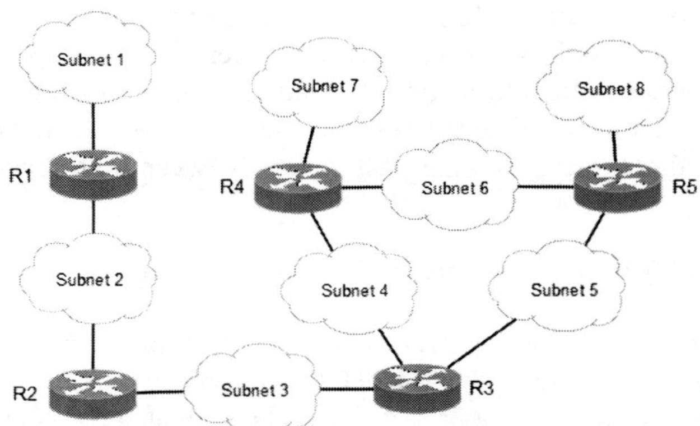

Figure 3.7 Network topology example.

network in the routing table. The IP address of the sender is used as the next hop.

3.6.1 The Count-to-Infinity Problem

RIP prevents routing loops by implementing a limit on the number of hops allowed in a path from the source to a destination. The maximum number of hops in a path is 15. If a router receives a routing update that contains a new or changed entry, and if increasing the metric value by 1 causes the metric to be infinity (that is, 16), the network destination is considered unreachable. This is the count-to-infinity problem. Referring to Figure 3.7, subnet 1 is 2 hops away from R3 and 1 hop away from R2. Therefore R2 advertises a cost of 1 for subnet 1 and R3 advertises a cost of 2, and traffic continues to be routed through R2. If R1 crashes, R2 waits for an update from R1 for 180 seconds. While waiting, R2 continues to send updates to R3 that keep the route to subnet 1 in R3's routing table. When R2's timer finally expires, it removes all routes through R1 from its routing table, including the route to subnet 1. It then receives an update from R3 advertising that R3 is 2 hops away from subnet 1. R2 installs this route and announces that it is 3 hops away from subnet 1. R3 receives this update, installs the route, and announces that it is 4 hops away from subnet 1. Things continue on in this manner until the cost of the route to subnet 1 reaches 16 in both routing tables. If the update interval is 30 seconds, this could take a long time!

RIP includes some important stability features to prevent incorrect routing information from being propagated: the split horizon, the poison reverse and the hold-down timers mechanisms.

RIP stability mechanisms

Split horizon, poison reverse and hold-down timers are important RIP mechanisms used to prevent incorrect routing information from being propagated.

With the split horizon feature, a router does not advertise routes on the link from which those routes were obtained. In Figure 3.7, for example, router R3 would not announce the route to subnet 1 on subnet 3 because it learned that route from the updates it received from R2 on subnet 3.

The poison reverse feature is an enhancement to split horizon. It uses the same idea of "not advertising routes on the link from which they were obtained", but it adds a positive action to that essentially negative rule. Poison reverse says that a router should advertise an infinite distance for routes on this link. With poison reverse, R3 would advertise subnet 1 with a cost of 16 to all systems on subnet 3. The cost of 16 means that subnet 1 cannot be reached through R3.

The hold-down timers ignore routing update information for a specified period of time (when network links are unstable). These timers can be reset when the timer expires, a routing update is received that has a better metric, or a routing update is received indicating that the original route to the network is valid.

3.6.2 Triggered Updates

Split horizon and poison reverse are able to solve the problem that was described above. But what happens if router R3 crashes? With split horizon, R4 and R5 do not advertise to R3 the route to subnet 3 because they learned that route from R3. They do, however, advertise the route to subnet 3 to each other. When R3 goes down, R4 and R5 perform their own count to infinity before they remove the route to subnet 3. Triggered updates address this problem: instead of waiting the normal 30 second update interval, a triggered update is sent immediately. Therefore, when an upstream router crashes or a local link goes down, immediately after the router updates its local routing table, it sends the changes to its neighbors. Without triggered updates counting to infinity can take almost 8 minutes, but with triggered updates neighbors are informed in a few seconds. Triggered updates also use network bandwidth efficiently, since they don't include the full routing table; they include only the routes that have changed.

Triggered updates take positive action to eliminate bad routes. Using triggered updates, a router advertises the routes deleted from its routing table with a infinite cost to force downstream routers to also remove them. So, if R3 crashes, R4 and R5 wait 180 seconds and remove the routes to subnets 1, 2, and 3 from their routing tables. They then send each other triggered updates with a metric of 16 for subnets 1, 2, and 3. Thus, they tell each other that they cannot reach these networks and no count to infinity occurs.

3.6.3 RIP Timers

RIP uses numerous timers to regulate its performance. These include a routing update timer, a route timeout timer, and a route flush timer. The routing update timer clocks the interval between periodic routing updates. Generally, it is set to 30 seconds, with a small random amount of time added whenever the timer is reset. This is done to help prevent congestion, which could result from all routers simultaneously attempting to update their neighbors. Each routing table entry has a route timeout timer associated with it. When the route timeout timer expires, the route is marked invalid but is retained in the table until the route flush timer expires.

3.6.4 RIP Messages

The left part of Figure 3.8 illustrates the general format of the RIPv1 messages. The following message types are supported:

- RIP Request (Optional) - sent by a router that was recently connected or when the aging time of a routing table entry expires ((timeout equal to 180 s). The Request can ask for information regarding a specific destination or all destinations.
- RIP Response - contains a distance vector and can be sent using split horizon or not. This message can be sent periodically (periodicity period of 30 seconds) to the broadcast address, optionally when information has changed (triggered update, sent to the broadcast address) or in response to a RIP Request (in this case, to the address that originated the Request).

3.6.5 RIPv2

RIP Version 2 (RIPv2) is a new version of RIP that basically defines extensions to the RIPv1 packet format (Figure 3.8): RIPv2 adds a network mask and a next hop address to the destination address and metric found in the original RIP packet. The network mask frees the RIPv2 router from the limitation of interpreting addresses based on strict address class rules. Using the mask, RIPv2 routers support variable-length subnets and Classless Inter-Domain Routing (CIDR) supernets. The next hop address is the IP address of the gateway that handles the route. If the address is 0.0.0.0, the source of the update packet is the gateway for the route. The next hop route permits a RIPv2 supplier to provide routing information about gateways that do not speak RIPv2. Its function is similar to an ICMP Redirect, pointing to the best gateway for a route and eliminating extra routing hops.

RIPv2 adds other new features to RIP. It transmits updates via the multicast address 224.0.0.9 to reduce the load on systems that are not capable of processing a RIPv2 packet. RIPv2 also introduces a packet authentication scheme to reduce the possibility of accepting erroneous updates from misconfigured systems.

Despite these changes, RIPv2 is compatible with RIPv1. The original RIP specification allowed for future versions of RIP. RIP has a version number in the packet header and includes several empty fields for extending the packet. The new values used by RIPv2 did not require any changes to the structure of the packet. The new values are simply placed in the empty fields that the original protocol reserved for future use.

Split horizon, poison reverse, triggered updates, and RIPv2 eliminate most of the problems with the original RIP protocol. However, RIPv2 is still a distance vector protocol and there are other routing technologies that are considered superior for large networks. In particular, link-state routing protocols provide rapid routing convergence and reduce the possibility of routing loops.

3.6.6 RIPng

Next generation RIP (RIPng), supporting IPv6 addressing, was defined in RFC 2080 [39] and has similar characteristics to IPv4 RIP: it is also a distance vector protocol, it uses the same timers, procedures, and message types as RIPv2, it uses the same hop count metric (with 16 indicating an unreachable value), it uses Request and Response messages in the same way that RIPv2 does, it supports the split-horizon feature, is multicast-based (using address

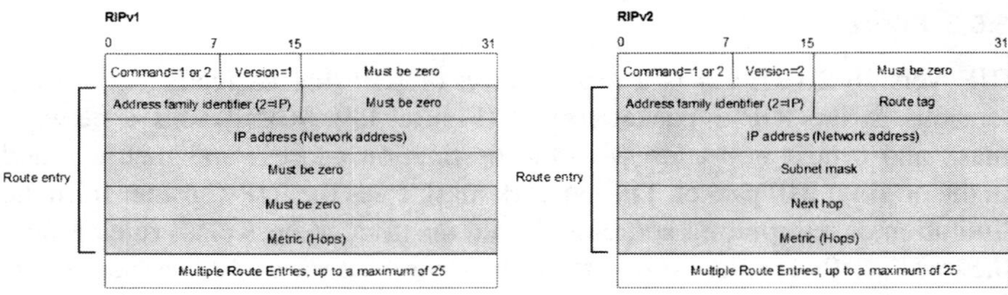

Figure 3.8 Comparison of the RIPv1 and RIPv2 packet formats.

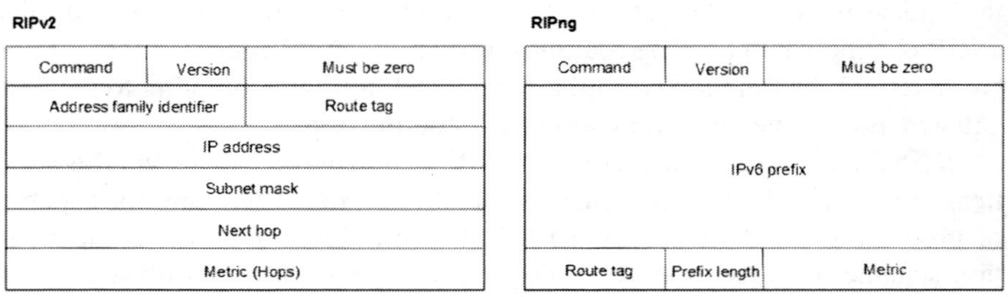

Figure 3.9 Comparing the RIPv2 and RIPng packet formats.

FF02::9) and runs over UDP on port 521. An exception to these parallel functions is authentication. RIPng does not have an authentication mechanism of its own, but instead relies on the authentication features built into IPv6.

There are, however, some updated features for IPv6 support. As can be seen from Figure 3.9, RIPng packets include the IPv6 prefix and prefix length and handles the next hop in a special way: the route tag and prefix length for next hop is all 0, the metric is 0xFF and the next hop must be link local.

The RIPng process is created automatically when RIPng is enabled on an interface with the command:

```
ipv6 rip <process-name> enable
```

This command is still needed when configuring optional features of RIPng: for example, command

```
maximum-paths 2
```

defines the maximum number of equal-cost routes that RIPng can support as 2 (a number from 1 to 64 can be used, with a default of 4).

Note that the process name is specified on each interface. This name is relevant only to the local router and is used to distinguish between multiple

```
# no ip route 0.0.0.0 0.0.0.0 10.3.1.4
# no ip route 10.4.1.0 255.255.255.0 10.3.1.4
# no ipv6 route 2001:A:4:1::/64 2001:A:3:1::4
# no ipv6 route ::/0 2001:A:3:1::4
```

Box 3.30 Static Route removal Router 1.

```
1 #router rip
2   #version 2
3   #network 10.0.0.0
```

Box 3.31 RIPv2 configuration for all Routers.

```
#interface FastEthernet 0/0
  #ipv6 rip 1 enable
```

Box 3.32 RIPng configuration for all Routers (and all interfaces).

RIPng processes that might be running on the router. Different interfaces can each run a unique RIPng process, without exchanging information between these interfaces, or multiple processes might be running on a single interface. Recall that only a single RIPv1 or RIPv2 process can run on a router, and an interface either belongs to this process or it does not run RIP.

Exercise - RIP Operation

Let us consider again the network illustrated in Figure 3.2. Fist of all, we have to remove all the static routes that were previously configured. Box 3.30 shows the commands that remove the static routes configured at Router 1.

Now, let us configure the RIPv2 and RIPng protocols. The configuration of the RIPv2 protocol is very straightforward, as shown in Box 3.31: the command in line 1 activates the RIP configuration mode, the line 2 command is needed to specify the intended version of the protocol and, finally, the line 3 command associates this classful network address to the RIP process: in this way, RIP is enabled on all interfaces that belong to this network, which will now send and receive RIP updates and the router will advertise this network in RIP routing updates sent to other routers every 30 seconds. This configuration should be replicated on all routers. RIPng configuration is made on a per interface basis, as shown in Box 3.32 for a particular FastEthernet interface.

After these configurations, the routing tables should now include all routes learned from RIP. Box 3.33 shows the IPv4 routing table of Router 1: besides the directly connected networks, the remaining routes are learned

```
1  #show ip route
2      10.0.0.0/24 is subnetted, 7 subnets
3  C      10.3.1.0 is directly connected, FastEthernet0/1
4  C      10.2.1.0 is directly connected, FastEthernet0/0
5  R      10.2.2.0 [120/1] via 10.2.1.2, 00:00:04, FastEthernet0/0
6  C      10.1.1.0 is directly connected, FastEthernet1/0
7  R      10.3.2.0 [120/1] via 10.3.1.4, 00:00:28, FastEthernet0/1
8  R      10.2.3.0 [120/2] via 10.3.1.4, 00:00:28, FastEthernet0/1
9                  [120/2] via 10.2.1.2, 00:00:04, FastEthernet0/0
10 R      10.4.1.0 [120/3] via 10.2.1.2, 00:00:04, FastEthernet0/0
```

Box 3.33 IPv4 routing table at Router 1 after RIP configuration.

```
1  #show ipv6 route
2  C    2001:A:1:1::/64 [0/0]
3         via ::, FastEthernet1/0
4  L    2001:A:1:1::1/128 [0/0]
5         via ::, FastEthernet1/0
6  C    2001:A:2:1::/64 [0/0]
7         via ::, FastEthernet0/0
8  L    2001:A:2:1::1/128 [0/0]
9         via ::, FastEthernet0/0
10 R    2001:A:2:2::/64 [120/2]
11        via FE80::C002:AAFF:FEAA:0, FastEthernet0/0
12 R    2001:A:2:3::/64 [120/3]
13        via FE80::C002:AAFF:FEAA:0, FastEthernet0/0
14        via FE80::C004:AAFF:FEAA:0, FastEthernet0/1
15 C    2001:A:3:1::/64 [0/0]
16        via ::, FastEthernet0/1
17 L    2001:A:3:1::1/128 [0/0]
18        via ::, FastEthernet0/1
19 R    2001:A:3:2::/64 [120/2]
20        via FE80::C004:AAFF:FEAA:0, FastEthernet0/1
21 R    2001:A:4:1::/64 [120/3]
22        via FE80::C004:AAFF:FEAA:0, FastEthernet0/1
23 L    FE80::/10 [0/0]
24        via ::, Null0
25 L    FF00::/8 [0/0]
26        via ::, Null0
```

Box 3.34 IPv6 routing table at Router 1 after RIPng configuration.

using RIP. The administrative distance associated to RIP is 120 and the route metric is the number of hops, as previously said. Note that there are two equal cost routes (metric equal to 2) to network 10.2.3.0/24, which are both displayed in the routing table (see Box 3.33 lines 8-9).

The IPv6 routing table of Router 1 is shown in Box 3.34: similarly to IPv4, there are two equal cost routes (with a cost of 2) to network 2001:A:2:3::/64, being both displayed in the routing table (see Box 3.34 lines 12-14). It is also relevant to notice that the next hop is represented by the link local address of the next router interface in the least cost route to the destination.

Figure 3.10 shows the RIPv2 Response packets that were captured in the link between routers 1 and 4, particularly one of the Response packets sent by

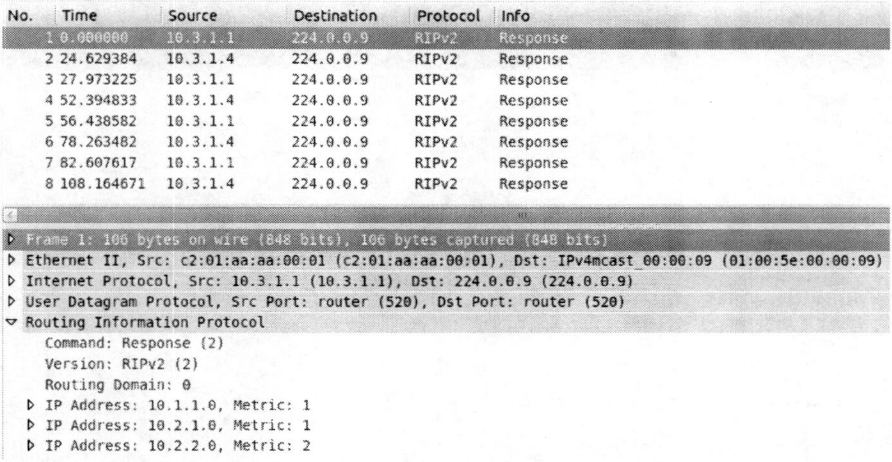

No.	Time	Source	Destination	Protocol	Info
1	0.000000	10.3.1.1	224.0.0.9	RIPv2	Response
2	24.629384	10.3.1.4	224.0.0.9	RIPv2	Response
3	27.973225	10.3.1.1	224.0.0.9	RIPv2	Response
4	52.394833	10.3.1.4	224.0.0.9	RIPv2	Response
5	56.438582	10.3.1.1	224.0.0.9	RIPv2	Response
6	78.263482	10.3.1.4	224.0.0.9	RIPv2	Response
7	82.607617	10.3.1.1	224.0.0.9	RIPv2	Response
8	108.164671	10.3.1.4	224.0.0.9	RIPv2	Response

```
▷ Frame 1: 106 bytes on wire (848 bits), 106 bytes captured (848 bits)
▷ Ethernet II, Src: c2:01:aa:aa:00:01 (c2:01:aa:aa:00:01), Dst: IPv4mcast_00:00:09 (01:00:5e:00:00:09)
▷ Internet Protocol, Src: 10.3.1.1 (10.3.1.1), Dst: 224.0.0.9 (224.0.0.9)
▷ User Datagram Protocol, Src Port: router (520), Dst Port: router (520)
▽ Routing Information Protocol
    Command: Response (2)
    Version: RIPv2 (2)
    Routing Domain: 0
  ▷ IP Address: 10.1.1.0, Metric: 1
  ▷ IP Address: 10.2.1.0, Metric: 1
  ▷ IP Address: 10.2.2.0, Metric: 2
```

Figure 3.10 RIPv2 response packet from Router 1 (at network 10.3.1.0/24).

Router 1. In this case, Router 1 announces three distance vectors to Router 4, which means that the split horizon mechanism is active: Router 1 only announces to Router 4 the routes whose minimum cost path does not pass through Router 4 (in other words, the routes that were not learned from Router 4); otherwise, the Response packet should include all distance vectors. So, we can conclude that split-horizon is active by default. Figure 3.11 shows the RIPv2 Response packet sent by Router 4.

Figure 3.12 shows the RIPng Response packets that were captured in the same link between routers 1 and 4, particularly one of the Response packets sent by Router 1. Router 1 announces three distance vectors to Router 4, so the split horizon mechanism is active. Figure 3.13 shows the RIPv2 Response packet sent by Router 4, also announcing three distance vectors. Note that the source address of these packets is the link local address of the sending interface.

Next, we shut down the interface of Router 1 to network 10.1.1.0/24 and captured the RIP packets that were exchanged. Figure 3.14 shows the RIPv2 Response packet that is sent by Router 1 to network 10.3.1.0/24, announcing network 10.1.1.0/24 with a metric of 16 (infinity). This packet illustrates the previously explained poison reverse mechanism in full operation: Router 1 advertises an infinite distance to network 10.1.1.0/24 on the link corresponding to network 10.3.1.0/24. After this announcement, Router 1 will remove this network from further Response packets, announcing only

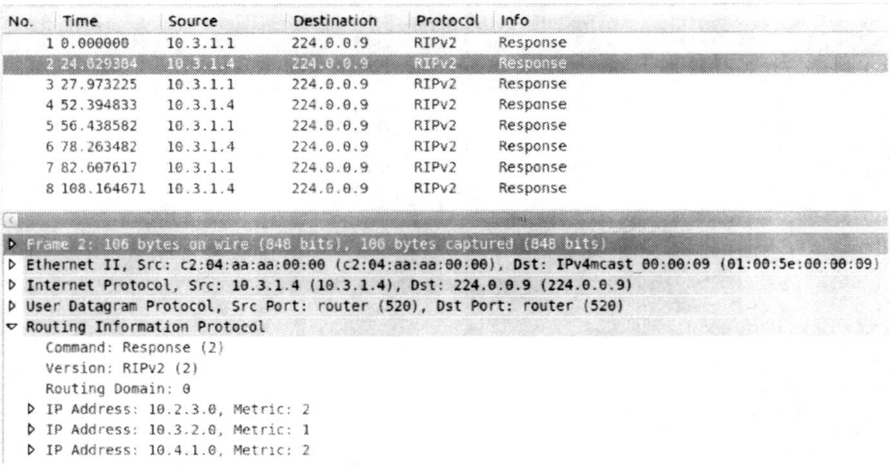

No.	Time	Source	Destination	Protocol	Info
1	0.000000	10.3.1.1	224.0.0.9	RIPv2	Response
2	24.029304	10.3.1.4	224.0.0.9	RIPv2	Response
3	27.973225	10.3.1.1	224.0.0.9	RIPv2	Response
4	52.394833	10.3.1.4	224.0.0.9	RIPv2	Response
5	56.438582	10.3.1.1	224.0.0.9	RIPv2	Response
6	78.263482	10.3.1.4	224.0.0.9	RIPv2	Response
7	82.607617	10.3.1.1	224.0.0.9	RIPv2	Response
8	108.164671	10.3.1.4	224.0.0.9	RIPv2	Response

```
▷ Frame 2: 106 bytes on wire (848 bits), 106 bytes captured (848 bits)
▷ Ethernet II, Src: c2:04:aa:aa:00:00 (c2:04:aa:aa:00:00), Dst: IPv4mcast_00:00:09 (01:00:5e:00:00:09)
▷ Internet Protocol, Src: 10.3.1.4 (10.3.1.4), Dst: 224.0.0.9 (224.0.0.9)
▷ User Datagram Protocol, Src Port: router (520), Dst Port: router (520)
▽ Routing Information Protocol
    Command: Response (2)
    Version: RIPv2 (2)
    Routing Domain: 0
  ▷ IP Address: 10.2.3.0, Metric: 2
  ▷ IP Address: 10.3.2.0, Metric: 1
  ▷ IP Address: 10.4.1.0, Metric: 2
```

Figure 3.11 RIPv2 response packet from Router 4 (at network 10.3.1.0/24).

the remaining networks (Figure 3.15). Regarding the RIPng protocol, the poison reverse mechanism operates exactly in the same way: as can be seen from Figure 3.16, Router 1 advertises an infinite distance to network 2001:A:1:1::/64 on the link corresponding to network 2001:A:3:1::/64 and, after that announcement, Router 1 will remove this network from further Response packets.

Finally, we disconnected Router 4 from network 10.3.2.0/24 (2001:A:3:2::/64) by shutting down its F0/1 interface. Obviously, the RIP processes will converge into new sets of alternative paths. Boxes 3.35 and 3.37 display the new Router 1 IPv4 and IPv6 routing tables, respectively: new paths have been calculated, passing through the alternative way, like for example the route to network 10.3.2.0/24 (2001:A:3:2::/64) that has a new metric of 3 and a new next hop corresponding to the interface of Router 2 (see line 7 in Box 3.35 and lines 12-13 in Box 3.37). A similar situation occurs if we look at Router 4 (Boxes 3.36 and 3.38) now Router R4 is able to reach network 10.3.2.0/24 (2001:A:3:2::/64) with a metric of 4 hops through Router 1 (see line 7 in Box 3.36 and lines 8-9 in Box 3.38). All other routes are also re-calculated using the alternative path.

3.6.7 Route Redistribution

As already said, multi-protocol environments are common in operational networks. Route Redistribution allows routes from one routing protocol to be

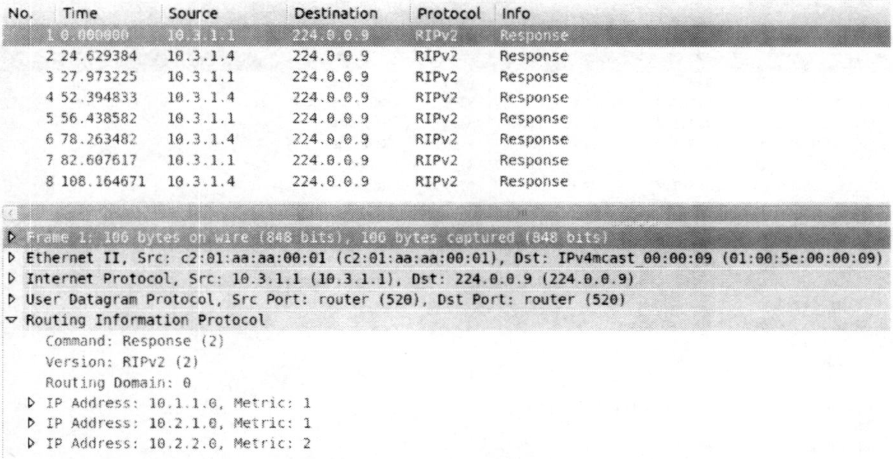

```
No.  Time        Source     Destination  Protocol  Info
   1 0.000000    10.3.1.1   224.0.0.9    RIPv2     Response
   2 24.629384   10.3.1.4   224.0.0.9    RIPv2     Response
   3 27.973225   10.3.1.1   224.0.0.9    RIPv2     Response
   4 52.394833   10.3.1.4   224.0.0.9    RIPv2     Response
   5 56.438582   10.3.1.1   224.0.0.9    RIPv2     Response
   6 78.263482   10.3.1.4   224.0.0.9    RIPv2     Response
   7 82.607617   10.3.1.1   224.0.0.9    RIPv2     Response
   8 108.164671  10.3.1.4   224.0.0.9    RIPv2     Response

▷ Frame 1: 106 bytes on wire (848 bits), 106 bytes captured (848 bits)
▷ Ethernet II, Src: c2:01:aa:aa:00:01 (c2:01:aa:aa:00:01), Dst: IPv4mcast_00:00:09 (01:00:5e:00:00:09)
▷ Internet Protocol, Src: 10.3.1.1 (10.3.1.1), Dst: 224.0.0.9 (224.0.0.9)
▷ User Datagram Protocol, Src Port: router (520), Dst Port: router (520)
▽ Routing Information Protocol
    Command: Response (2)
    Version: RIPv2 (2)
    Routing Domain: 0
  ▷ IP Address: 10.1.1.0, Metric: 1
  ▷ IP Address: 10.2.1.0, Metric: 1
  ▷ IP Address: 10.2.2.0, Metric: 2
```

Figure 3.12 RIPng response packet from Router 1 (at network 2001:A:3:1::/64).

```
#show ip route
   10.0.0.0/24 is subnetted, 7 subnets
C     10.3.1.0 is directly connected, FastEthernet0/1
C     10.2.1.0 is directly connected, FastEthernet0/0
R     10.2.2.0 [120/1] via 10.2.1.2, 00:00:24, FastEthernet0/0
C     10.1.1.0 is directly connected, FastEthernet1/0
R     10.3.2.0 [120/3] via 10.2.1.2, 00:00:24, FastEthernet0/0
R     10.2.3.0 [120/2] via 10.2.1.2, 00:00:24, FastEthernet0/0
R     10.4.1.0 [120/3] via 10.2.1.2, 00:00:24, FastEthernet0/0
```

Box 3.35 IPv4 routing table at Router 1 after network 10.3.2.0/24 removal.

advertised into another routing protocol. The routing protocol receiving these redistributed routes usually marks the routes as external. External routes are usually less preferred than locally-originated routes. At least one redistribution point (router) needs to exist between the two routing domains. This routing device will actually run both routing protocols.

It is possible to redistribute from one routing protocol to the same routing protocol. Static routes and connected interfaces can also be redistributed into a routing protocol. Routes will only be redistributed if they exist in the routing table. Routing metrics are a key consideration when performing route redistribution. Each routing protocol usually utilizes a unique (and thus incompatible) metric. Routes redistributed from the injecting protocol must be manually (or globally) stamped with a metric that is understood by the receiving protocol.

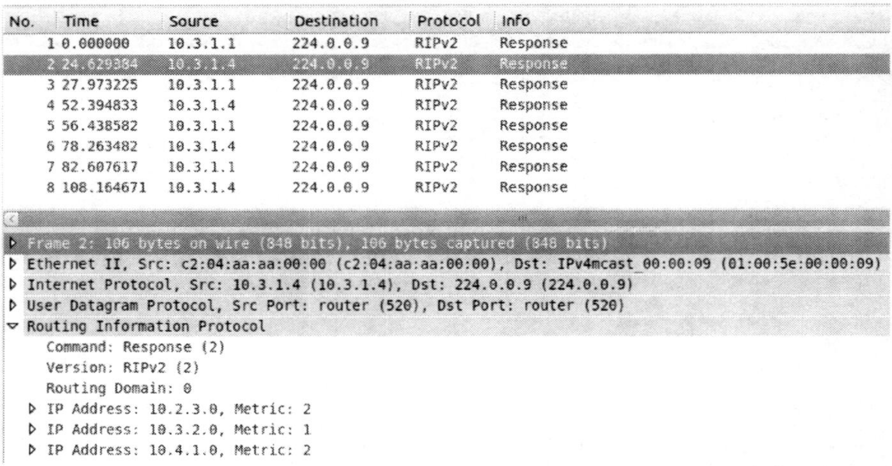

No.	Time	Source	Destination	Protocol	Info
1	0.000000	10.3.1.1	224.0.0.9	RIPv2	Response
2	24.629384	10.3.1.4	224.0.0.9	RIPv2	Response
3	27.973225	10.3.1.1	224.0.0.9	RIPv2	Response
4	52.394833	10.3.1.4	224.0.0.9	RIPv2	Response
5	56.438582	10.3.1.1	224.0.0.9	RIPv2	Response
6	78.263482	10.3.1.4	224.0.0.9	RIPv2	Response
7	82.607617	10.3.1.1	224.0.0.9	RIPv2	Response
8	108.164671	10.3.1.4	224.0.0.9	RIPv2	Response

```
▷ Frame 2: 106 bytes on wire (848 bits), 106 bytes captured (848 bits)
▷ Ethernet II, Src: c2:04:aa:aa:00:00 (c2:04:aa:aa:00:00), Dst: IPv4mcast_00:00:09 (01:00:5e:00:00:09)
▷ Internet Protocol, Src: 10.3.1.4 (10.3.1.4), Dst: 224.0.0.9 (224.0.0.9)
▷ User Datagram Protocol, Src Port: router (520), Dst Port: router (520)
▽ Routing Information Protocol
    Command: Response (2)
    Version: RIPv2 (2)
    Routing Domain: 0
  ▷ IP Address: 10.2.3.0, Metric: 2
  ▷ IP Address: 10.3.2.0, Metric: 1
  ▷ IP Address: 10.4.1.0, Metric: 2
```

Figure 3.13 RIPng response packet from Router 4 (at network 2001:A:3:1::/64).

```
#show ip route
    10.0.0.0/24 is subnetted, 7 subnets
C     10.3.1.0 is directly connected, FastEthernet0/0
R     10.2.1.0 [120/1] via 10.3.1.1, 00:00:19, FastEthernet0/0
R     10.2.2.0 [120/2] via 10.3.1.1, 00:00:19, FastEthernet0/0
R     10.1.1.0 [120/1] via 10.3.1.1, 00:00:19, FastEthernet0/0
R     10.3.2.0 [120/4] via 10.3.1.1, 00:00:19, FastEthernet0/0
R     10.2.3.0 [120/3] via 10.3.1.1, 00:00:19, FastEthernet0/0
R     10.4.1.0 [120/4] via 10.3.1.1, 00:00:19, FastEthernet0/0
```

Box 3.36 IPv4 routing table at Router 4 after network 10.3.2.0/24 removal.

Just as an example, redistributing other protocols into RIP can be done using the following possible commands:

```
# router rip
  # redistribute static
  # redistribute igrp <AS_Number>
  # redistribute eigrp <AS_Number>
  # redistribute ospf <Process_Number> default-metric <value>
```

When redistributing a protocol into RIP, a low metric (such as 1) should be used because a high metric limits RIP: defining a metric of 10 for redistributed routes will only enable these routes to be advertised to routers up to 5 hops away, at which point the metric (hop count) exceeds 15. However, doing this increases the possibility of routing loops if there are multiple redistribution points and a router learns about the network with a better metric from the redistribution point than from the original source. Therefore, the metric should not be neither too high, preventing it from being advertised to

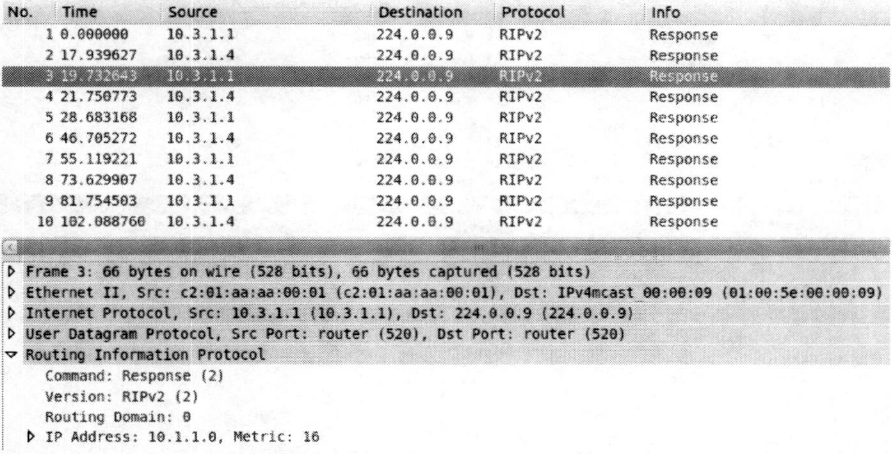

No.	Time	Source	Destination	Protocol	Info
1	0.000000	10.3.1.1	224.0.0.9	RIPv2	Response
2	17.939627	10.3.1.4	224.0.0.9	RIPv2	Response
3	19.732643	10.3.1.1	224.0.0.9	RIPv2	Response
4	21.750773	10.3.1.4	224.0.0.9	RIPv2	Response
5	28.683168	10.3.1.1	224.0.0.9	RIPv2	Response
6	46.705272	10.3.1.4	224.0.0.9	RIPv2	Response
7	55.119221	10.3.1.1	224.0.0.9	RIPv2	Response
8	73.629907	10.3.1.4	224.0.0.9	RIPv2	Response
9	81.754503	10.3.1.1	224.0.0.9	RIPv2	Response
10	102.088760	10.3.1.4	224.0.0.9	RIPv2	Response

```
▷ Frame 3: 66 bytes on wire (528 bits), 66 bytes captured (528 bits)
▷ Ethernet II, Src: c2:01:aa:aa:00:01 (c2:01:aa:aa:00:01), Dst: IPv4mcast_00:00:09 (01:00:5e:00:00:09)
▷ Internet Protocol, Src: 10.3.1.1 (10.3.1.1), Dst: 224.0.0.9 (224.0.0.9)
▷ User Datagram Protocol, Src Port: router (520), Dst Port: router (520)
▽ Routing Information Protocol
    Command: Response (2)
    Version: RIPv2 (2)
    Routing Domain: 0
  ▷ IP Address: 10.1.1.0, Metric: 16
```

Figure 3.14 RIPv2 response (reverse poisoning) packet from Router 1 (at network 10.3.1.0/24).

all the routers, or too low, leading to routing loops when there are multiple redistribution points.

3.7 Open Shortest Path First

The Open Shortest Path First (OSPF) protocol, defined in RFC 2328 [41], is a link-state routing protocol that addresses most of the pitfalls presented by RIP:

- Has no limitation on the hop count.
- Supports variable length subnet masks (VLSM).
- Uses IP multicast to send link-state updates, which ensures less processing on routers that are not listening to OSPF packets and a better use of bandwidth since updates are only sent in case routing changes occur.
- Has better convergence than RIP, because routing changes are propagated instantaneously and not periodically.
- Allows for a better load balancing.
- Allows for a logical definition of networks, dividing routers into areas, which limits the explosion of link state updates over the whole network and provides a mechanism for aggregating routes and cutting down on the unnecessary propagation of subnet information.
- Allows for routing authentication.

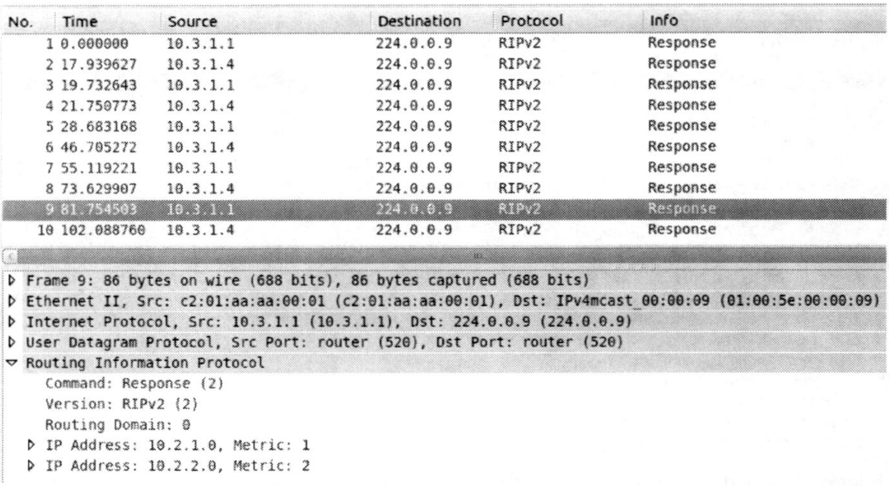

No.	Time	Source	Destination	Protocol	Info
1	0.000000	10.3.1.1	224.0.0.9	RIPv2	Response
2	17.939627	10.3.1.4	224.0.0.9	RIPv2	Response
3	19.732643	10.3.1.1	224.0.0.9	RIPv2	Response
4	21.750773	10.3.1.4	224.0.0.9	RIPv2	Response
5	28.683168	10.3.1.1	224.0.0.9	RIPv2	Response
6	46.705272	10.3.1.4	224.0.0.9	RIPv2	Response
7	55.119221	10.3.1.1	224.0.0.9	RIPv2	Response
8	73.629907	10.3.1.4	224.0.0.9	RIPv2	Response
9	81.754503	10.3.1.1	224.0.0.9	RIPv2	Response
10	102.088760	10.3.1.4	224.0.0.9	RIPv2	Response

```
▷ Frame 9: 86 bytes on wire (688 bits), 86 bytes captured (688 bits)
▷ Ethernet II, Src: c2:01:aa:aa:00:01 (c2:01:aa:aa:00:01), Dst: IPv4mcast_00:00:09 (01:00:5e:00:00:09)
▷ Internet Protocol, Src: 10.3.1.1 (10.3.1.1), Dst: 224.0.0.9 (224.0.0.9)
▷ User Datagram Protocol, Src Port: router (520), Dst Port: router (520)
▽ Routing Information Protocol
    Command: Response (2)
    Version: RIPv2 (2)
    Routing Domain: 0
  ▷ IP Address: 10.2.1.0, Metric: 1
  ▷ IP Address: 10.2.2.0, Metric: 2
```

Figure 3.15 RIPv2 response (after reverse poisoning) packet from Router 1 (at network 10.3.1.0/24).

- Allows for the transfer and tagging of external routes injected into an Autonomous System, keeping track of external routes injected by exterior protocols.

Routers running OSPF collect routing information from all other routers in the network (or from within a defined area of the network) and, then, each router independently calculates its best paths to all destinations in the network, using the Dijkstra Shortest Path First (SPF) algorithm [18]. For all the routers in the network to make consistent routing decisions, each link-state router must keep a record of the following information:

- Its immediate neighbor routers - If the router loses contact with a neighbor router it invalidates all paths through that router and recalculates its paths through the network. In OSPF, adjacency information about neighbors is stored in the OSPF neighbor table, also known as an adjacency database.
- All the other routers in the network, or in its area of the network, and their attached networks - The router recognizes other routers and networks through Link State Advertisements (LSAs), which are flooded through the network. LSAs are stored in a topology table or database, also called Link State Database (LSDB).

No.	Time	Source	Destination	Protocol	Info
1	0.000000	fe80::c001:aaff:feaa:1	ff02::9	RIPng version 1	Response
2	10.447328	fe80::c004:aaff:feaa:0	ff02::9	RIPng version 1	Response
3	14.902862	fe80::c001:aaff:feaa:1	ff02::9	RIPng version 1	Response
4	26.909832	fe80::c001:aaff:feaa:1	ff02::9	RIPng version 1	Response
5	36.856324	fe80::c004:aaff:feaa:0	ff02::9	RIPng version 1	Response
6	53.628455	fe80::c001:aaff:feaa:1	ff02::9	RIPng version 1	Response
7	62.437606	fe80::c004:aaff:feaa:0	ff02::9	RIPng version 1	Response

```
▷ Frame 3: 86 bytes on wire (688 bits), 86 bytes captured (688 bits)
▷ Ethernet II, Src: c2:01:aa:aa:00:01 (c2:01:aa:aa:00:01), Dst: IPv6mcast_00:00:00:09 (33:33:00:00:00:09)
▷ Internet Protocol Version 6, Src: fe80::c001:aaff:feaa:1 (fe80::c001:aaff:feaa:1), Dst: ff02::9 (ff02::9)
▷ User Datagram Protocol, Src Port: ripng (521), Dst Port: ripng (521)
▽ RIPng
    Command: Response (2)
    Version: 1
  ▷ IP Address: 2001:a:1:1::/64, Metric: 16
```

Figure 3.16 RIPng response (reverse poisoning) packet from Router 1 (at network 2001:A:3:1::/64).

```
#show ipv6 route
C    2001:A:1:1::/64 [0/0]
      via ::, FastEthernet1/0
L    2001:A:1:1::1/128 [0/0]
      via ::, FastEthernet1/0
C    2001:A:2:1::/64 [0/0]
      via ::, FastEthernet0/0
L    2001:A:2:1::1/128 [0/0]
      via ::, FastEthernet0/0
R    2001:A:2:2::/64 [120/2]
      via FE80::C002:AAFF:FEAA:0, FastEthernet0/0
R    2001:A:2:3::/64 [120/3]
      via FE80::C002:AAFF:FEAA:0, FastEthernet0/0
C    2001:A:3:1::/64 [0/0]
      via ::, FastEthernet0/1
L    2001:A:3:1::1/128 [0/0]
      via ::, FastEthernet0/1
R    2001:A:3:2::/64 [120/4]
      via FE80::C002:AAFF:FEAA:0, FastEthernet0/0
L    FE80::/10 [0/0]
      via ::, Null0
L    FF00::/8 [0/0]
      via ::, Null0
```

Box 3.37 IPv6 routing table at Router 1 after network 10.3.2.0/24 removal.

- The best paths to each destination - Each router independently calculates the best paths to each destination in the network using Dijkstra's SPF algorithm. All paths are kept in the LSDB. The best paths are then put in the routing table (also called the forwarding database). Packets arriving at the router are forwarded based on the information held in the routing table.

```
#show ipv6 route
R    2001:A:1:1::/64 [120/2]
      via FE80::C001:AAFF:FEAA:1, FastEthernet0/0
R    2001:A:2:1::/64 [120/2]
      via FE80::C001:AAFF:FEAA:1, FastEthernet0/0
R    2001:A:2:2::/64 [120/3]
      via FE80::C001:AAFF:FEAA:1, FastEthernet0/0
R    2001:A:2:3::/64 [120/4]
      via FE80::C001:AAFF:FEAA:1, FastEthernet0/0
C    2001:A:3:1::/64 [0/0]
      via ::, FastEthernet0/0
L    2001:A:3:1::4/128 [0/0]
      via ::, FastEthernet0/0
R    2001:A:3:2::/64 [120/5]
      via FE80::C001:AAFF:FEAA:1, FastEthernet0/0
R    2001:A:4:1::/64 [120/5]
      via FE80::C001:AAFF:FEAA:1, FastEthernet0/0
L    FE80::/10 [0/0]
      via ::, Null0
L    FF00::/8 [0/0]
      via ::, Null0
```

Box 3.38 IPv6 routing table at Router 4 after network 10.3.2.0/24 removal.

3.7.1 Route Computation

OSPF uses a SPF algorithm to calculate the shortest path to all known destinations, forming a Shortest Path Tree (SPT). The shortest path is calculated using the Dijkstra algorithm, and these are the basic steps to build the SPT:

- Upon initialization or due to any change in routing information, a router generates a Link State Advertisement. This advertisement represents the collection of all link-states on that router.
- All routers exchange link-states by flooding. Each router that receives a link-state update should store a copy in its link-state database and propagate the update to other routers.
- After completing its database, each router calculates a SPT to all destinations based on the Dijkstra algorithm (Figure 3.17). The destinations, their associated cost and the next hop to reach them form the IP routing table.
- In case no changes in the OSPF network occur, such as cost of a link or a network being added or deleted, OSPF remains quiet. Any changes that occur are communicated through link-state packets, and the Dijkstra algorithm is re-run in order to find the shortest path.
- The algorithm places each router at the root of a tree and calculates the shortest path to each destination based on the cumulative cost required to reach that destination. Each router will have its own view of the topology even though all the routers will build a SPT using the same LSDB. A

Dijkstra's Algorithm:

```
s       source node.
Dn      cost of the least-cost path from node s to node n

M = {s};
for each  n ∉ M
        Dn  = dsn;
while (M ≠ all nodes) do
        Find w ∉ M for which Dw = min{Dj ; j ∉ M};
        Add w to M;
        for each  n ∉ M
                Dn = minw [ Dn, Dw + dwn ];
                Update route;
enddo
```

Figure 3.17 The Dijkstra algorithm: calculating the shortest path for node s.

routing table is derived from the SPF tree, containing the best route to each router.

To run OSPF, a router must have a Router ID, which is a 32-bit unsigned integer, the unique identifier of the router in the AS. A Router ID can be assigned to an OSPF router manually. If no Router ID is specified, the system automatically selects one for the router as follows: if loopback interfaces are configured, the system selects the highest IP address among them; if no loopback interface is configured, the highest IP address among addresses of active interfaces on the router is selected.

3.7.2 Neighboring and Adjacencies

Routers that share a common segment become neighbors on that segment. Neighbors are elected via the Hello protocol. Hello packets are sent periodically out of each interface using IP multicast. Routers become neighbors as soon as they see themselves listed in the neighbor's Hello packet. In order to become neighbors, two routers have to agree on the following:

- Area ID - their interfaces have to belong to the same area on that segment;
- Authentication - routers have to exchange the same password on a particular segment;
- Hello and Dead Intervals - the Hello interval specifies the length of time, in seconds, between the hello packets that a router sends on an OSPF interface, while the dead interval is the number of seconds that a router's

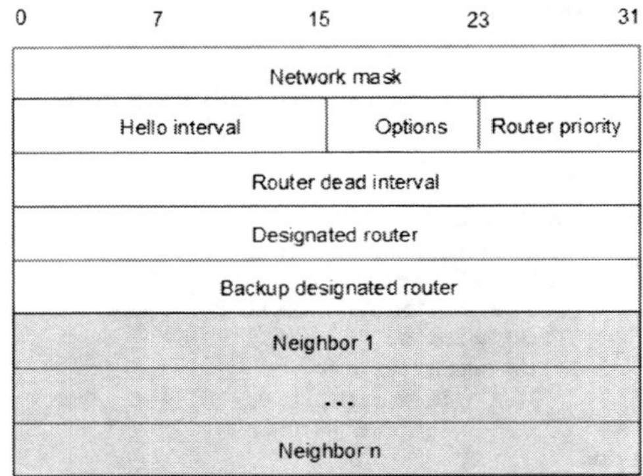

Figure 3.18 The OSPF Hello message format.

Hello packets have not been seen before its neighbors declare the OSPF router down;

- Stub area flag.

After becoming neighbors, routers have to establish adjacency relationships between them. Adjacent routers are routers that go beyond the simple Hello exchange and proceed into the database exchange process. In order to minimize the amount of information exchange on a particular segment, OSPF elects one router to be a Designated Router (DR), and one router to be a Backup Designated Router (BDR), on each multi-access segment. The BDR is elected as a backup mechanism in case the DR goes down.

DR and BDR election is done via the Hello protocol. Hello packets, whose general format is illustrated in Figure 3.18, are exchanged via IP multicast packets on each segment. The router with the highest OSPF priority on a segment will become the DR for that segment. The same process is repeated for the BDR. In case of a tie, the router with the highest Router ID will win. The default for the interface OSPF priority is one. Remember that the DR and BDR concepts are per multiaccess segment.

OSPF will always form an adjacency with the neighbor on the other side of a point-to-point interface, such as point-to-point serial lines. There is no concept of DR or BDR. The state of the serial interfaces is point-to-point.

Special care should be taken when configuring OSPF over multi-access non-broadcast medias such as Frame Relay, X.25, ATM. The protocol considers these media like any other broadcast media such as Ethernet. NBMA (Non-Broadcast Multiple Access) clouds are usually built in a hub and spoke topology. The selection of the DR becomes an issue because the DR and BDR need to have full physical connectivity with all routers that exist on the cloud. Because of the lack of broadcast capabilities, the DR and BDR need to have a static list of all other routers attached to the cloud.

If a router with a higher priority is added to the network later on it does not take over the DR and no re-election takes place. It is possible for a router to be a DR in one network and a normal router in another at the same time.

After election, the routers are in the Exstart state as the DR and BDR create an adjacency with each other and the router with the highest priority acts as the master and they begin creating their link-state databases using Database Description Packets.

The process of discovering routes by exchanging Database Description Packets (DBD) is known as Exchange. These packets contain details such as the link-state type, the address of the advertising router, the cost of the link and the sequence number that identifies how recent the link information is. Unicasts are used to compare link state databases to see which Link State Advertisements (LSAs) are missing or out of date.

Once a DBD has been received, a Link State ACK is sent containing the link-state entry sequence number. The slave router compares the information and if it is newer it sends a request to update. In order to update its LSDB the slave router sends a Link State Request. This is known as the Loading state.

A Link State Update is sent in response to a Link State Request and it contains the requested LSAs. Once a Link State Update has been received, a Link State ACK is sent again and the adjacency has been formed. At this point the databases are considered to be synchronous.

In the Full state, routers can route traffic and continue sending hello packets to each other in order to maintain the adjacency and the routing information.

Figure 3.19 illustrates the link state database synchronization process.

All OSPF messages share a similar packet structure, beginning with a 24-byte header, as shown in Figure 3.20. Version is the version number, Type indicates the type of OSPF packet (according to Table 3.1), Packet Length is the length of the message, in bytes, including the 24 bytes of the header, Router ID is the ID of the router that generated this message (generally its IP address on the interface over which the message was sent), Area ID is an identification of the OSPF area to which this message belongs, when areas are used, Checksum is a 16-bit checksum computed in a manner similar to a standard IP checksum (the entire message is included in the calculation except the Authentication field), AuType is the type of authentication used (0 means No Authentication, 1 is for Simple Password Authentication and 2

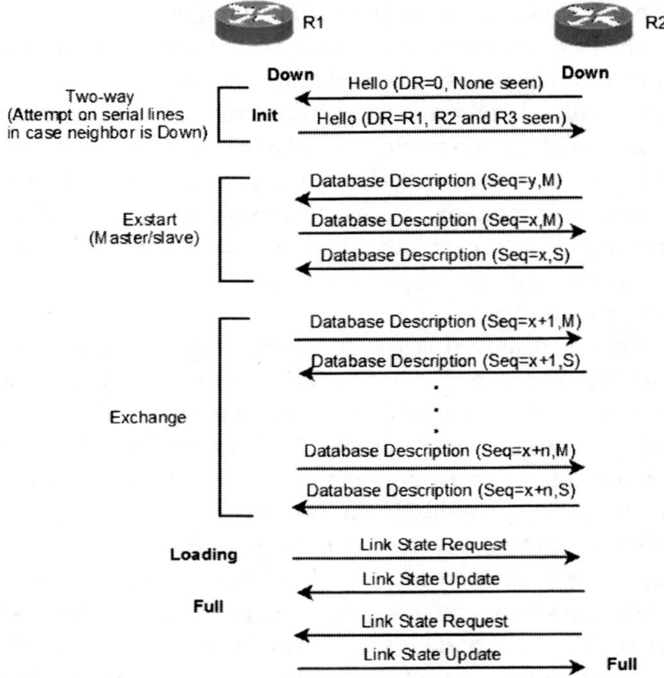

Figure 3.19 Link state database synchronization.

for Cryptographic Authentication), Authentication is a 64-bit field used for authentication of the message, as needed.

The Hello message is used to discover neighbors and build adjacencies between them. Database Description (DBD) messages are used to check for database synchronization between routers. Link-State Request (LSR) packets request for specific link-state records from another router. Link-State Update (LSU) packets send specifically requested link-state records, while Link State Acknowledgment (LSAck) messages are used to acknowledge the other packet types.

3.7.3 OSPF Database

The OSPF database is organized in two tables:

- Router Link States - This is the routers related information table. Routers are identified by theirs Router IDs.

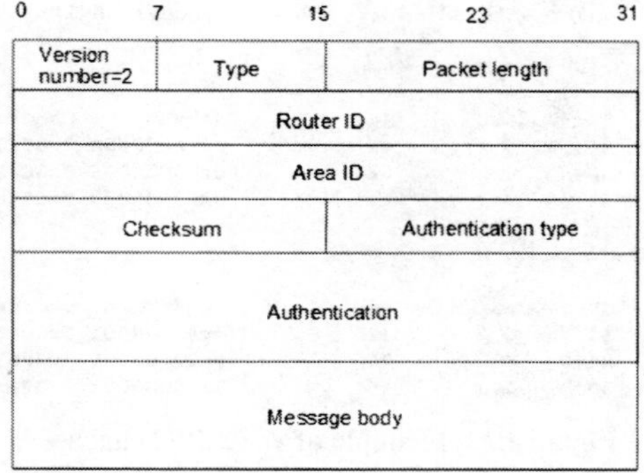

Figure 3.20 Common OSPF packet format.

Table 3.1 OSPF message types.

Type Value	OSPF Message Type
1	Hello
2	Database Description
3	Link State Request
4	Link State Update
5	Link State Acknowledgment

- Net Link States - Networks/Links related information table. Networks are also identified by their IDs.

Figure 3.21 shows an example of an OSPF database, corresponding to a brief output of the database contents.

Figure 3.22 shows an example of a Router link State: for each router, this table contains information about the directly connected networks.

Figure 3.23 shows an example of a Network link State: for each network, it contains information about directly attached routers.

3.7.4 Enabling OSPF on a Router

Enabling OSPF on a router involves the following two steps in configuration mode:

- Enable the OSPF process by using the command

```
# router ospf <process-id>
```

- Assign areas to interfaces using the command

```
        OSPF Router with ID (20.20.20.1) (Process ID 1)

              Router Link States (Area 0)

Link ID         ADV Router      Age         Seq#       Checksum Link count
20.20.20.1      20.20.20.1      40          0x8000000A 0x00E7FB 2
30.30.30.2      30.30.30.2      69          0x80000006 0x002906 2
30.30.30.3      30.30.30.3      41          0x80000007 0x00283D 2

              Net Link States (Area 0)

Link ID         ADV Router      Age         Seq#       Checksum
10.10.10.3      30.30.30.3      41          0x80000001 0x00051C
20.20.20.2      30.30.30.2      70          0x80000001 0x00A164
30.30.30.3      30.30.30.3      154         0x80000001 0x00A91C
```

Figure 3.21 Example of an OSPF Database.

```
LS age: 321
Options: (No TOS-capability, DC)
LS Type: Router Links
Link State ID: 20.20.20.1 <---------------- Router ID
Advertising Router: 20.20.20.1
LS Seq Number: 8000000A
Checksum: 0xE7FB
Length: 48
Number of Links: 2 <---------------- Number of Links

  Link connected to: a Transit Network <---------------- Network Type
   (Link ID) Designated Router address: 20.20.20.2 <---- Network ID
   (Link Data) Router Interface address: 20.20.20.1 <---- Interface IP Address
    Number of TOS metrics: 0
     TOS 0 Metrics: 1 <---------------- Interface Cost

  Link connected to: a Transit Network
   (Link ID) Designated Router address: 10.10.10.3
   (Link Data) Router Interface address: 10.10.10.1
    Number of TOS metrics: 0
     TOS 0 Metrics: 1
```

Figure 3.22 Example of a Router Link State.

```
# network <network-id> <wildcard-mask> <area-id>
```

where `process-id` is a numeric value local to the router. It does not have to match process-ids on other routers. It is possible to run multiple OSPF processes on the same router, but is not recommended as it creates multiple database instances that add extra overhead to the router. The `network` command is a way of assigning an interface of a specific network (`network-id`) to a certain area (`area-id`). The `wildcard-mask` is used as a shortcut and it helps putting a list of interfaces in the same area with one configuration line. The mask contains wild card bits where 0 is a match and 1 is a "do not care" bit, e.g. 0.0.255.255 indicates a match in the first two bytes of the network number.

```
Routing Bit Set on this LSA
LS age: 483
Options: (No TOS-capability, DC)
LS Type: Network Links
Link State ID: 10.10.10.3 (address of Designated Router) <---- Network ID
Advertising Router: 30.30.30.3
LS Seq Number: 80000001
Checksum: 0x51C
Length: 32
Network Mask: /24
  Attached Router: 30.30.30.3 }
  Attached Router: 20.20.20.1 } <---- Attached routers (RID)
```

Figure 3.23 Example of a Network Link State.

3.7.5 Authentication

It is possible to authenticate OSPF packets such that routers can participate in routing domains based on predefined passwords. By default, a router uses a Null authentication, which means that routing exchanges over a network are not authenticated. Two authentication methods are available for OSPF: Simple password authentication and Message Digest authentication (MD-5).

Simple password authentication allows a password (key) to be configured per area. Routers in the same area that want to participate in the routing domain will have to be configured with the same key. The drawback of this method is that it is vulnerable to passive attacks. Anybody with a link analyzer could easily get the password off the wire.

Message Digest authentication is a cryptographic authentication. A key (password) and a key-id are configured on each router. The router uses an algorithm based on the OSPF packet, the key, and the key-id to generate a "message digest" that gets appended to the packet. Unlike the simple authentication, the key is not exchanged over the wire. A non-decreasing sequence number is also included in each OSPF packet to protect against replay attacks.

3.7.6 Hierarchical Routing

In small networks, the web of router links is not complex, and paths to individual destinations are easily deduced. In large networks, the resulting web is highly complex, and the number of potential paths to each destination is large. Dijkstra calculations comparing all possible routes can be very complex and take significant time. Since the Link State Database covers the topology of the entire network, each router must maintain an entry for every network in the area, even if not every route is selected for the routing table. Besides, in a large network changes are inevitable, so routers spend many

CPU cycles recalculating the SPF algorithm and updating the routing table. OSPF does not perform route summarization by default. If routes are not summarized, routing tables can become very large, depending on the size of the network. Link-state routing protocols usually reduce the size of the Dijkstra calculations by partitioning the network into areas.

Multiple areas simplify Dijkstra calculations

Dijkstra calculations are simplified by partitioning the network into areas.

OSPF supports the definition of multiple Areas to optimize processing in SPF calculations, memory use and minimize the number of LSAs being transmitted. Areas are firstly defined on the routers and then interfaces are assigned to areas. The default area is 0.0.0.0 and should exist even if there is only one area in the whole network. As more areas are added, 0.0.0.0 becomes the "backbone area".

Separate LSDBs are maintained per area and networks outside of an area are advertised into that area, routers internal to an area have less work to do as only topology changes within an area affect a modification of the SPF specific to that area. Another benefit of implementing areas is that networks within an area can be advertised as a summary so reducing the size of the routing table and the processing on routers external to this area. Creating summaries is made easier if addresses within an area are contiguous.

OSPF uses a two-layer area hierarchy:

- Backbone area - An OSPF area whose primary function is the fast and efficient movement of IP packets. Generally, end users are not found within a backbone area. The backbone area is also called OSPF area 0, which is defined as the core to which all other areas connect (directly or virtually).
- Regular (non backbone) or normal area - An OSPF area whose primary function is to connect users and resources. Regular areas are usually set up along functional or geographic groupings. By default, a regular area does not allow traffic from another area to use its links to reach other areas. Regular areas can have several subtypes, including standard area, stub area, totally stubby area, not-so-stubby area (NSSA), and totally stubby NSSA.

In a multiple area environment there are four types of routers:

- Internal Router - All its directly connected networks are within the same area as itself. It is only concerned with the LSDB for that area.

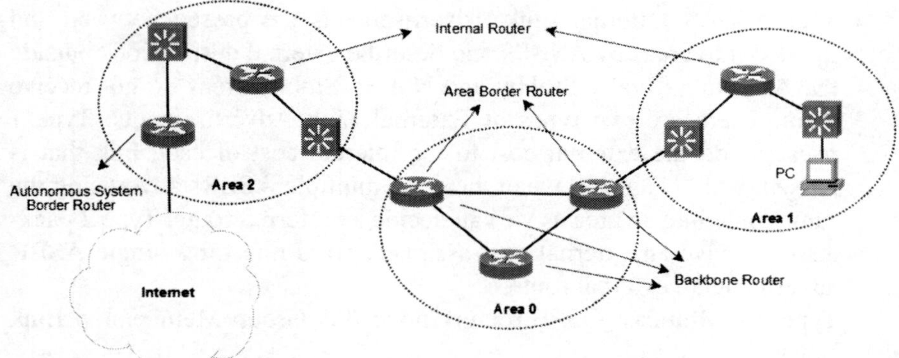

Figure 3.24 OSPF hierarchical routing and different types of routers.

- Area Border Router - This type of router has interfaces in multiple areas and so has to maintain multiple LSDBs as well as be connected to the backbone. It sends and receives Summary Links Advertisements from the backbone area and they describe one network or a range of networks within the area.
- Backbone Router - Has an interface connected to the backbone.
- AS Boundary Router - Has an interface connected to a non-OSPF network which is considered to be outside its Autonomous System (AS). The router holds AS external routes which are advertised throughout the OSPF network and each router within the OSPF network knows the path to each ASBR.

There are seven types of Link State Advertisements (LSAs) that are usually exchanged between OSPF routers:

- Type 1 - Router Links Advertisements are passed within an area by all OSPF routers and describe the router links to the network. These are only flooded within a particular area.
- Type 2 - Network Links Advertisements are flooded within an area by the DR and describe the routers attached to particular networks.
- Type 3 - Summary Link Advertisements are passed between areas by ABRs and describe networks within an area.
- Type 4 - AS (Autonomous System) Summary Link Advertisements are passed between areas and describe the path to the AS Boundary Router (ASBR). These LSAs are not flooded into Totally Stubby Areas.

- Type 5 - AS External Link Advertisements are passed between and flooded into areas by ASBRs and describe external destinations outside the AS. Stub, Totally Stubby and Not So Stubby areas do not receive them. There are two types of External Link Advertisements: Type 1 packets add the external cost to the internal cost of each link that is passed, which is useful when there are multiple ASBRs advertising the same route into an area as we can decide a preferred route; Type 2 packets only have an external cost assigned, so is fine for a single ASBR advertising an external route.
- Type 6 - Multicast OSPF routers flood this Group Membership Link Entry.
- Type 7 - NSSA AS external routes flooded by the ASBR. The ABR converts these into Type 5 LSAs before flooding them into the Backbone. The difference between Type 7 and Type 5 LSAs is that Type 5 LSAs ones are flooded into multiple areas, whereas Type 7 are only flooded into NSSAs.

A stub area is an area which is out on a limb with no routers or areas beyond it. A stub area is configured to prevent AS External Link Advertisements (Type 5) being flooded into the Stub area. The benefits of configuring a Stub area are that the size of the LSDB is reduced along with the routing table and less CPU cycles are used to process LSAs. Any router wanting access to a network outside the area sends the packets to the default route (0.0.0.0).

A Totally Stubby Area is a Stub Area with the addition that Summary Link Advertisements (Type 3/4) are not sent into the area, as well as External routes. A default route is advertised instead.

The Not So Stubby Area (NSSA) accepts Type 7 LSAs, which are external route advertisements like Type 5 but they are only flooded within the NSSA. This type of area is usually used by an ISP when connecting to a branch office running an IGP. If it was a standard area linking the ISP to the branch office, then the ISP would receive all the Type 5 LSAs from the branch, which it does not want. Because Type 7 LSAs are only flooded to the NSSA, the ISP is saved from the external routes whereas the NSSA can still receive them.

Besides these seven common LSAs, other special types have been proposed for future use:

- Type 8: These LSAs will be used for internetworking between the Border Gateway Protocol (BGP) and OSPF (mainly, to carry BGP attributes into OSPF).

- Type 9, 10 and 11: These LSAs are called "Opaque LSAs" and will used on future upgrades of OSPF for application specific purposes [12, 2].

3.7.6.1 Virtual Links

Each AS has a backbone area, which is responsible for distributing routing information between none-backbone areas. Routing information between non-backbone areas must be forwarded by the backbone area. Therefore, OSPF requires that all non-backbone areas must maintain connectivity to the backbone area and the backbone area itself must maintain connectivity. In practice, due to physical limitations, the requirements may not be satisfied. In this case, the solution is to configure OSPF virtual links.

A virtual link is established between two area border routers via a non-backbone area and is configured on both ABRs to take effect. The area that provides the non-backbone area internal route for the virtual link is a "transit area".

This situation is illustrated in Figure 3.25, where Area 2 has no direct physical link to the backbone area 0. Configuring a virtual link between ABRs R2 and R3 can connect Area 2 to the backbone area. Area 1 is to be used as a transit area and R3 is the entry point into area 0. In this way R2 and area 2 will have a logical connection to the backbone.

In order to configure a virtual link, use the following sub-command (of the router ospf process configuration) on Router R2 and R3:

```
# area <area-id> virtual-link <RID>
```

where `<area-id>` is the transit area (Area 1) and `<RID>` is, of course, the router ID.

3.7.6.2 Route Summarization

Summary Links Advertisements are sent by Area Border Routers and by default they advertise every individual network within each area to which it is connected. Networks can be condensed into a network summary, reducing the number of Summary Links Advertisements being sent and the LSDBs of the routers outside the area. In addition, if there is a network change then this will not be propagated into the backbone and other areas, minimizing the recalculation of SPF.

There are two types of summarizations: Inter-Area Route Summarization and External Route Summarization. Inter-area route summarization is done on ABRs and it applies to routes from within the AS. It does not apply to external routes injected into OSPF via redistribution.

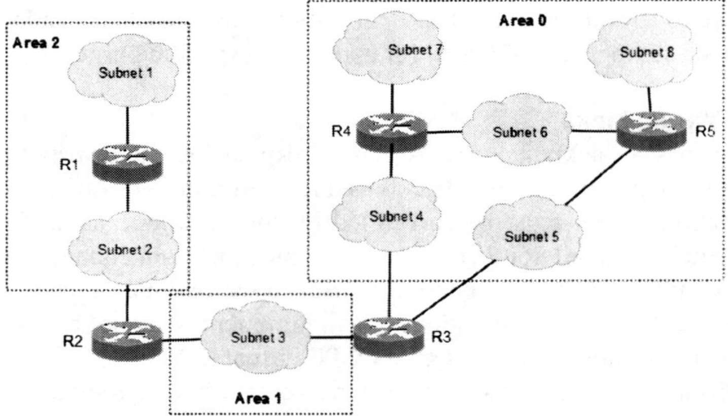

Figure 3.25 Network scenario that must include the configuration of a Virtual Link.

In order to take advantage of summarization, network numbers in areas should be assigned in a contiguous way in order to be able to aggregate them into one range. To specify an address range, enter the following command in router configuration mode:

```
# area <area-id> range <address> <mask>
```

where <area-id> identifies the area containing networks to be summarized and <address> and <mask will specify the range of addresses to be summarized in one range.

External route summarization is specific to external routes that are injected into OSPF via redistribution. The external ranges that are being summarized should be contiguous. Summarization overlapping ranges from two different routers can cause packets to be sent to the wrong destination. Summarization is done via the following router OSPF subcommand:

```
# summary-address <ip-address> <mask>
```

This command is effective only on ASBRs doing redistribution into OSPF.

3.7.6.3 External Routes

In order to make non-OSPF networks available to routers within an OSPF network, the router connected to the non-OSPF network needs to be configured as an AS Boundary Router (ASBR). AS External Link Advertisements (one for each external route) are flooded into the OSPF network (except Stub networks). There are two types of metric for external destinations:

- Type-1 destination networks - The cost to an external network directly connected to the ASBR (close) plus the internal path cost within the OSPF network gives the total cost.
- Type-2 destination networks - The cost to a "far away" network (i.e. not directly connected to the ASBR) is merely the number of hops from the ASBR to the external network.

If a number of routes to a network are advertised to an internal OSPF router, then the router picks the Type 1 route rather than the Type 2 route. If this router learns the route via different protocols then it decides which route to use firstly based on the preference value (configurable) and then on route weight (non-configurable).

3.7.6.4 Redistributing Routes

Domains with difference routing protocols can exchange routes. This is called route redistribution. There are two types of route redistribution:

- One-way redistribution - Redistributes only the networks learned from one routing protocol into the other routing protocol. Uses a default or static route so that devices in that other part of the network can reach the first part of the network.
- Two-way redistribution - Redistributes routes between the two routing processes in both directions.

Redistributing routes into OSPF from other routing protocols or from static will cause these routes to become OSPF external routes. To redistribute routes into OSPF, use the following command in router configuration mode:

```
# redistribute <protocol> <process-id> [metric <value>]
        [metric-type <type-value>] [route-map <map-name>] [subnets]
```

where `<protocol>` and `<process-id>` are the protocol that we are injecting into OSPF and its process-id, if it exits, respectively. The `<metric>` is the cost we are assigning to the external route. If no metric is specified, OSPF puts a default value of 20 when redistributing routes from all protocols except BGP routes, (which get a metric of 1). The `<type-value>` is 1 or 2, corresponding to the categories of the external routes' types. A route-map is the method used to control the redistribution of routes between routing domains, so `<map-name>` is the identifier of a specific route-map. When redistributing routes into OSPF, only routes that are not subnetted are redistributed if the

```
1 # router ospf 1
2   # redistribute static
3   # redistribute rip
```

Box 3.39 Examples of OSPF redistribution commands.

`subnets` keyword is not specified. Box 3.39 contains examples of OSPF redistribution commands: (line 2) for redistribution of static routes into OSPF process 1 and (line 3) for redistribution of RIP routes into OSPF process 1.

Generally, mutual redistribution between protocols should be done in a controlled manner, since incorrect configuration could lead to potential looping of routing information. Information learned from a protocol should not be injected back into the same protocol. Passive interfaces and distribute lists should be applied on the redistributing routers.

The following commands

```
# router ospf <process-id>
  # passive-interface <interface>
```

are used in all routing protocols to disable sending updates out from a specific interface. In OSPF the passive-interface command suppresses Hello packets and hence neighbor relationships.

3.7.6.5 Injecting Defaults into OSPF

An autonomous system boundary router (ASBR) can be forced to generate a default route into the OSPF domain. A router becomes an ASBR whenever routes are redistributed into an OSPF domain. However, an ASBR does not, by default, generate a default route into the OSPF routing domain. To have OSPF generate a default route the following command should be used:

```
# default-information originate [always] [metric <metric-value>]
[metric-type <type-value>] [route-map <map-name>]
```

There are two ways to generate a default. The first is to advertise 0.0.0.0 inside the domain, but only if the ASBR itself already has a default route. The second is to advertise 0.0.0.0 regardless whether the ASBR has a default route. The latter can be set by adding the keyword `always`. Special care should be taken when using the `always` keyword. If the router advertises a default (0.0.0.0) inside the domain and does not have a default itself or a path to reach the destinations, routing will be broken. The `<metric-value>` and `<type-value>` are the cost and type (E1 or E2) assigned to the default route. The route map specifies the set of conditions that need to be satisfied in order for the default to be generated.

3.7.7 OSPFv3

OSPFv3 is specified in RFC 2740 [43] and uses the same fundamental mechanisms as OSPFv2: the SPF algorithm, the flooding mechanism, the DR election, support for areas, and so on. Constants and variables such as timers and metrics are also the same. OSPFv3 is not backward-compatible with OSPFv2. So if we want to use OSPF to route both IPv4 and IPv6, we must run both OSPFv2 and OSPFv3.

3.7.7.1 OSPFv3 Differences from OSPFv2

Besides LSAs, OSPFv3 introduces several changes when compared to OSPFv2:

- Per link protocol processing - An interface to a link can have more than one IPv6 address. In fact, a single link can belong to multiple subnets, and two interfaces attached to the same link but belonging to different IPv6 subnets can still communicate. OSPFv3 changes the OSPFv2 language of "subnet" to "link" and allows the exchange of packets between two neighbors on the same link but belonging to different IPv6 subnets.
- Removal of addressing semantics - OSPFv3 Router and Network LSAs do not carry IP addresses. A new LSA is defined for that purpose. This has some scaling advantages. However, 32-bit RIDs, Area IDs, and LSA IDs are maintained in IPv6. These OSPFv3 IDs are still expressed in dotted decimal, allowing easy overlay of an OSPFv3 network on an existing OSPFv2 network.
- Neighbors are always identified by Router ID - All neighbors on all link types are identified by a RID.
- Addition of a link-local flooding scope - OSPFv3 retains the domain (or AS) and area flooding scopes of OSPFv2, but adds a link-local flooding scope. Note that IPv6 makes much use of the link-local scope. A new LSA, the Link LSA, has been added for carrying information that is only relevant to neighbors on a single link. This LSA has link-local flooding scope, meaning that it cannot be flooded beyond any attached router.
- Use of link-local addresses - OSPFv2 packets have a link-local scope, that is, they are not forwarded by any router. OSPFv3 uses the routers' link-local IPv6 addresses (these addresses always begin with FF80::/10) as the source and next-hop addresses.
- Support for multiple instances per link - There are applications in which multiple OSPF routers can be attached to a single broadcast link but should not form a single adjacency among them. An example of this

is a shared Network Access Point (NAP). OSPFv3 allows for multiple instances per link by adding an *Instance ID* to the OSPF packet header to distinguish instances. An interface assigned to a given *Instance ID* will drop OSPF packets whose *Instance ID* does not match.

- Removal of OSPF specific authentication - IPv6 has, using the Authentication extension header, a standard authentication procedure. Because of this, OSPFv3 has no need for its own authentication of OSPFv3 packets; it just uses IPv6 authentication.
- More flexible handling of unknown LSA types - While OSPFv2 always discards unknown LSA types, OSPFv3 can either treat them as having link-local flooding scope, or store and flood them as if they are understood, while ignoring them in their own SPF algorithms. This can result in easier network changes and easier integration of new capabilities in OSPFv3 than in OSPFv2.

3.7.7.2 OSPFv3 Messages

OSPFv2 and OSPFv3 both have the same protocol number of 89. Like OSPFv2, OSPFv3 uses multicast whenever possible. The IPv6 AllSPFRouters multicast address is FF02::5, and the AllDRouters multicast address is FF02::6. Both have link-local scope. There is an obvious similarity in the last bits with the OSPFv2 addresses of 224.0.0.5 and 224.0.0.6.

OSPFv3 uses the same five message types as OSPFv2: Hello, Database Description, LS Database Request, LS Database Update, and LS Acknowledgment. The message header, shown in Figure 3.26, is somewhat different from the OSPFv2 message header. The version number is, of course, 3 rather than 2. But, more important, there are no fields for authentication. As discussed in the previous section, OSPFv3 uses the Authentication extension header of the IPv6 packet itself rather than its own authentication process. There is also an *Instance ID*, which allows multiple OSPFv3 instances to run on the same link. The *Instance ID* has local link significance only, because the OSPFv3 message is not forwarded beyond the link on which it is originated.

OSPFv3 routers use their link-local addresses as the source of Hello packets. No IPv6 prefix information is contained in Hello packets. Two routers will become adjacent even if no IPv6 prefix is common between the neighbors except the link-local address. This is different from OSPFv2: OSPFv2 neighbors will only become adjacent if the neighbors belong to the same IP subnet, and the common subnet is configured as the primary IP address on the neighboring interfaces.

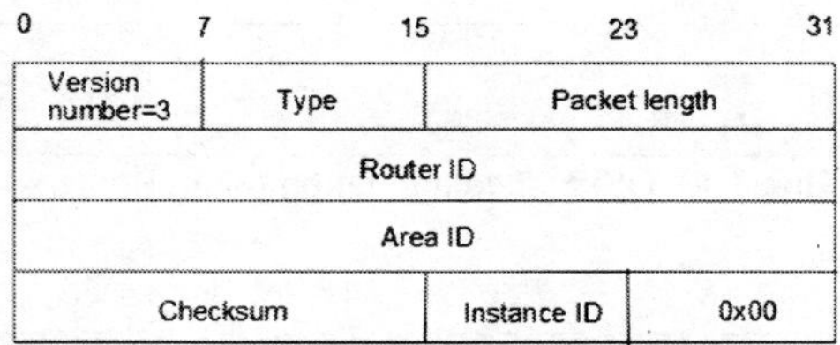

Figure 3.26 OSPFv3 packet header.

Other parameters must match between two neighbors before they will become adjacent. These parameters are the same as IPv4: neighbors must share the same *Area ID*, must have the same *Hello Interval* and *Dead Time*, and both must have the same value in the E-bit, indicating whether the area is a stub area or not. An OSPFv3 packet must also have the same *Instance ID* as the receiving interface, or the OSPFv3 packets will be dropped.

OSPFv3 uses a 32-bit number for a *Router ID*. If IPv4 is configured on the router, by default, the *Router ID* is chosen in the same way it is by OSPFv2 for IPv4. The highest IPv4 address configured on a loopback interface will become the *Router ID*, or if no loopback interfaces are configured, the highest address on any other interface will become the *Router ID*. IPv6 neighbors are always known by their *Router IDs*, unlike IPv4, where point-to-point network neighbors are known by *Router IDs* and broadcast, NBMA and point-to-multipoint network neighbors are known by their interface IP addresses. If IPv4 is not configured in the network, the *Router ID* must be configured using the IPv6 OSPF routing process command *router-id* before the OSPF process starts.

When OSPFv3 is configured on an interface, the routing process is created. Interface parameters, such as the cost of a link, are modified at the interface configuration, but global parameters are modified at the OSPF process level.

Exercise - OSPF Operation

Let us consider again the network illustrated in Figure 3.2. In order to configure OSPFv2, the sequence of commands shown in Box 3.40 should be used on all routers, where `x.x.x.x` should be replaced by the appropriate *router-id* (i.e. a different value for each router). Only a single area is considered.

The configuration of OSPFv3 is similar to OSPFv2 with two exceptions. OSPFv3 for IPv6 is enabled by specifying an OSPF *Process ID* and an area

```
#router ospf 1
  #router-id x.x.x.x
  #network 10.0.0.0 0.255.255.255 area 0
```

Box 3.40 OSPFv2 configuration for all Routers.

```
1 #ipv6 router ospf 1
2   #router-id x.x.x.x
3 #interface F0/0
4   #ipv6 ospf 1 area 0
5 ...
```

Box 3.41 OSPFv3 configuration for all Routers (and all interfaces).

```
1  #show ip route
2       10.0.0.0/24 is subnetted, 7 subnets
3  C       10.3.1.0 is directly connected, FastEthernet0/1
4  C       10.2.1.0 is directly connected, FastEthernet0/0
5  O       10.2.2.0 [110/20] via 10.2.1.2, 00:06:22, FastEthernet0/0
6  C       10.1.1.0 is directly connected, FastEthernet1/0
7  O       10.3.2.0 [110/20] via 10.3.1.4, 00:06:22, FastEthernet0/1
8  O       10.2.3.0 [110/30] via 10.3.1.4, 00:06:22, FastEthernet0/1
9                   [110/30] via 10.2.1.2, 00:06:22, FastEthernet0/0
10 O       10.4.1.0 [110/21] via 10.3.1.4, 00:06:22, FastEthernet0/1
```

Box 3.42 IPv4 routing table at Router 1 after OSPFv2 configuration.

at the interface configuration level. If an OSPFv3 process has not yet been created, it is automatically created. All IPv6 addresses configured on the interface are included in the specified OSPF process. Box 3.41 shows the commands used to configure OSPFv3, where `x.x.x.x` should also be replaced by the appropriate *router-id*. The line 4 command places the routers' F0/0 interface in area 0 and creates the OSPFv3 process with ID "1" on the router. Remember that the `network area` command can enable OSPFv2 on multiple interfaces with the single command, whereas the OSPFv3 equivalent command has to be configured on each interface that will be running OSPFv3.

After the correct configuration of these protocols, the routing tables should include all routes, learned using OSPF. Box 3.42 shows the OSPFv2 routing table of Router 1: knowing that the OSPF costs of all interfaces are equal to 10, except for the interfaces that are connected to stub networks (which are interfaces F1/0 of routers 1 and 5), the metrics of the different routes are obtained by summing the costs of the outgoing interfaces in the minimum cost paths. So, for example there are two equal cost (cost value of 30) routes to network 10.2.3.0/24 (see lines 8-9), one pointing to IP address 10.3.1.4 (the cost is calculated by summing the costs of the F0/1 interfaces of

```
#show ipv6 route
C    2001:A:1:1::/64 [0/0]
      via ::, FastEthernet1/0
L    2001:A:1:1::1/128 [0/0]
      via ::, FastEthernet1/0
C    2001:A:2:1::/64 [0/0]
      via ::, FastEthernet0/0
L    2001:A:2:1::1/128 [0/0]
      via ::, FastEthernet0/0
O    2001:A:2:2::/64 [110/20]
      via FE80::C002:AAFF:FEAA:0, FastEthernet0/0
O    2001:A:2:3::/64 [110/30]
      via FE80::C002:AAFF:FEAA:0, FastEthernet0/0
      via FE80::C004:AAFF:FEAA:0, FastEthernet0/1
C    2001:A:3:1::/64 [0/0]
      via ::, FastEthernet0/1
L    2001:A:3:1::1/128 [0/0]
      via ::, FastEthernet0/1
O    2001:A:3:2::/64 [110/20]
      via FE80::C004:AAFF:FEAA:0, FastEthernet0/1
O    2001:A:4:1::/64 [110/21]
      via FE80::C004:AAFF:FEAA:0, FastEthernet0/1
L    FE80::/10 [0/0]
      via ::, Null0
L    FF00::/8 [0/0]
      via ::, Null0
```

Box 3.43 IPv6 routing table at Router 1 after OSPFv3 configuration.

routers 1 and 4 and the cost of the F0/0 interface of Router 5) and the other pointing to IP address 10.2.1.1 (the cost is calculated by summing the cost of the F0/0 interface of Router 1 and the costs of the F0/1 interfaces of routers 2 and 3).

Box 3.43 shows the OSPFv3 routing table of Router 1: the routes are similar to the ones corresponding to OSPFv2, since they were calculated using the same algorithm. In this case, the next hop address is always a link local address.

Next, we want to check the contents of the OSPF database. Box 3.44 shows the brief output of the OSPF database corresponding to Router 1, which is divided in two parts: the summary of the Router Link States, showing the different Router Link Advertisements, and the summary of the Net Link States, showing the Network Link Advertisements. Note that the considered network has only area 0.

From the output, we can see that there are 5 devices running OSPF in area 0, because we got 5 different Type 1 LSAs. Each router generates one Type 1 LSA. For each LSA, we can also see information about its age, sequence number, checksum and the number of links it carries (from the sixth column).

Each LSA maintains a sequence number. Each time OSPF floods an LSA, which occurs every 30 minutes to maintain proper database synchronization, the sequence number is incremented by one. When a router encounters two

```
#sh ip ospf database

            OSPF Router with ID (1.1.1.1) (Process ID 1)

        Router Link States (Area 0)

Link ID       ADV Router    Age    Seq#          Checksum Link count
1.1.1.1       1.1.1.1       674    0x80000003 0x00CBE3 3
2.2.2.2       2.2.2.2       646    0x80000002 0x002F92 2
3.3.3.3       3.3.3.3       621    0x80000003 0x008D21 2
4.4.4.4       4.4.4.4       620    0x80000002 0x003B6C 2
5.5.5.5       5.5.5.5       620    0x80000002 0x005126 3

        Net Link States (Area 0)

Link ID       ADV Router    Age    Seq#          Checksum
10.2.1.1      1.1.1.1       703    0x80000001 0x0058C3
10.2.2.3      3.3.3.3       647    0x80000001 0x0041C7
10.2.3.5      5.5.5.5       621    0x80000001 0x005C95
10.3.1.1      1.1.1.1       674    0x80000001 0x00B062
10.3.2.5      5.5.5.5       621    0x80000001 0x008D60
```

Box 3.44 OSPFv2 database (brief output).

instances of an LSA, it must determine which is the most recent one, which is of course the one that has the higher sequence number.

A Type 2 LSA is present every time there is an OSPF multi-access network (transit network). As previously said, in this kind of network a Designated Router (DR) and a Backup DR are chosen. All other routers build a neighborship with DR and BDR routers in the transit network, and they send their Type 1 LSA announcements only to them (using multicast address 224.0.0.6). Then, the DR retransmits these Type 1 LSAs to all its neighbors, inside and outside the transit network. It also builds a new Type 2 LSA with information about the transit network. Only the DR is in charge of generating this Type 2 LSA and sending it to all its neighbors, even the ones in the transit network, putting itself as the virtual owner of this LSA. In our example, there are of course five transit networks.

Box 3.45 shows the details of the Router Link State database (only a partial view, corresponding to Router 1). OSPF differentiate between Transit and Stub Networks: a Transit network is a network where multiple routers can exist, therefore where multiple neighborships can be built, whereas a Stub network is a network where no router or at most one other router can exist (such as loopbacks or point-to-point networks). From this output, we can see that R1 is connected to 3 links, two corresponding to transit networks and one to a stub network. By looking at the available info for the stub network (lines 12-16), we see the IP and mask of the interface and the cost that R1 has to reach it. For the transit networks (lines 18-22 and lines 24-28), we have the address of the DR, the address of the router interface that is connected to that

```
#sh ip ospf database router
LS age: 799
  Options: (No TOS-capability, DC)
  LS Type: Router Links
  Link State ID: 1.1.1.1
  Advertising Router: 1.1.1.1
  LS Seq Number: 80000003
  Checksum: 0xCBE3
  Length: 60
  Number of Links: 3

    Link connected to: a Stub Network
     (Link ID) Network/subnet number: 10.1.1.0
     (Link Data) Network Mask: 255.255.255.0
      Number of TOS metrics: 0
       TOS 0 Metrics: 1

    Link connected to: a Transit Network
     (Link ID) Designated Router address: 10.3.1.1
     (Link Data) Router Interface address: 10.3.1.1
      Number of TOS metrics: 0
       TOS 0 Metrics: 10

    Link connected to: a Transit Network
     (Link ID) Designated Router address: 10.2.1.1
     (Link Data) Router Interface address: 10.2.1.1
      Number of TOS metrics: 0
       TOS 0 Metrics: 10
...
```

Box 3.45 OSPFv2 database - example of a Router Link State (corresponding to Router 1).

network and the associated OSPF cost. All this information is very valuable because it allows to build a complete picture of the network, knowing how routers are connected and the costs associated to all links.

Box 3.46 shows the details of the Network Link State database (only a partial view, corresponding to Network 10.2.1.0/24). IP address 10.2.1.0 is the address that the DR has on this link (see line 6); the advertising router is 1.1.1.1 (line 7) and corresponds to the router in charge of gathering all the info of the transit network and building the Type 2 LSA. We can also see that there are 2 routers in this transit network: routers 1.1.1.1 and 2.2.2.2 (lines 12-13). Note that there is no cost associated to this network, because it is announced by each router in a Type 1 LSA.

Box 3.47 shows the details of the OSPFv3 Router Link State database (only a partial view, corresponding to Router 1). Unlike OSPFv2, stub networks are not included in this database: they are included in the Prefix Link States database, shown in Box 3.49, corresponding to all entries with *Referenced Link State ID* of 0. As can be seen from this database, each stub network entry identifies the advertising router (in this case, with RID 1.1.1.1), the number of prefixes (one, in this case), the prefix address (only the global

```
#sh ip ospf database router
  Routing Bit Set on this LSA
  LS age: 1515
  Options: (No TOS-capability, DC)
  LS Type: Network Links
  Link State ID: 10.2.1.1 (address of Designated Router)
  Advertising Router: 1.1.1.1
  LS Seq Number: 80000001
  Checksum: 0x58C3
  Length: 32
  Network Mask: /24
  Attached Router: 1.1.1.1
  Attached Router: 2.2.2.2
...
```

Box 3.46 OSPFv2 database - example of a Network Link State (corresponding to network 10.2.1.0/24).

```
1  #sh ipv6 ospf database router
2    LS age: 897
3    Options: (V6-Bit E-Bit R-bit DC-Bit)
4    LS Type: Router Links
5    Link State ID: 0
6    Advertising Router: 1.1.1.1
7    LS Seq Number: 80000004
8    Checksum: 0xAEFA
9    Length: 56
10   Number of Links: 2
11
12     Link connected to: a Transit Network
13       Link Metric: 10
14       Local Interface ID: 5
15       Neighbor (DR) Interface ID: 4
16       Neighbor (DR) Router ID: 4.4.4.4
17
18     Link connected to: a Transit Network
19       Link Metric: 10
20       Local Interface ID: 4
21       Neighbor (DR) Interface ID: 4
22       Neighbor (DR) Router ID: 2.2.2.2
23 ...
```

Box 3.47 OSPFv3 database - example of a Router Link State (corresponding to Router 1).

address) and the metric value (1 in this case, which is the OSPF cost of this interface, as already said).

Returning again to Box 3.47, we see that the Link State ID of this entry is equal to zero (line 5), which indicates that this LSA was originated exactly in the router (1.1.1.1) that is being considered, and there are two transit networks: for each one, the entry presents the OSPF metric of the router interface, the *Local Interface ID* (corresponding to the ID of the link local address of this router), the *Neighbor Interface ID* (corresponding to the ID of the link local address of the DR - these IDs are announced by each router to its neighbors) and the *Neighbor (DR) Router ID* (which is the Router ID of the Designated Router).

```
1  #sh ipv6 ospf database network
2    LS age: 938
3    Options: (V6-Bit E-Bit R-bit DC-Bit)
4    LS Type: Network Links
5    Link State ID: 4 (Interface ID of Designated Router)
6    Advertising Router: 2.2.2.2
7    LS Seq Number: 80000001
8    Checksum: 0x3ED
9    Length: 32
10   Attached Router: 2.2.2.2
11   Attached Router: 1.1.1.1
12 ...
```

Box 3.48 OSPFv3 database - example of a Network Link State (corresponding to network 2001:A:2:1::/64).

The Network Link State, partially represented in Box 3.48, is very similar to the one corresponding to OSPFv2. The only difference lies on the Link State ID (line 5), which is now the Interface ID of the Designated Router (which was previously announced).

Returning again to the Prefix Link States database, partially shown in Box 3.49, the second entry (lines 17-30) corresponds to a transit network presenting a *Referenced Link State ID* of 4. The remaining information is analogous to the information contained in the first entry (lines 2-15), corresponding to a stub network: the advertising router (in this case, 2.2.2.2), the number of prefixes (one, in this case, since there is only one global network address configured), the prefix address (the global network address) and the metric value (0 because the true metric is included in the Router Link State).

In order to understand the OSPF database synchronization process, we stopped all routers in our GNS3 emulation and re-started only routers 1 and 2, capturing all exchanged packets in network 10.2.1.0/24 (2001:A:2:1::/64). The sequence of OSPFv2 packets is shown in Figure 3.27, while the complete sequence of OSPFv3 packets is shown in Figure 3.33: obviously, all packets are sent by the connected router interfaces, in this case 10.2.1.1 (with link local address fe80::c001:aaff:feaa:0) and 10.2.1.2 (with link local address fe80::c002:aaff:feaa:0).

The Hello packets, sent in IPv4 to the broadcast address 224.0.0.5 and in IPv6 to the broadcast address ff02::5, only contain the identification of the sending router and the already known neighbors (Figure 3.28).

After exchanging Hello packets, routers enter the EXstart state, a three-way handshake process where routers agree on the master and slave roles. The Database description packet number 3, illustrated in Figure 3.29, is the first packet of this sequence and basically shows the Master/Slave bit set.

```
#sh ipv6 ospf database prefix
  Routing Bit Set on this LSA
  LS age: 698
  LS Type: Intra-Area-Prefix-LSA
  Link State ID: 0
  Advertising Router: 1.1.1.1
  LS Seq Number: 80000002
  Checksum: 0x8B04
  Length: 44
  Referenced LSA Type: 2001
  Referenced Link State ID: 0
  Referenced Advertising Router: 1.1.1.1
  Number of Prefixes: 1
  Prefix Address: 2001:A:1:1::
  Prefix Length: 64, Options: None, Metric: 1

  Routing Bit Set on this LSA
  LS age: 699
  LS Type: Intra-Area-Prefix-LSA
  Link State ID: 4096
  Advertising Router: 2.2.2.2
  LS Seq Number: 80000001
  Checksum: 0x254E
  Length: 44
  Referenced LSA Type: 2002
  Referenced Link State ID: 4
  Referenced Advertising Router: 2.2.2.2
  Number of Prefixes: 1
  Prefix Address: 2001:A:2:1::
  Prefix Length: 64, Options: None, Metric: 0
...
```

Box 3.49 OSPFv3 database sample Prefix Link States.

```
#show ip route
     10.0.0.0/24 is subnetted, 7 subnets
C       10.3.1.0 is directly connected, FastEthernet0/1
C       10.2.1.0 is directly connected, FastEthernet0/0
O       10.2.2.0 [110/20] via 10.2.1.2, 00:10:20, FastEthernet0/0
C       10.1.1.0 is directly connected, FastEthernet1/0
O       10.3.2.0 [110/40] via 10.2.1.2, 00:10:20, FastEthernet0/0
O       10.2.3.0 [110/30] via 10.2.1.2, 00:10:20, FastEthernet0/0
O       10.4.1.0 [110/31] via 10.2.1.2, 00:10:20, FastEthernet0/0
```

Box 3.50 IPv4 routing table at Router 1 after disconnection.

The next phase is the Exchange state, where routers discover routes by exchanging Database Description Packets: the packet shown in Figure 3.30 is one of them and includes a Router LSA with information corresponding to Router 1.

Once a slave router receives a Database Description Packet, it compares its information (through the included sequence numbers) with already known information and, if it is newer, it sends a Link State Request in order to update existing information. The packet shown in Figure 3.31 is a Link State Request sent to Router 1.

No.	Time	Source	Destination	Protocol	Info
1	0.000000	10.2.1.1	224.0.0.5	OSPF	Hello Packet
2	9.081021	10.2.1.2	224.0.0.5	OSPF	Hello Packet
3	9.982701	10.2.1.1	10.2.1.2	OSPF	DB Description
4	10.026730	10.2.1.1	224.0.0.5	OSPF	Hello Packet
5	14.979632	10.2.1.1	10.2.1.2	OSPF	DB Description
6	19.106908	10.2.1.2	10.2.1.1	OSPF	DB Description
7	19.108997	10.2.1.1	10.2.1.2	OSPF	DB Description
8	19.109008	10.2.1.2	224.0.0.5	OSPF	Hello Packet
9	19.111122	10.2.1.2	10.2.1.1	OSPF	DB Description
10	19.113168	10.2.1.1	10.2.1.2	OSPF	DB Description
11	19.115284	10.2.1.2	10.2.1.1	OSPF	DB Description
12	19.117342	10.2.1.1	10.2.1.2	OSPF	LS Request
13	19.117371	10.2.1.2	10.2.1.1	OSPF	LS Request
14	19.119434	10.2.1.1	10.2.1.2	OSPF	DB Description
15	19.119467	10.2.1.2	10.2.1.1	OSPF	LS Update
16	19.121514	10.2.1.1	10.2.1.2	OSPF	LS Update
17	19.123652	10.2.1.2	224.0.0.5	OSPF	LS Update
18	19.125698	10.2.1.1	10.2.1.2	OSPF	LS Acknowledge
19	19.583361	10.2.1.2	224.0.0.5	OSPF	LS Update
20	19.620567	10.2.1.2	224.0.0.5	OSPF	LS Update
21	19.622689	10.2.1.1	224.0.0.5	OSPF	LS Update
22	19.662397	10.2.1.2	224.0.0.5	OSPF	LS Update
23	20.000948	10.2.1.1	224.0.0.5	OSPF	Hello Packet
24	21.627593	10.2.1.2	224.0.0.5	OSPF	LS Acknowledge
25	21.629713	10.2.1.1	224.0.0.5	OSPF	LS Acknowledge
26	29.103389	10.2.1.2	224.0.0.5	OSPF	Hello Packet
27	30.000229	10.2.1.1	224.0.0.5	OSPF	Hello Packet

Figure 3.27 OSPFv2 bootstrap sequence (in network 10.2.1.0/24).

A Link State Update is sent in response to this Link State Request, containing the requested LSAs. One of these packets is illustrated in Figure 3.32, containing a Router LSA. Of course, information sent in these LSAs is the most recent one. Figure 3.34 presents a Link State Update in OSPFv3 sent by Router 1 to Router 2: this update includes several (3) LSAs, specifically a Router LSA, a Link LSA and a Prefix LSA.

Finally, a last experiment was conducted: Router 4 is disconnected from network 10.3.2.0/24 (2001:A:3:2::/64) in order to verify how the OSPF processes converge to a new set of alternative paths. Some seconds after the disconnection, the OSPFv2 routing tables of routers 1 and 4 are the ones shown in Boxes 3.50 and 3.51, while the OSPFv3 routing tables are shown in Boxes 3.52 and 3.53. Now, Router 1 is able to reach network 10.3.2.0/24 (2001:A:3:2::/64) with a cost of 40 via Router 2, while Router 4 is able to reach that network with a cost of 50 via Router 1. With this topology, there is only one possible path to any IP network of the scenario.

No.	Time	Source	Destination	Protocol	Info
1	0.000000	10.2.1.1	224.0.0.5	OSPF	Hello Packet

```
▷ Frame 1: 94 bytes on wire (752 bits), 94 bytes captured (752 bits)
▷ Ethernet II, Src: c2:01:aa:aa:00:00 (c2:01:aa:aa:00:00), Dst: IPv4mcast_00:00:05 (01:00:5e:00:00:05)
▷ Internet Protocol, Src: 10.2.1.1 (10.2.1.1), Dst: 224.0.0.5 (224.0.0.5)
▽ Open Shortest Path First
  ▽ OSPF Header
      OSPF Version: 2
      Message Type: Hello Packet (1)
      Packet Length: 48
      Source OSPF Router: 1.1.1.1 (1.1.1.1)
      Area ID: 0.0.0.0 (Backbone)
      Packet Checksum: 0xe694 [correct]
      Auth Type: Null
      Auth Data (none)
  ▽ OSPF Hello Packet
      Network Mask: 255.255.255.0
      Hello Interval: 10 seconds
    ▷ Options: 0x12 (L, E)
      Router Priority: 1
      Router Dead Interval: 40 seconds
      Designated Router: 0.0.0.0
      Backup Designated Router: 0.0.0.0
      Active Neighbor: 2.2.2.2
  ▷ OSPF LLS Data Block
```

Figure 3.28 OSPFv2 Hello packet (in network 10.2.1.0/24).

```
#show ip route
     10.0.0.0/24 is subnetted, 7 subnets
C        10.3.1.0 is directly connected, FastEthernet0/0
O        10.2.1.0 [110/20] via 10.3.1.1, 00:11:01, FastEthernet0/0
O        10.2.2.0 [110/30] via 10.3.1.1, 00:11:01, FastEthernet0/0
O        10.1.1.0 [110/11] via 10.3.1.1, 00:11:01, FastEthernet0/0
O        10.3.2.0 [110/50] via 10.3.1.1, 00:11:01, FastEthernet0/0
O        10.2.3.0 [110/40] via 10.3.1.1, 00:11:01, FastEthernet0/0
O        10.4.1.0 [110/41] via 10.3.1.1, 00:11:01, FastEthernet0/0
```

Box 3.51 IPv4 routing table at Router 4 after disconnection.

When the disconnection takes place, OSPF databases have to synchronize again, in order to reflect the new network topology. Figures 3.35 and 3.36 show the convergence sequence captured in network 10.2.1.0/24 (2001:A:2:1::/64) and corresponding to OSPFv2 and OSPFv3, respectively. Router 1, for example, forwards a Router LSA that was sent by Router 4 stating that it is only connected to a transit network, the one located between routers 1 and 4 (network 10.3.1.0/24 or 2001:A:3:1::/64). This means that Router 4 is not connected to network 10.3.2.0/24 (2001:A:3:2::/64) anymore.

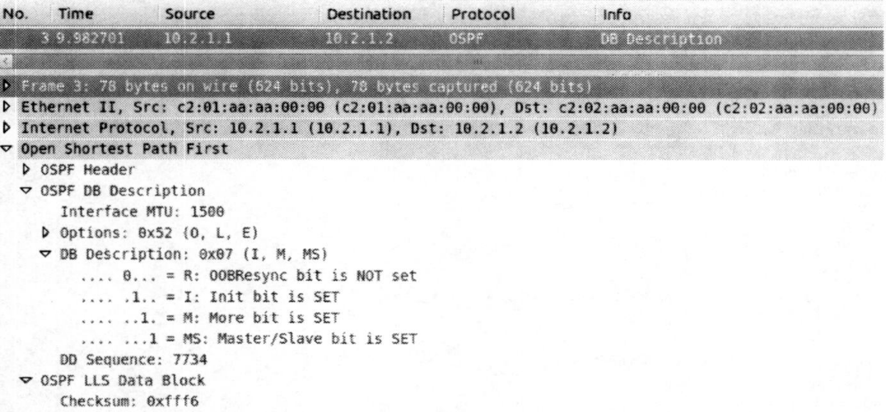
No. Time Source Destination Protocol Info
3 9.982701 10.2.1.1 10.2.1.2 OSPF DB Description
Frame 3: 78 bytes on wire (624 bits), 78 bytes captured (624 bits)
Ethernet II, Src: c2:01:aa:aa:00:00 (c2:01:aa:aa:00:00), Dst: c2:02:aa:aa:00:00 (c2:02:aa:aa:00:00)
Internet Protocol, Src: 10.2.1.1 (10.2.1.1), Dst: 10.2.1.2 (10.2.1.2)
Open Shortest Path First
OSPF Header
OSPF DB Description
Interface MTU: 1500
Options: 0x52 (O, L, E)
DB Description: 0x07 (I, M, MS)
.... 0... = R: OOBResync bit is NOT set
.... .1.. = I: Init bit is SET
.... ..1. = M: More bit is SET
....1 = MS: Master/Slave bit is SET
DD Sequence: 7734
OSPF LLS Data Block
Checksum: 0xfff6

Figure 3.29 OSPFv2 Database Description packet during the EXstart state (in network 10.2.1.0/24).

3.7.8 OSPF Design Issues

The number of routers connected to the same LAN is important. Each LAN has a DR and BDR that build adjacencies with all other routers. The fewer neighbors that exist on the LAN, the smaller the number of adjacencies a DR or BDR have to build. That depends on how much power the router has. The OSPF priority can be changed to select the DR. If possible, we should avoid having the same router to be the DR on more than one segment.

The maximum number of routers per area depends on several factors: the type of area, the available CPU power, the king of media, if OSPF is being run NBMA mode or not, if there are a lot of external LSAs in the network or if other areas are well summarized. So, it is difficult to specify a maximum number of routers per area.

ABRs will keep a copy of the database for all areas they service. The number of areas per ABR is a number that is dependent on many factors, including type of area, ABR CPU power, number of routes per area and number of external routes per area. For this reason, a specific number of areas per ABR cannot be recommended. Of course, it is better not to overload an ABR when it is possible to spread the areas over other routers.

A partial mesh topology has proven to behave much better than a full mesh. A carefully laid out point-to-point or point-to-multipoint network works much better than multipoint networks that have to deal with DR issues.

```
No.   Time           Source      Destination   Protocol   Info
    7 19.108997      10.2.1.1    10.2.1.2      OSPF       DB Description

▷ Frame 7: 98 bytes on wire (784 bits), 98 bytes captured (784 bits)
▷ Ethernet II, Src: c2:01:aa:aa:00:00 (c2:01:aa:aa:00:00), Dst: c2:02:aa:aa:00:00 (c2:02:aa:aa:00:00)
▷ Internet Protocol, Src: 10.2.1.1 (10.2.1.1), Dst: 10.2.1.2 (10.2.1.2)
▽ Open Shortest Path First
  ▷ OSPF Header
  ▷ OSPF DB Description
  ▽ LSA Header
      LS Age: 40 seconds
      Do Not Age: False
    ▷ Options: 0x22 (DC, E)
      Link-State Advertisement Type: Router-LSA (1)
      Link State ID: 1.1.1.1
      Advertising Router: 1.1.1.1 (1.1.1.1)
      LS Sequence Number: 0x80000002
      LS Checksum: 0x824b
      Length: 60
  ▷ OSPF LLS Data Block
```

Figure 3.30 OSPFv2 Database Description packet during the Exchange state (in network 10.2.1.0/24).

```
No.   Time           Source      Destination   Protocol   Info
   13 19.117371      10.2.1.2    10.2.1.1      OSPF       LS Request

▷ Frame 13: 70 bytes on wire (560 bits), 70 bytes captured (560 bits)
▷ Ethernet II, Src: c2:02:aa:aa:00:00 (c2:02:aa:aa:00:00), Dst: c2:01:aa:aa:00:00 (c2:01:aa:aa:00:00)
▷ Internet Protocol, Src: 10.2.1.2 (10.2.1.2), Dst: 10.2.1.1 (10.2.1.1)
▽ Open Shortest Path First
  ▷ OSPF Header
  ▽ Link State Request
      Link-State Advertisement Type: Router-LSA (1)
      Link State ID: 1.1.1.1
      Advertising Router: 1.1.1.1 (1.1.1.1)
```

Figure 3.31 OSPFv2 Link State Request packet (in network 10.2.1.0/24).

Router memory issues usually come up when too many external routes are injected in the OSPF domain. Memory can be conserved by using a good OSPF design. Summarization at the area border routers and use of stub areas could further minimize the number of routes exchanged. The total memory used by OSPF is the sum of the memory used in the routing table and the memory used in the link-state database.

3.8 Multicast Routing

Unicast protocols deliver packets from a single source to a single destination. However, several network applications (like software upgrading solutions,

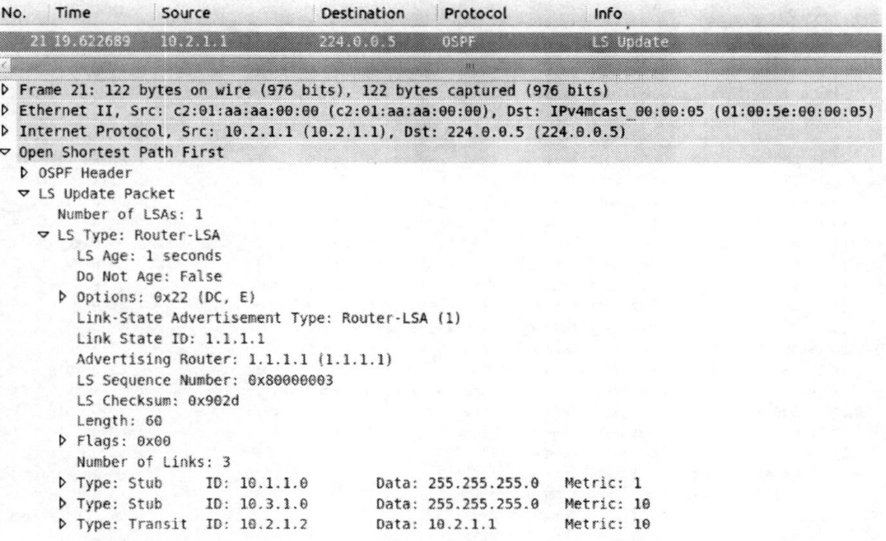
No. Time Source Destination Protocol Info
21 19.622689 10.2.1.1 224.0.0.5 OSPF LS Update
Frame 21: 122 bytes on wire (976 bits), 122 bytes captured (976 bits)
Ethernet II, Src: c2:01:aa:aa:00:00 (c2:01:aa:aa:00:00), Dst: IPv4mcast_00:00:05 (01:00:5e:00:00:05)
Internet Protocol, Src: 10.2.1.1 (10.2.1.1), Dst: 224.0.0.5 (224.0.0.5)
Open Shortest Path First
OSPF Header
LS Update Packet
Number of LSAs: 1
LS Type: Router-LSA
LS Age: 1 seconds
Do Not Age: False
Options: 0x22 (DC, E)
Link-State Advertisement Type: Router-LSA (1)
Link State ID: 1.1.1.1
Advertising Router: 1.1.1.1 (1.1.1.1)
LS Sequence Number: 0x80000003
LS Checksum: 0x902d
Length: 60
Flags: 0x00
Number of Links: 3
Type: Stub ID: 10.1.1.0 Data: 255.255.255.0 Metric: 1
Type: Stub ID: 10.3.1.0 Data: 255.255.255.0 Metric: 10
Type: Transit ID: 10.2.1.2 Data: 10.2.1.1 Metric: 10

Figure 3.32 OSPFv2 Link State Update packet (in network 10.2.1.0/24).

continuous media streaming, shared data applications or interactive gaming) require the delivery of packets from one or more senders to a group of receivers.

The multicast abstraction - a single send operation that results in copies of the sent data being delivered to many receivers - can be implemented in many ways. One possibility is for the sender to use a separate unicast transport connection to each of the receivers. An application-level data unit that is passed to the transport layer is then duplicated at the sender and transmitted over each of the individual connections. This approach requires no explicit multicast support from the network layer to implement the multicast abstraction; multicast is emulated using multiple point-to-point unicast connections. A second alternative is to provide explicit multicast support at the network layer. In this approach, a single datagram is transmitted from the sending host. This datagram (or a copy of this datagram) is then replicated at a network router whenever it must be forwarded on multiple outgoing links in order to reach the receivers.

Clearly, the second approach towards multicast makes more efficient use of network bandwidth in that only a single copy of a datagram will ever traverse a link. Other the other hand, considerable network layer support is needed to implement a multicast-aware network layer.

No.	Time	Source	Destination	Protocol	Info
1	0.000000	fe80::c001:aaff:feaa:0	ff02::5	OSPF	Hello Packet
2	9.072837	fe80::c002:aaff:feaa:0	ff02::5	OSPF	Hello Packet
3	9.990008	fe80::c001:aaff:feaa:0	fe80::c002:aaff:feaa:0	OSPF	DB Description
4	9.994600	fe80::c001:aaff:feaa:0	ff02::5	OSPF	Hello Packet
5	14.978787	fe80::c001:aaff:feaa:0	fe80::c002:aaff:feaa:0	OSPF	DB Description
6	19.078017	fe80::c002:aaff:feaa:0	fe80::c001:aaff:feaa:0	OSPF	DB Description
7	19.080111	fe80::c001:aaff:feaa:0	fe80::c002:aaff:feaa:0	OSPF	DB Description
8	19.080182	fe80::c002:aaff:feaa:0	ff02::5	OSPF	Hello Packet
9	19.084385	fe80::c002:aaff:feaa:0	fe80::c001:aaff:feaa:0	OSPF	DB Description
10	19.086488	fe80::c001:aaff:feaa:0	fe80::c002:aaff:feaa:0	OSPF	DB Description
11	19.088601	fe80::c002:aaff:feaa:0	fe80::c001:aaff:feaa:0	OSPF	DB Description
12	19.090719	fe80::c002:aaff:feaa:0	fe80::c001:aaff:feaa:0	OSPF	LS Request
13	19.090741	fe80::c001:aaff:feaa:0	fe80::c002:aaff:feaa:0	OSPF	LS Request
14	19.092844	fe80::c002:aaff:feaa:0	fe80::c001:aaff:feaa:0	OSPF	LS Update
15	19.092882	fe80::c001:aaff:feaa:0	fe80::c002:aaff:feaa:0	OSPF	DB Description
16	19.094987	fe80::c001:aaff:feaa:0	fe80::c002:aaff:feaa:0	OSPF	LS Update
17	19.134629	fe80::c002:aaff:feaa:0	ff02::5	OSPF	LS Update
18	19.138804	fe80::c001:aaff:feaa:0	fe80::c002:aaff:feaa:0	OSPF	LS Acknowledge
19	19.589751	fe80::c002:aaff:feaa:0	ff02::5	OSPF	LS Update
20	19.627272	fe80::c001:aaff:feaa:0	ff02::5	OSPF	LS Update
21	19.669164	fe80::c002:aaff:feaa:0	ff02::5	OSPF	LS Update
22	20.005414	fe80::c001:aaff:feaa:0	ff02::5	OSPF	Hello Packet
23	21.594845	fe80::c002:aaff:feaa:0	ff02::5	OSPF	LS Acknowledge
24	21.594906	fe80::c001:aaff:feaa:0	ff02::5	OSPF	LS Acknowledge
25	25.056338	fe80::c002:aaff:feaa:0	ff02::5	OSPF	LS Update
26	27.566241	fe80::c001:aaff:feaa:0	ff02::5	OSPF	LS Acknowledge
27	29.062257	fe80::c002:aaff:feaa:0	ff02::5	OSPF	Hello Packet
28	30.009344	fe80::c001:aaff:feaa:0	ff02::5	OSPF	Hello Packet

Figure 3.33 OSPFv3 bootstrap sequence (in network 2001:A:2:1::/64).

3.8.1 Multicast Addressing

In the multicast paradigm, a single "identifier" is used for a group of receivers, and a copy of the datagram that is addressed to the group using this single "identifier" is delivered to all of the multicast receivers associated with that group. The single "identifier" that represents a group of receivers is a Class D multicast address (Figure 3.37). The group of receivers associated with a class D address is referred to as a multicast group.

Multicast addresses are dynamically assigned. An IP datagram sent to a multicast address is forwarded to everyone who has joined the multicast group. If an application is terminated, the multicast address is (implicitly) released.

The range of addresses between 224.0.0.0 and 224.0.0.255, inclusive, is reserved for the use of routing protocols and other low-level topology discovery or maintenance protocols. Multicast routers should not forward any multicast datagram with destination addresses in this range. Examples of special and reserved Class D addresses are, for example: 224.0.0.1 - All

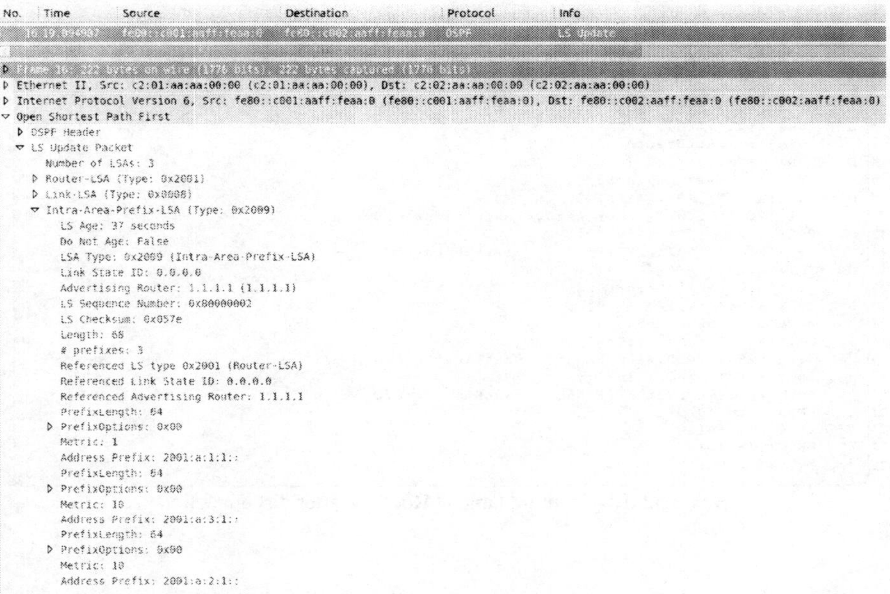

Figure 3.34 OSPFv3 Link State Update packet (in network 2001:A:2:1::/64).

systems on this subnet; 224.0.0.2 - All routers on this subnet; 224.0.1.1 - NTP (Network Time Protocol); 224.0.0.9 - RIPv2.

Figure 3.38 illustrates the mapping between multicast addresses at the Ethernet and IP levels.

3.8.2 The IGMP Protocol

The Internet Group Management protocol (IGMP), version 2, was defined in RFC 2236 [40] and operates between a host and its directly attached router. This protocol provides the means for a host to inform its attached router that an application running on the host wants to join a specific multicast group. Given that the scope of IGMP interaction is limited to a host and its attached router, another protocol is clearly required to coordinate the multicast routers (including the attached routers) throughout the Internet, so that multicast datagrams are routed to their final destinations. This functionality is accomplished by network layer multicast routing algorithms such as PIM (Protocol Independent Multicast), DVMRP (Distance Vector Multicast Routing Protocol), MOSFP (Multicast Open Shortest Path) and BGP. Network layer

```
#show ipv6 route
C    2001:A:1:1::/64 [0/0]
      via ::, FastEthernet1/0
L    2001:A:1:1::1/128 [0/0]
      via ::, FastEthernet1/0
C    2001:A:2:1::/64 [0/0]
      via ::, FastEthernet0/0
L    2001:A:2:1::1/128 [0/0]
      via ::, FastEthernet0/0
O    2001:A:2:2::/64 [110/20]
      via FE80::C002:AAFF:FEAA:0, FastEthernet0/0
O    2001:A:2:3::/64 [110/30]
      via FE80::C002:AAFF:FEAA:0, FastEthernet0/0
C    2001:A:3:1::/64 [0/0]
      via ::, FastEthernet0/1
L    2001:A:3:1::1/128 [0/0]
      via ::, FastEthernet0/1
O    2001:A:3:2::/64 [110/40]
      via FE80::C002:AAFF:FEAA:0, FastEthernet0/0
O    2001:A:4:1::/64 [110/31]
      via FE80::C002:AAFF:FEAA:0, FastEthernet0/0
L    FE80::/10 [0/0]
      via ::, Null0
L    FF00::/8 [0/0]
      via ::, Null0
```

Box 3.52 IPv6 routing table at Router 1 after disconnection.

multicast in the Internet thus consists of two complementary components: IGMP and multicast routing protocols.

3.8.2.1 IGMP Messages

Figure 3.39 illustrates the IGMP packet format. This protocol supports the following messages:

- General Membership Query - Message sent by a router to all hosts on an attached interface to determine the set of all multicast groups that have been joined by the hosts on that interface;
- Specific Membership Query - Message sent by a router to query if a specific multicast group has been joined by hosts on an attached interface (this message includes the multicast address of the group being queried in the multicast group address field);
- Membership Report - Message sent by a host to report it wants to join or is joined to a given multicast group;
- Leave Group - Message sent by a host to report that it is leaving a given multicast group.

Membership Report messages can be generated by a host when it joins a multicast group or as a response to a Membership Query message sent by the router. Membership Report messages are received by the router and all hosts on the attached interface. Each Membership Report contains the

```
#show ipv6 route
O    2001:A:1:1::/64 [110/11]
      via FE80::C001:AAFF:FEAA:1, FastEthernet0/0
O    2001:A:2:1::/64 [110/20]
      via FE80::C001:AAFF:FEAA:1, FastEthernet0/0
O    2001:A:2:2::/64 [110/30]
      via FE80::C001:AAFF:FEAA:1, FastEthernet0/0
O    2001:A:2:3::/64 [110/40]
      via FE80::C001:AAFF:FEAA:1, FastEthernet0/0
C    2001:A:3:1::/64 [0/0]
      via ::, FastEthernet0/0
L    2001:A:3:1::4/128 [0/0]
      via ::, FastEthernet0/0
O    2001:A:3:2::/64 [110/50]
      via FE80::C001:AAFF:FEAA:1, FastEthernet0/0
O    2001:A:4:1::/64 [110/41]
      via FE80::C001:AAFF:FEAA:1, FastEthernet0/0
L    FE80::/10 [0/0]
      via ::, Null0
L    FF00::/8 [0/0]
      via ::, Null0
```

Box 3.53 IPv6 routing table at Router 4 after disconnection.

multicast address of a single group that the responding host has joined. Note that an attached router does not really care which hosts have joined a given multicast group or even how many hosts on the same LAN have joined the same group. Each Membership Query message sent by a router also includes a "Maximum Response Time" (MRT) value field: after receiving a Membership Query message and before sending a Membership Report message for a given multicast group, a host waits a random amount of time between zero and the MRT. If the host observes a Membership Report message from some other attached host for that given multicast group, it discards its own pending Membership Report message, since the host now knows that the attached router already knows that one or more hosts are joined to that multicast group.

On a given LAN, only one router sends IGMP queries (the so-called Querier Router): the router with the smallest IP address becomes the querier on a network. Besides, only one router forwards multicast packets to the network (the Forwarder).

Note that a malicious sender can inject datagrams into the multicast group datagram flow. Even with benign senders, since there is no network layer coordination of the use of multicast addresses, it is possible that two different multicast groups will choose to use the same multicast address. From a multicast application viewpoint, this will result in interleaved strange multicast traffic. These problems may seem to be insurmountable drawbacks for developing multicast applications. Although the network layer does not provide for filtering, ordering or privacy of multicast datagrams, these mechanisms can all be implemented at the application layer.

No.	Time	Source	Destination	Protocol	Info
1	0.000000	10.2.1.1	224.0.0.5	OSPF	LS Update
2	0.203886	10.2.1.2	224.0.0.5	OSPF	Hello Packet
3	1.435329	10.2.1.1	224.0.0.5	OSPF	Hello Packet
4	2.512306	10.2.1.2	224.0.0.5	OSPF	LS Acknowledge
5	10.202849	10.2.1.2	224.0.0.5	OSPF	Hello Packet
6	11.461858	10.2.1.1	224.0.0.5	OSPF	Hello Packet
7	20.233455	10.2.1.2	224.0.0.5	OSPF	Hello Packet
8	21.460268	10.2.1.1	224.0.0.5	OSPF	Hello Packet
9	30.231955	10.2.1.2	224.0.0.5	OSPF	Hello Packet
10	31.439672	10.2.1.1	224.0.0.5	OSPF	Hello Packet
11	32.251778	10.2.1.2	224.0.0.5	OSPF	LS Update
12	32.293315	10.2.1.2	224.0.0.5	OSPF	LS Update
13	34.766736	10.2.1.1	224.0.0.5	OSPF	LS Acknowledge
14	40.205505	10.2.1.2	224.0.0.5	OSPF	Hello Packet

```
▷ Frame 1: 98 bytes on wire (784 bits), 98 bytes captured (784 bits)
▷ Ethernet II, Src: c2:01:aa:aa:00:00 (c2:01:aa:aa:00:00), Dst: IPv4mcast_00:00:05 (01:00:5e:00:00:05)
▷ Internet Protocol, Src: 10.2.1.1 (10.2.1.1), Dst: 224.0.0.5 (224.0.0.5)
▽ Open Shortest Path First
  ▷ OSPF Header
  ▽ LS Update Packet
      Number of LSAs: 1
    ▽ LS Type: Router-LSA
        LS Age: 2 seconds
        Do Not Age: False
      ▷ Options: 0x22 (DC, E)
        Link-State Advertisement Type: Router-LSA (1)
        Link State ID: 4.4.4.4
        Advertising Router: 4.4.4.4 (4.4.4.4)
        LS Sequence Number: 0x80000005
        LS Checksum: 0x9e46
        Length: 36
      ▷ Flags: 0x00
        Number of Links: 1
      ▷ Type: Transit   ID: 10.3.1.1        Data: 10.3.1.4        Metric: 10
```

Figure 3.35 OSPFv2 convergence sequence (in network 10.2.1.0/24).

3.8.2.2 Different Versions of the IGMP Protocol

After studying IGMPv2 in detail, let us briefly discuss the other versions os the IGMP protocol. The first version of IGMP, IGMPv1, presented in [15], defined two types of IGMP messages:

- Membership Query, with IP destination 224.0.0.1.
- Membership Report, with IP destination equal to the multicast group address.

A host responds to a query message with a report message, reporting each multicast group to which it belongs on the network interface from which the query was received. When a host joins a new group, it should immediately transmit a report message for that group (in case it is the first member of that group on the network), rather than waiting for a query message.

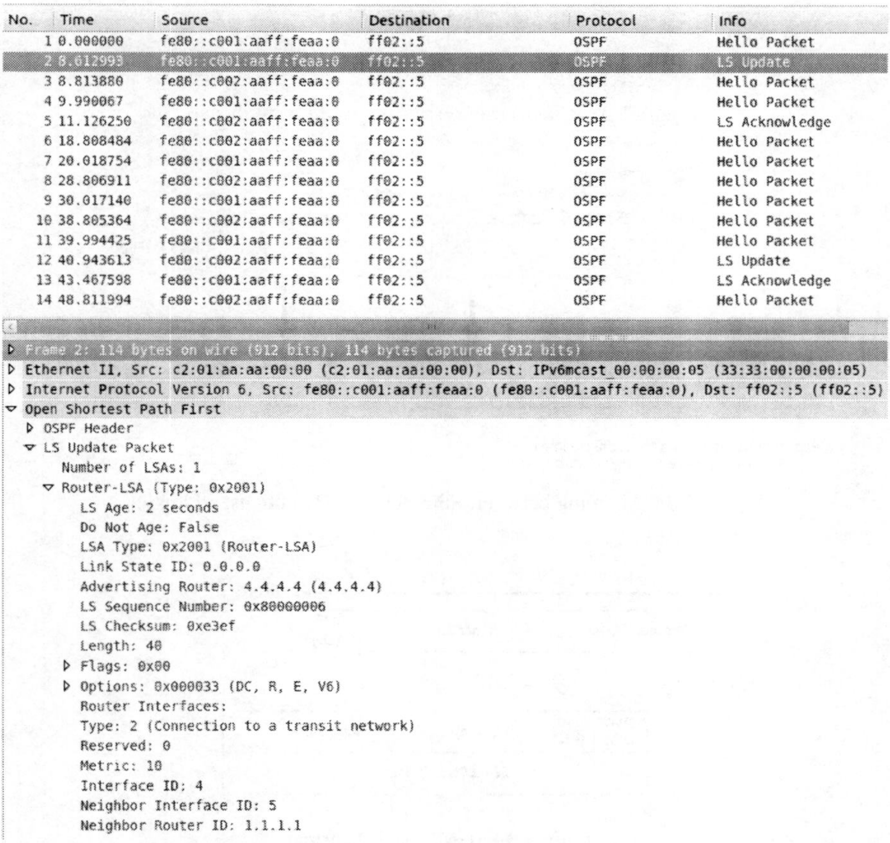

No.	Time	Source	Destination	Protocol	Info
1	0.000000	fe80::c001:aaff:feaa:0	ff02::5	OSPF	Hello Packet
2	8.612993	fe80::c001:aaff:feaa:0	ff02::5	OSPF	LS Update
3	8.813880	fe80::c002:aaff:feaa:0	ff02::5	OSPF	Hello Packet
4	9.990067	fe80::c001:aaff:feaa:0	ff02::5	OSPF	Hello Packet
5	11.126250	fe80::c002:aaff:feaa:0	ff02::5	OSPF	LS Acknowledge
6	18.808484	fe80::c002:aaff:feaa:0	ff02::5	OSPF	Hello Packet
7	20.018754	fe80::c001:aaff:feaa:0	ff02::5	OSPF	Hello Packet
8	28.806911	fe80::c002:aaff:feaa:0	ff02::5	OSPF	Hello Packet
9	30.017140	fe80::c001:aaff:feaa:0	ff02::5	OSPF	Hello Packet
10	38.805364	fe80::c002:aaff:feaa:0	ff02::5	OSPF	Hello Packet
11	39.994425	fe80::c001:aaff:feaa:0	ff02::5	OSPF	Hello Packet
12	40.943613	fe80::c002:aaff:feaa:0	ff02::5	OSPF	LS Update
13	43.467598	fe80::c001:aaff:feaa:0	ff02::5	OSPF	LS Acknowledge
14	48.811994	fe80::c002:aaff:feaa:0	ff02::5	OSPF	Hello Packet

```
▷ Frame 2: 114 bytes on wire (912 bits), 114 bytes captured (912 bits)
▷ Ethernet II, Src: c2:01:aa:aa:00:00 (c2:01:aa:aa:00:00), Dst: IPv6mcast_00:00:00:05 (33:33:00:00:00:05)
▷ Internet Protocol Version 6, Src: fe80::c001:aaff:feaa:0 (fe80::c001:aaff:feaa:0), Dst: ff02::5 (ff02::5)
▽ Open Shortest Path First
  ▷ OSPF Header
  ▽ LS Update Packet
      Number of LSAs: 1
    ▽ Router-LSA (Type: 0x2001)
        LS Age: 2 seconds
        Do Not Age: False
        LSA Type: 0x2001 (Router-LSA)
        Link State ID: 0.0.0.0
        Advertising Router: 4.4.4.4 (4.4.4.4)
        LS Sequence Number: 0x80000006
        LS Checksum: 0xe3ef
        Length: 40
      ▷ Flags: 0x00
      ▷ Options: 0x000033 (DC, R, E, V6)
        Router Interfaces:
        Type: 2 (Connection to a transit network)
        Reserved: 0
        Metric: 10
        Interface ID: 4
        Neighbor Interface ID: 5
        Neighbor Router ID: 1.1.1.1
```

Figure 3.36 OSPFv3 convergence sequence (in network 2001:A:2:1::/64).

In IGMPv3, which was initially defined by RFC 3376 [9] and has been updated by RFC 4604 [50], besides the multicast session, the receiver specifies from which sender he wants to receive multicast packets (this mechanism is known as Source Specific Multicast). IGMPv3 defines the following types of messages:

- General Membership Query - This query is used to learn which groups have members on an attached network. IP destination is equal to 224.0.0.1, the group address is equal to 0 and the number of sources is equal to 0.
- Group-specific Membership Query - This query is used to learn if a particular group has any members on an attached network. IP destination

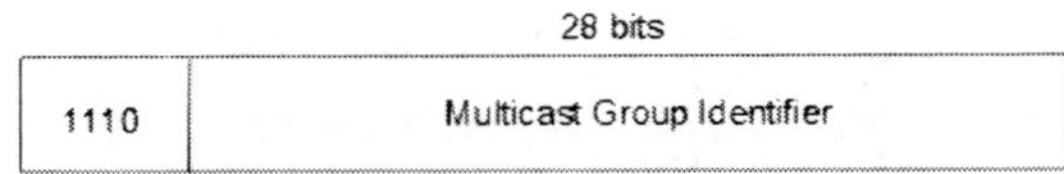

Figure 3.37 Class D addresses are used for multicast.

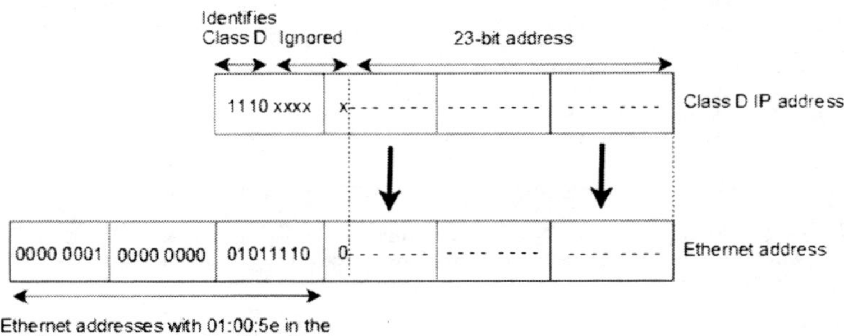

Figure 3.38 Mapping between Ethernet and IP multicast addresses.

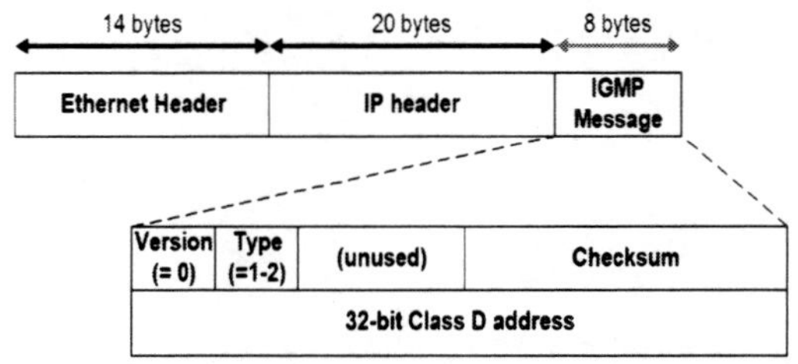

Figure 3.39 IGMP packet format.

is equal to the multicast group address; the group address is equal to the multicast group address and the number of sources is equal to 0.

- Group-and-source-specific Membership Query - This query is used to learn if any neighboring interface wants to receive packets sent to a specified group address, from any of a specified list of sources. IP destination is equal to the multicast group address; group address is equal to the multicast group address and the source field is equal to the source address of interest.
- Membership Report - This message is used to report each multicast group to which the host belongs. IP destination is equal to 224.0.0.22;
- V.1 Membership report - This message is supported to be compatible with IGMPv1.

- V.2 Membership report - This message is supported to be compatible with IGMPv2.
- V.2 Leave group - This message is supported to be compatible with IGMPv2.

IGMPv3 adds support for "Source Filtering", i.e. the ability for a system to report interest in receiving packets only from specific source addresses, as required to support Source-Specific Multicast (SSM), or from all but specific source addresses, sent to a particular multicast group address.

We have seen that traditional multicast forwarding is performed using multicast group addresses, taken from a reserved range (224.0.0.0/4 for IPv4, FF00::/8 for IPv6) to uniquely identify a group of hosts desiring to receive certain traffic. Since any host can act as a source, this multicast implementation is known as Any Source Multicast (ASM).

Source-Specific Multicast extends the ASM concept to identify a set of multicast hosts not only by their group address but also by source. An SSM group, called a channel, is identified as (S,G) where S is the source address and G is the group address. This is in contrast to the definition of an ASM multicast route written as (*,G). IANA has reserved the IPv4 address range 232.0.0.0/8 and the IPv6 range FF3x::/32 for SSM.

SSM brings several important benefits over ASM. Because an SSM channel is defined by both a source and a group address, group addresses can be re-used by multiple sources while keeping channels unique. For instance, the SSM channel (192.168.45.7, 232.7.8.9) is different from (192.168.3.104, 232.7.8.9), and hosts subscribed to one will not receive traffic from the other. This allows for greater flexibility in choosing a multicast group while also protecting against denial of service attacks; hosts will only receive traffic from explicitly requested sources.

One of the biggest advantages SSM holds over ASM is that it does not rely on the designation of a Rendezvous Point (RP) to establish a multicast tree (we will look at this later in this chapter). Because the source of an SSM channel is always known in advance, multicast trees are efficiently built from the channel hosts towards the source (based on the unicast routing topology), without the need for an RP to join a source and shared multicast tree. The corollary of this, which may be undesirable in some multicast implementations, is that the multicast source(s) must be learned in advance via some external method (e.g. manual configuration).

3.8.3 The MLD Protocol

The Multicast Listener Discovery (MLD) protocol was defined in RFC 2710 [42] and is used to exchange membership status information between IPv6 routers that support multicasting and members of multicast groups on a network segment. The version 1 of MLD has a similar operation to IGMPv2, while MLDv2 is similar to IGMPv3.

MLD supports the following message types:

- Multicast Listener Query - Sent by a multicast router to poll a network segment for group members. Queries can be general (requesting group membership for all groups), or specific (requesting group membership for a specific group).
- Multicast Listener Report - Sent by a host when it joins a multicast group, or in response to an MLD Multicast Listener Query sent by a router.
- Multicast Listener Done - Sent by a host when it leaves a host group and might be the last member of that group on the network segment.

Although being similar to IGMPv2, MLDv1 presents some distinct features:

- All the multicast devices on a subnet use a special IPv6 link-local address as their source address in their communication to other multicast devices. The use of the link-local source address prevents the MLD packet from traveling beyond the local link.
- In MLD, when a host wants to leave a group, it sends a Done message. The Done message is similar to the IGMPv2 Leave message. It is addressed to the all-routers IPv6 link-local scope address, FF02::2.
- In MLD, the router Queries are called Multicast Listener Queries. The General Queries are addressed to the all-nodes IPv6 link-local scope address, FF02::1. When a router receives a Done message, it sends a Multicast-Address-Specific Query. Its function is similar to IGMPv2 Group-Specific Query.

3.8.4 Multicast Routing Solutions

How multicast routers can route packets amongst themselves in order to ensure that each router receives the multicast group traffic that it needs? A spanning tree should be constructed connecting all members of the multicast group. Of course, the tree may contain routers that do not have attached hosts belonging to the multicast group.

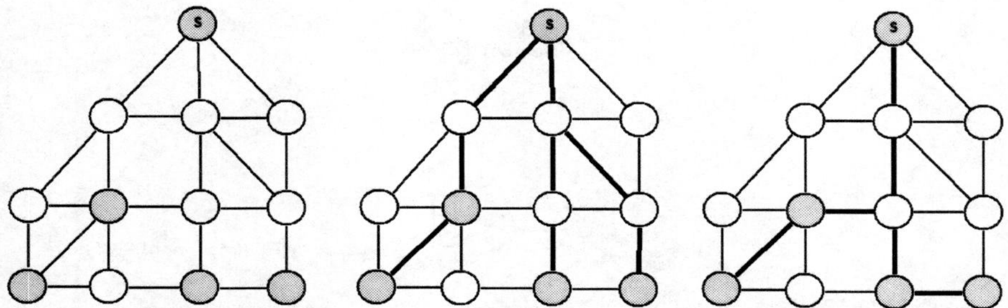

Figure 3.40 Multicast routing as a graph problem. (Left) multicast nodes; (center) building a least cost path tree; (right) - building a center-based tree.

Let us consider the network scenario illustrated in Figure 3.40, where each router is represented as a node and each link is represented by an edge. There is a unitary cost for traversing each edge in each direction. Dark routers are connected to a multicast group and thus need to receive multicast traffic. There is one multicast source (S), which is also a member of the multicast group. The problem is how to construct a tree that connects all multicast group members?

There are two solutions to this graph problem:

- Build a Shortest Path Tree or Source-Based Tree (center figure) - A tree that minimizes the path cost from the source to each receiver. It is a good solution if there is a single sender; if there are multiple senders, one tree per sender is needed. This tree is easy to compute.
- Build a Minimum-Cost Tree (right figure) - A tree that minimizes the total cost of the edges. It is a good solution if there are multiple senders, but it is very expensive to compute (not practical for more than 30 nodes, since this is known to be an NP-complete problem).

In practice, routing protocols implement two approaches for determining the multicast routing tree:

- Source-Based Tree - Essentially implements solution 1. Builds one shortest path tree for each sender and the tree is built from receiver to the sender.
- Core-Based Tree - Builds a single distribution tree that is shared by all senders. Does not implement solution 2 because it is too expensive. Instead, it selects one router as a "core" (called "Rendezvous Point") and all receivers build a shortest path to the core.

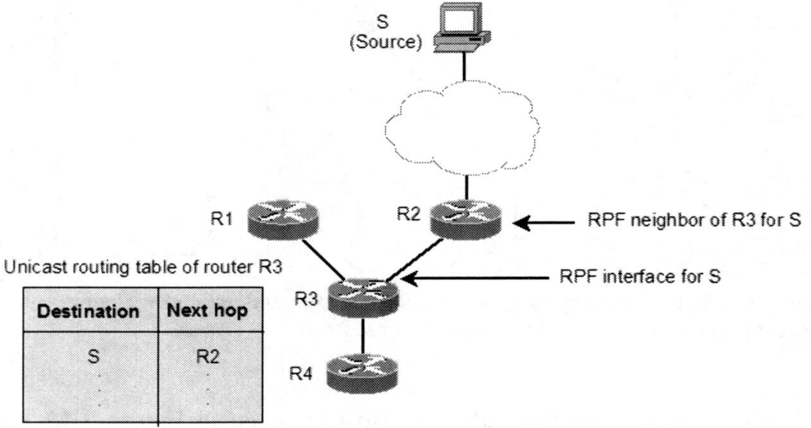

Figure 3.41 Illustration of the reverse path forwarding algorithm.

3.8.4.1 Source-Based Tree Approach

The least cost path multicast routing algorithm is a link-state algorithm. It requires that each router knows the state of each link in the network in order to be able to compute the least cost path tree from the source to all destinations. A simpler multicast routing algorithm, one which requires much less link state information than the least cost path routing algorithm, is the Reverse Path Forwarding (RPF) algorithm.

The idea behind reverse path forwarding is simple [64]. When a router receives a multicast packet with a given source address, it transmits the packet on all of its outgoing links (except the one on which it was received) only if the packet arrived on the link that is on its own shortest path back to the sender. Otherwise the router simply discards the incoming packet without forwarding it on any of its outgoing links. Such a packet can be dropped because the router knows it either will receive, or has already received, a copy of this packet on the link that is on its own shortest path back to the sender. Note that reverse path forwarding does not require that a router know the complete shortest path from itself to the source; it only needs to know the next hop on its unicast shortest path to the sender.

Figure 3.41 illustrates RPF. Router R3 will forward the source packet it has received from R2 (since R2 is on its least cost path to S) but will ignore (drop, without forwarding) any source packets it receives from any other routers.

Note however that even if router R4 has no attached hosts that have joined to the multicast group, it will still receive unwanted multicast packets from router R3. The solution to this problem is known as pruning. A multicast router that receives multicast packets and has no attached hosts joined to that group will send a prune message to its upstream router. If a router receives prune messages from each of its downstream routers, then it disables the routing table entry and can forward a prune message upstream.

So, a prune message temporarily disables a routing table entry, removing a link from the multicast tree. No multicast messages are sent on a pruned link. Prune messages are sent by (i) a router with no group members in its local network and no connection to other routers (sent on RPF interface); (ii) a router with no group members in its local network which has received a prune message on all non-RPF interfaces (sent on RPF interface); (iii) a router with group members which has received a packet from a non-RPF neighbor (to non-RPF neighbor).

If a router sends a prune message upstream and, later, it needs to join the same multicast group, some mechanism is needed to restore the branch of the spanning tree that was previously pruned. One possibility is to add a graft message that allows a router to "unprune" a branch. Another option is to allow pruned branches to time-out and be added again to the multicast RPF tree; a router can then re-prune the added branch if the multicast traffic is still not wanted.

Figures 3.42 and 3.43 illustrate the construction of a least cost path tree for a particular network scenario. First of all, routing tables are set according to RPF forwarding (initially, outgoing interfaces are all interfaces other than the RPF interface). Then each router forwards packets that arrive on its RPF interface on all non-RPF interfaces (the arrows that are shown in Figure 3.42). Whenever a packet is received on a non-RPF interface, it is discarded, as shown by the crosses in Figure 3.42. Note that, due to the flooding of messages, many routers receive multiple copies of the same packet.

Whenever a router receives a multicast packet on an interface that is not the RPF interface, the router sends a prune message on that interface. Besides, a router also sends a prune message on its RPF interface if it is not connected to a LAN with group members, which happens when the router does not have a downstream neighbor in the distribution tree or when it has received prune messages from all downstream neighbors. This situation is illustrated in Figure 3.43, where prune messages correspond to the dotted arrows. Note that, in this case, R2 sends a prune message to R4 since the interface connecting to R4 is not the RPF for source S; R8 sends a prune message because

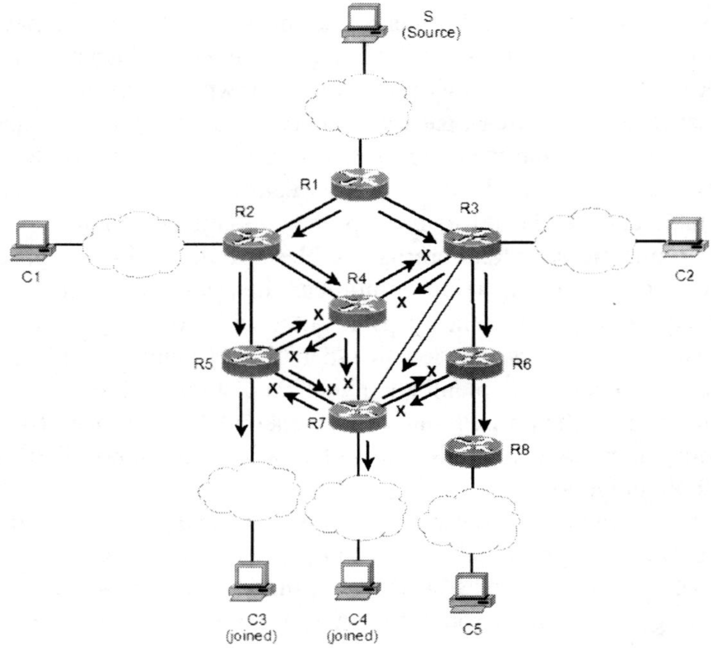

Figure 3.42 Building a source based tree - flooding and packet discarding.

it does not have a group member in its local network and it also does not have downstream neighbors; R6 sends a prune message to R3 because R6 has received prune messages from all downstream neighbors.

When a receiver joins, a pruned routing table entry must be re-activated. A Graft message is sent, disabling the previous prune state and re-activating the routing table entry. A router that receives a graft message forwards the graft message on its RPF interface, until a router that is part of the distribution tree is reached (Figure 3.44, considering that host C5 joins the multicast group).

3.8.4.2 Core-Based Tree Approach

In the core-based tree approach two cases can be considered. In the first case, all packets sent to a multicast group are routed along the same singe multicast tree, regardless of the sender. In this situation, the multicast routing problem appears quite simple: find a tree within the network that connects all routers having an attached host belonging to that multicast group. One might also want the tree to have minimal "cost", obtaining an optimal multicast routing

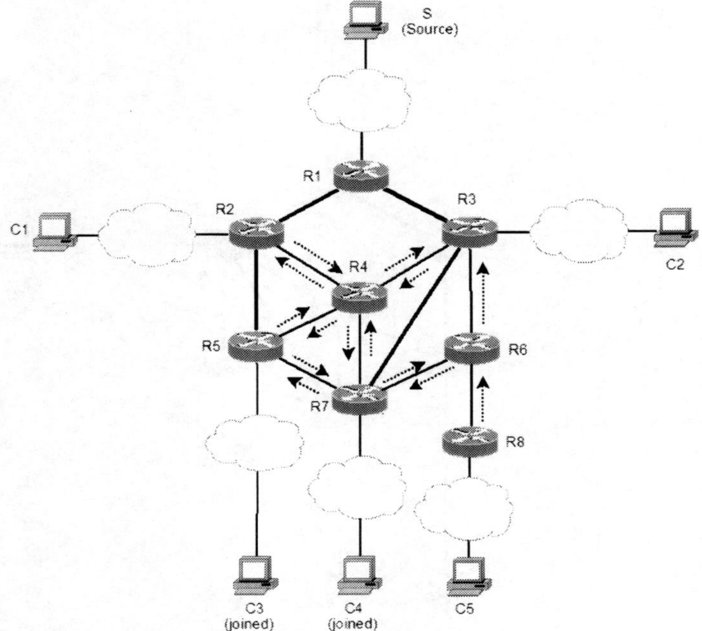

Figure 3.43 Building a source based tree - pruning.

tree (the one having the smallest sum of the tree link costs). The problem of finding a minimum cost tree is known as the Steiner Tree problem [47]. Solving this problem has been shown to be NP-complete [31], but there are some approximation algorithms that perform quite well in practice [104].

Even though good heuristics exist for the Steiner tree problem, none of the existing Internet multicast routing algorithms have been based on this approach. One reason for that is that information is needed about all links in the network. Another reason is that in order for a minimum cost tree to be maintained, the algorithm needs to be re-run whenever link costs change. Finally, we will see that other considerations, such as the ability to leverage the routing tables that have already been computed for unicast routing, play an important role in judging the suitability of a multicast routing algorithm. Performance and optimality are other important concerns [64].

An alternate approach towards determining the group-shared multicast tree, one that is used in practice by several Internet multicast routing algorithms, is based on the notion of defining a center node (also known as a rendezvous point or a core) in the single shared multicast routing tree. In the

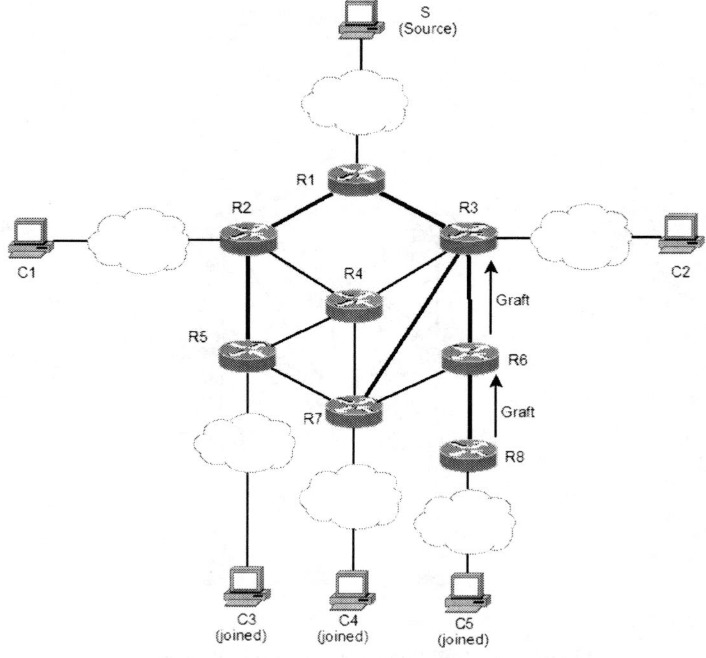

Figure 3.44 Building a source based tree - grafting.

center-based approach, a center node is first identified for the multicast group. Routers with attached hosts belonging to the multicast group then unicast so-called "join" messages addressed to the center node. A join message is forwarded using unicast routing towards the center until it either arrives at a router that already belongs to the multicast tree or arrives at the center. In either case, the path that the join message has followed defines the branch of the routing tree between the edge router that initiated the join message and the center. One can think of this new path as being "grafted" onto the existing multicast tree for the group.

Figures 3.45 and 3.46 illustrate the construction of a core-based multicast routing tree. Suppose that router R4 is selected as the center of the tree. Each router that wants to receive multicast packets sends a Join message through its RPF interface with respect to core (Figure 3.45). If the join message arrives to a router that is not part of the core-based tree, an (*, G) entry is added to the router multicast routing table (assuming that is multicast group is G), the interface where the join message was received is added to the outgoing interface list and the incoming interface is set to the RPF interface. Finally, this router forwards the join message on its RPF interface in the direction of the core. If the router is already part of the shared tree for group G, the router adds the interface where the join message arrived to the outgoing interface list of the corresponding (*, G) entry, but does not forward the join message. Note that the distribution tree that is formed is a reverse shortest-path tree rooted at the core.

When the source starts sending data to the core, this node will forward data according to the routing table entry. This situation is illustrated in Figure 3.46. When source S transmits a multicast packet, it is forwarded on the unicast shortest path to the core (possible routes are from S to R2 and to R4 or from S to R3 and then to R4).When the packet reaches the tree at the core, it is sent downstream in the core-based tree. Note that sending packets always to the core and then forwarding them to the distribution tree only at the core can be inefficient; sometimes using the core-based tree could result in a shortest-path packet delivery [64].

An important issue on center-based tree multicast routing is the center selection process. Several algorithms were proposed in the literature to solve this problem [100].

3.8.5 Multicast Routing Protocols

The three currently standardized Internet multicast routing protocls are DVMRP, MOSPF, and PIM.

3.8.5.1 DVMRP

The distance vector multicast routing protocol (DVMRP) was proposed in [103] and implements source-based trees with reverse path forwarding, pruning and grafting. DVMRP uses a distance vector algorithm that allows each router to compute the outgoing link (next hop) that is on its shortest path back to each possible source. In addition to computing next hop information, DVMRP also computes a list of dependent downstream routers for pruning purposes.

When a router has received a prune message from all of its dependent downstream routers for a given group, it will propagate a prune message upstream to the router from which it receives its multicast traffic for that group. A DVMRP prune message contains a prune lifetime (with a default value of two hours) that indicates how long a pruned branch will remain pruned before being automatically restored. DVMRP graft messages are sent by a router to

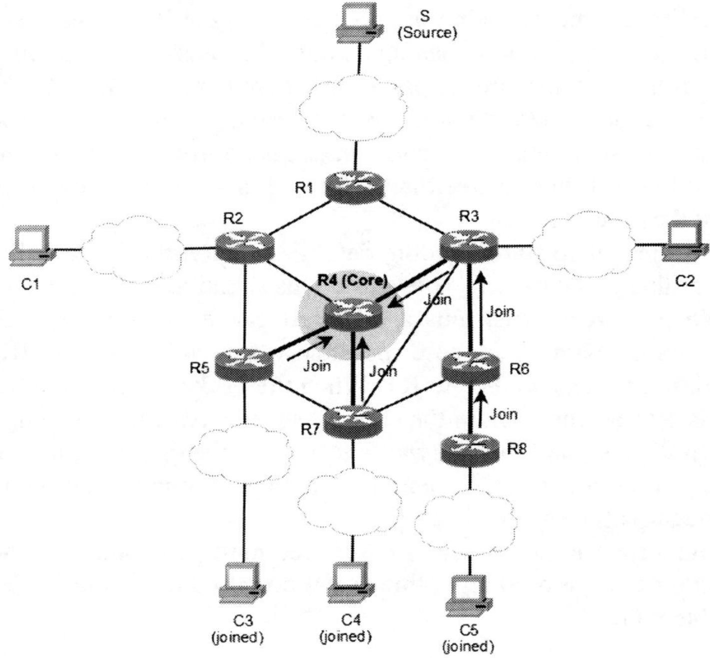

Figure 3.45 Building a core based tree - Joining.

its upstream neighbor to force a previously-pruned branch to be added back on to the multicast tree.

The DVMRP routing table has no information about multicast groups. The forwarding table represents the perception that a given router has about the minimum cost tree rooted on the source node, for each pair (source, group). This table is built based on the combination of the multicast routing table, known multicast groups and received prune messages. Box 3.54 represents an example of a DVMRP forwarding table. The InPort is the upstream port for the pair (source, group), where "Pr" indicates that a prune message was sent to the upstream router; OutPorts are downstream ports to where multicast packets of the (source, group) pair will be forwarded, where "p" indicates that this router has received a prune message from a router that is located downstream.

DVMRP supports tunnels between multicast routers: tunnels are administratively configured and the tunnel end points act as they were neighbor routers. This is illustrated in Figure 3.47. Suppose that the left multicast router

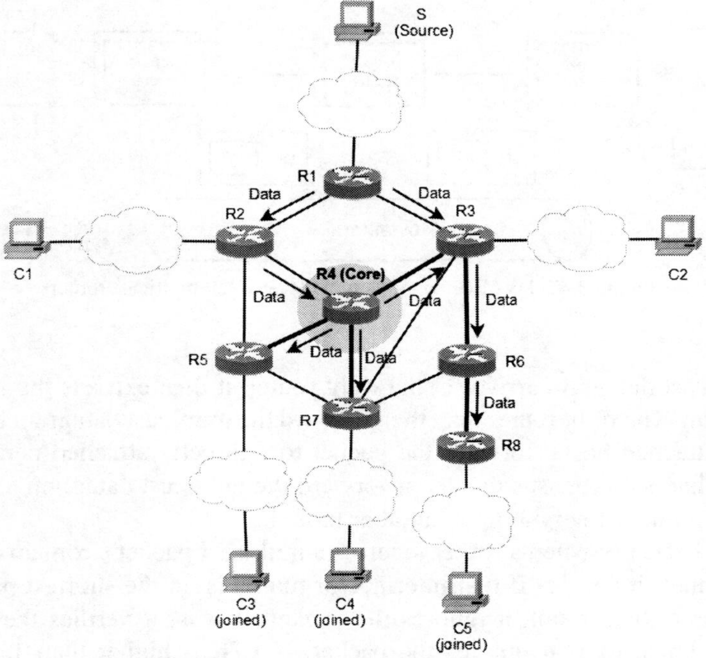

Figure 3.46 Building a core based tree - flooding.

Source Subnet	Multicast Group	TTL	InPort	OutPorts
128.1.0.0	224.1.1.1	200	1 Pr	2p 3p
	224.2.2.2	100	1	2p 3
	224.3.3.3	250	1	2
128.2.0.0	224.1.1.1	150	2	2p 3

Box 3.54 Example of a DVMRP forwarding table

wants to forward a multicast datagram to the multicast right router. Also suppose that these routers are not physically connected to each other and that the intervening router is not multicast capable. To implement tunneling, the left router takes the multicast datagram and "encapsulates" it inside a standard unicast datagram. That is, the entire multicast datagram (including source and multicast address fields) is carried as the payload of an IP unicast datagram. The unicast datagram is then addressed to the unicast address of the right router and forwarded towards the right router by the left router. The unicast router between them dutifully forward the unicast packet to the right, being unaware that the unicast datagram itself contains a multicast datagram. When

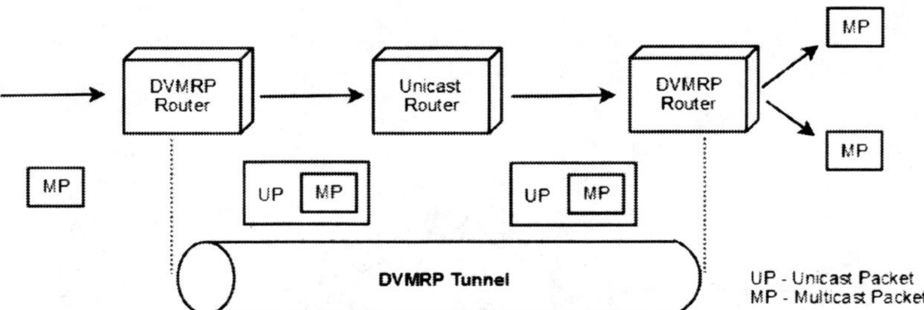

Figure 3.47 DVMRP supports tunnels between multicast routers.

the unicast datagram arrives at the right router, it then extracts the multicast datagram. The right router may then forward the multicast datagram on to one of its attached hosts, forward the packet to a directly attached neighboring router that is multicast capable, or forward the multicast datagram to another logical multicast neighbor via another tunnel.

So, basically when a router receives a multicast packet (from an interface or a tunnel) it verifies if that interface or tunnel is on the shortest path from the source. It it is not, it ignores the packet. If it is, it verifies the TTL. If TTL is equal to 1, it ignores the packet. If TTL is higher than 1, it sends through all interfaces and tunnels where (i) the router is the previous hop on the shortest path; (ii) the "pruning" process for this multicast session has not been applied.

3.8.5.2 MOSPF

The Multicast Open Shortest Path First protocol (MOSPF), defined in RFC [71], operates in an AS that uses the OSPF protocol for unicast routing. MOSPF extends OSPF by having routers add their multicast group membership to the link state advertisements that are broadcast by routers as part of the OSPF protocol. With this extension, all routers have not only complete topology information, but also know which edge routers have attached hosts belonging to various multicast groups. With this information, the routers within the AS can build source-specific, pre-pruned, shortest path trees for each multicast group.

MOSPF is a source-based tree algorithm, implementing a RPF strategy with pruning. Since each router knows the complete network topology, it can calculate the shortest path tree for each active multicast session. Each MOSPF router has a local database of the multicast groups, together with a list of all hosts that are connected to each group. The router will deliver

```
Destination   Source      Upstream   Downstream   TTL
224.1.1.1     128.1.0.2   !1         !2 !3        5
224.1.1.1     128.4.1.1   !1         !2 !3        3
224.1.1.1     128.5.2.1   !1         !2 !3        2
224.2.2.2     128.2.0.4   !2         !1           8
```

Box 3.55 Example of a MOSPF forwarding cache

datagrams to the group's members based on this knowledge. When an initial datagram arrives, each router builds the shortest-path tree with source located on the source address of the datagram using the Dijkstra algorithm. This operation is repeated for each (source, group) pair. After this step, Group-Membership LSAs are used to prune the paths that do not contain groups members. Each router calculates its position in the shortest-path distribution tree corresponding to this datagram and an entry is created in the forwarding cache containing the (source, group) pair, the upstream node and the downstream interfaces. At this moment, the shortest-path tree is discarded, freeing all resources associated to its creation. The forwarding cache entry is used to forward subsequent datagrams for that (source, group) pair.

The forwarding cache, illustrated in Box 3.55 is not periodically updated. It only changes when the network topology changes or the Membership Groups change. These changes are announced through LSAs that are sent using a flooding process.

MOSPF doe not support tunnels: in the OSPF router LSA messages there is a flag that indicates if the router supports multicast or not; routers that do not support multicast are unable to belong to the shortest path tree.

3.8.5.3 PIM

The Protocol Independent Multicast (PIM) routing protocol explicitly envisions two different multicast distribution scenarios: dense mode [1], where multicast group members are densely located, that is, many or most of the routers in the area need to be involved in routing multicast datagrams; sparse mode [24, 25], where the number of routers with attached group members is small with respect to the total number of routers.

In dense mode, since most routers will be involved in multicast (by having group members attached to them), it is reasonable to assume that each and every router should be involved in multicast. Thus, an approach like RPF, which floods datagrams to every multicast router (unless a router explicitly prunes itself) is well suited to this scenario. On the other hand, in sparse mode, the routers that need to be involved in multicast forwarding are few and far be-

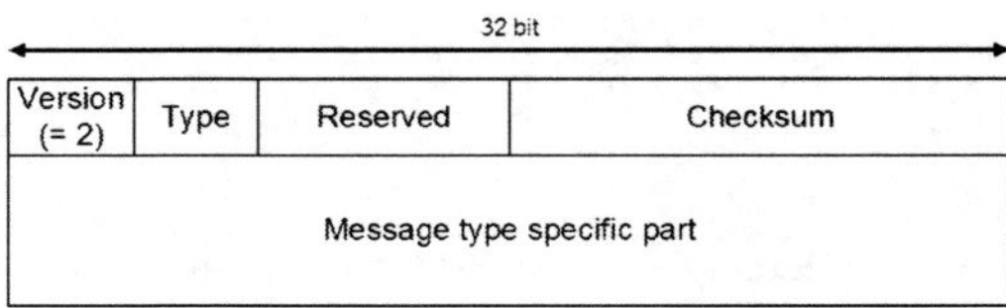

Figure 3.48 PIMv2 packet format.

tween. In this case, a data-driven multicast technique like RPF, which forces a router to constantly send prune messages simply to avoid receiving multicast traffic is much less satisfactory. So, in sparse mode the default assumption should be that a router is not involved in a multicast distribution; the router should not have to do any work unless it wants to join a multicast group. This argues for a center-based approach, where routers send explicit join messages, but are otherwise uninvolved in multicast forwarding. The sparse mode approach can be considered receiver-driven (that is, nothing happens until a receiver explicitly joins a group), while the dense mode approach can be considered data-driven (that is, datagrams are multicast everywhere, unless explicitly pruned). PIM accommodates this dense versus sparse dichotomy by offering two explicit modes of operation: dense mode and sparse mode.

PIM means "protocol independent", so it is independent of the underlying unicast routing protocol. PIM can inter operate with any underlying unicast routing protocol. Figure 3.48 shows the general format of the PIMv2 messages (the current used version of PIM), while Figure 3.49 shows which message types are supported in each operation mode. PIM messages are encapsulated in IP datagrams with protocol number equal to 103 and can be sent as unicast or multicast packets. The 224.0.0.13 address is reserved as the ALL-PIM-Routers group.

PIM Dense Mode is a flood-and-prune reverse path forwarding technique, similar to DVMRP, and its basic operation will be explained using Figure 3.50. PIM Dense Mode starts by flooding the multicast traffic, and then stopping it at each link where it is not needed, using a Prune message. In the most recent Cisco IOS versions, the Prune state is refreshed periodically in order to recognize topology changes more quickly.

In Figure 3.50, if Router R5 choses R3 as its RPF neighbor based on unicast routing back to the source, then when it receives a multicast packet from R2 it will send a rate-limited Prune to R2. Routers R4 and R5 are connected to the same LAN, so they will see the multicast packets from each other. In order to avoid the forwarding of packets from both routers, they will send

PIM-DM messages	Type	PIM-DM	PIM-SM
Hello	0	✓	✓
Register	1		✓
Register-Stop	2		✓
Join/Prune	3	✓	✓
Bootstrap	4		✓
Assert	5	✓	✓
Graft	6	✓	
Graft-Ack	7	✓	
Candidate-RP-Advertisement	8		✓

Figure 3.49 PIM messages for the dense and sparse operation modes.

Assert messages to LAN B. The router with the best routing metric wins, with higher IP address being the tie-breaker. If, for example, R5 wins, R4 knows it does not have to forward multicasts on LAN B. This non-forwarder router then send a Prune on its RPF interface.

Supposing that Router R8 has no receivers downstream, it can send a Prune to Router R5. However, if Router R7 has a downstream receiver, it will send a PIM Join message in order to continue receiving multicast data packets. If Router R6 has no downstream or local receivers, it will send a Prune to R3. If later on one of its clients connects to the multicast group, Router R6 will send a PIM Graft message to R3, asking it to resume sending the specified multicast group traffic. Router R3 will send a Graft-Acknowledge message back to R6.

PIM Sparse Mode is a center-based approach. The core is the Rendezvous-Point (RP) and receivers know who is the RP (is statically configured or dynamically elected). Let us consider the network illustrated in Figure 3.51, where router R5 is the RP.

When receiver C1 joins, a Join message is sent to the RP on the RPF interface. Note that intermediate routers set up multicast state and forward the Join message towards the RP. There is no acknowledgment to this Join message. Join messages are periodically sent upstream to refresh/maintain the PIM routing tree. When host C2 joins, a Join message is forwarded only to the first router that is part of the core-based tree (R2).

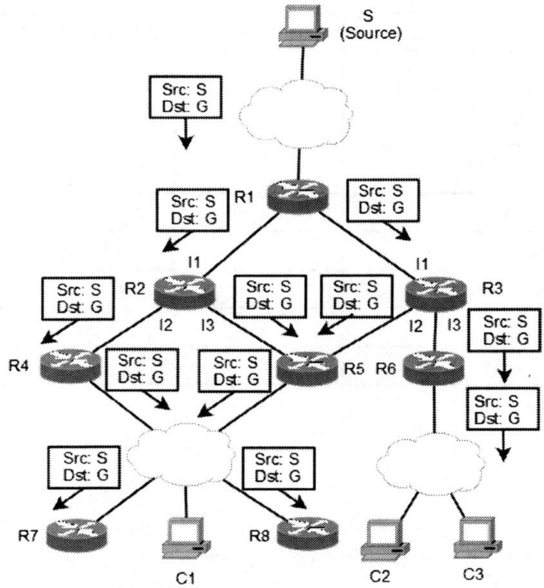

Figure 3.50 Illustration of the PIM dense mode operation.

When the source sends the first multicast packet to the RP, the packet is attached to an RP Register message (Figure 3.52). When the packet reaches the RP, it is forwarded in the tree and the RP sends a Join message on the reverse path to the source. When the Join message reaches R1, this router sends a native multicast packet to the RP (in addition to the packet attached to the Register message).

When the RP receives the native multicast packet it sends a Register Stop message to R1. This message stops the transmission of Register messages from R1 (Figure 3.53).

One novel feature of PIM Sparse Mode is the ability to switch from a group-shared tree to a source-specific tree after joining the RP. In the context of PIM Sparse Mode, the source-based tree is called shortest path tree (SPT). A source-specific tree may be preferred due to the decreased traffic concentration that occurs when multiple source-specific trees are used. In this example, when data to receivers exceeds a threshold, routers switch to a source-based tree. This is done by sending an explicit Join message to the source on the RPF interface: router R3 initiates the switch to the SPT for source S by creating an (S,G) routing table entry and by sending a Join(S,G) message

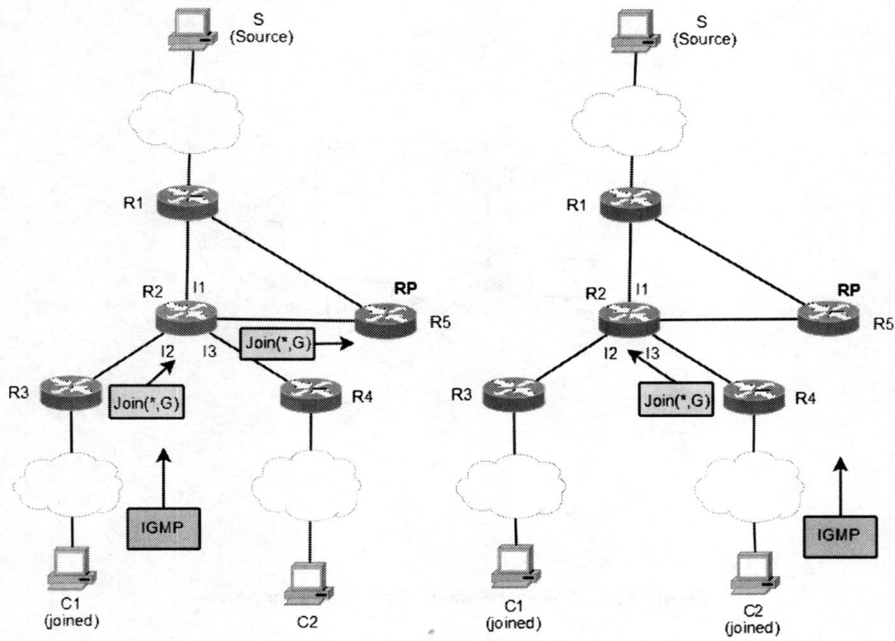

Figure 3.51 PIM sparse mode - receiver joining: (left) C1 joins; (right) C2 joins.

on its RPF interface for S (left part of Figure 3.54). When R2 receives the Join message, it creates an (S,G) routing table entry and sends a Join(S,G) to R1. R1 adds the interface to R2 to is outgoing interface list and starts forwarding multicast packets to R2, as well as to the RP. When the SPT is established, there may be duplicate packets being sent for some time (one from the RPT and one from the SPT). When data arrives from the source through the SPT, a Prune(S1,G,RP) message is sent to the RP in order to stop duplicate transmissions. This special Prune message removes the router from the Rendezvous-Point Tree (RPT). From now on, data will be forwarded only along the shortest-path tree.

We saw that there are multiple multicast routing protocols, with different characteristics and behaviors. The following factors should be considered when evaluating a multicast protocol:

- Scalability - The ability of the protocol to easily accommodate higher numbers of groups or senders in a group should be evaluated. The degradation of the routers' performance under that increase should also be evaluated.

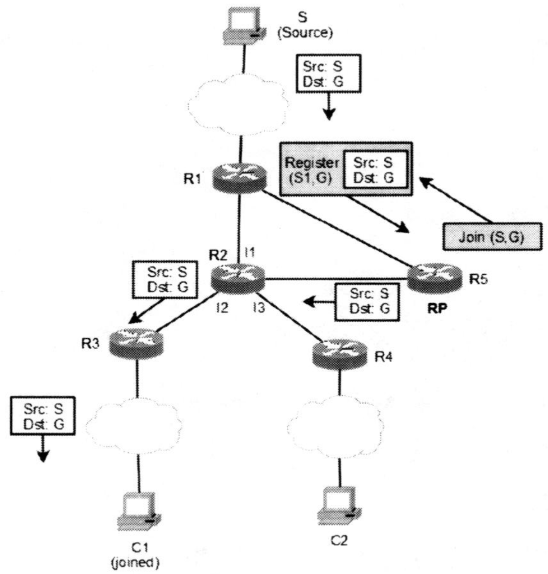

Figure 3.52 PIM sparse mode - the Register message.

- Reliance on the underlying unicast routing - Some multicast protocols rely on specific underlying unicast routing protocols, while others are completely independent of the unicast routing.
- Quantity of unwanted traffic received - In protocols like MOSPF or PIM-Sparse Mode routers receive data only if they have an attached host in the multicast group, while for other protocols (like DVMRP or PIM Dense Mode) the default is to receive all traffic for all multicast groups.
- Traffic concentration - The group-shared tree algorithm tends to concentrate traffic on a smaller number of links, whereas the source-specific tree approach tends to distribute multicast traffic more evenly.
- Optimality of forwarding paths - Since determining the minimum cost multicast tree is hard to implement in practice, heuristic approaches have been adopted in real scenarios.

Exercise - PIM Sparse Mode

Considering again the network of Figure 3.2, with OSPF configured as the unicast routing protocol, let us configure PIM sparse mode: this can be done by using the commands shown in Box 3.56 on every router. The Rendez-vous

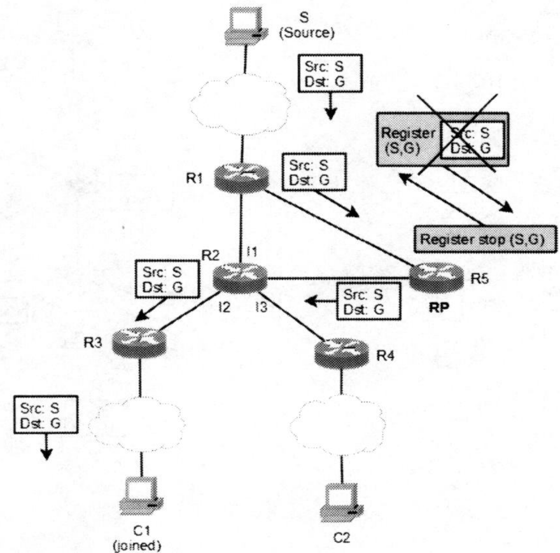

Figure 3.53 PIM sparse mode - the Register Stop message.

```
# ip multicast-routing
# ip pim rp-address 10.3.1.4
# interface Fx/x
  # ip pim sparse-mode
```

Box 3.56 Multicast configuration.

```
# interface F1/0
  # ip igmp join-group 234.234.234.234
```

Box 3.57 Multicast client configuration at Router 1 for session 234.234.234.234.

Point (RP) was selected as interface F0/0 of Router 4. Let us suppose that the multicast traffic generator is located at Router 5 and there is a multicast client at interface F1/0 of Router 1. Considering that the IP address of the multicast session is 234.234.234.234, the F0/1 interface of Router 1 can become a multicast client of this session if the commands presented in Box 3.57 are inserted. In this experiment, multicast traffic will be emulated by using the *ping* command to address 234.234.234.234. We will consider only the IPv4 implementation, but in IPv6 the differences are not too significant.

Figure 3.55 shows the IGMP packets that circulate at network 10.1.1.0/24, illustrating the possible interactions between a multicast client and the multi-

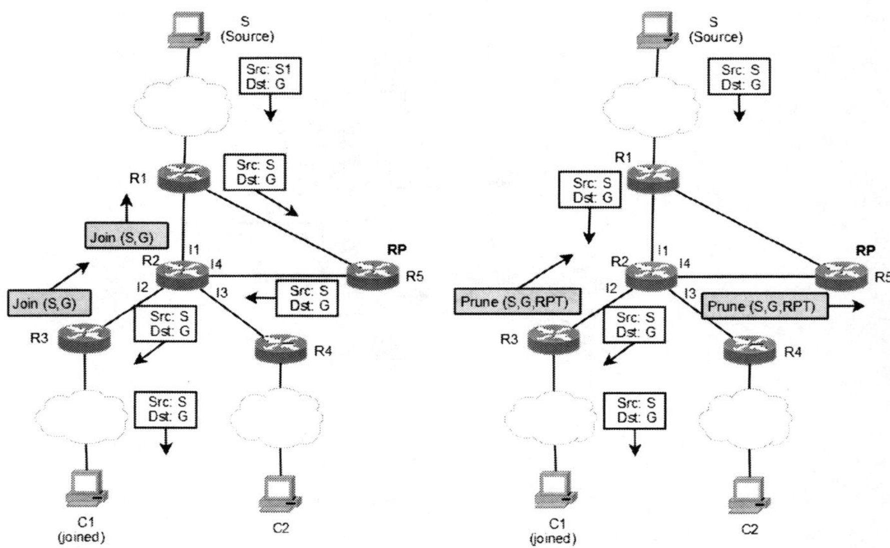

Figure 3.54 PIM sparse mode - switching from group-shared tree to source-based tree: (left) R3 switches to a SPT; (right) Data follows a SPT.

cast querier router of the network it is attached to (for simplicity, in this case the client and the querier router reside in the same router). Initially, when the client adheres to a multicast session, it sends a Membership Report message, reporting it wants to adhere to the specific multicast session. Periodically, the querier router sends a General Membership Query message, asking for any client of any multicast session. Of course, active clients immediately answer with a Membership Report message indicating the group(s) they have adhered to. When a client leaves a multicast session, it sends a Leave Group message specifying the multicast address of that session. Then, the querier router immediately sends a specific Membership Query message in order to verify if there is any other active client for that multicast session. Of course, the querier router will continue to send periodic General Membership Query messages in order to verify if there are any active clients for any multicast session.

Once Router R1 receives the IGMP Membership Report message indicating that the client has adhered to multicast session 234.234.234.234, Router R1 sends a PIM Join message to the RP (10.1.3.4), as illustrated in Figure 3.56.

No.	Time	Source	Destination	Protocol	Info
1	0.000000	10.1.1.1	234.234.234.234	IGMP	V2 Membership Report / Join group 234.234.234.234
2	23.371458	10.1.1.1	224.0.0.1	IGMP	V2 Membership Query, general
3	33.350570	10.1.1.1	234.234.234.234	IGMP	V2 Membership Report / Join group 234.234.234.234
4	60.352042	10.1.1.1	224.0.0.2	IGMP	V2 Leave Group 234.234.234.234
5	61.339006	10.1.1.1	234.234.234.234	IGMP	V2 Membership Query / Join group 234.234.234.234
6	84.338621	10.1.1.1	224.0.0.1	IGMP	V2 Membership Query, general

Figure 3.55 IGMP protocol interactions (in network 10.1.1.0/24).

No.	Time	Source	Destination	Protocol	Info
48	43.510771	10.3.1.1	224.0.0.13	PIMv2	Join/Prune

```
▷ Frame 48: 68 bytes on wire (544 bits), 68 bytes captured (544 bits)
▷ Ethernet II, Src: c2:01:aa:aa:00:01 (c2:01:aa:aa:00:01), Dst: 01:00:5e:00:00:0d (01:00:5e:00:00:0d)
▷ Internet Protocol, Src: 10.3.1.1 (10.3.1.1), Dst: 224.0.0.13 (224.0.0.13)
▽ Protocol Independent Multicast
    0010 .... = Version: 2
    .... 0011 = Type: Join/Prune (3)
    Reserved byte(s): 00
    Checksum: 0xe607 [correct]
  ▽ PIM parameters
      Upstream-neighbor: 10.3.1.4
      Groups: 1
      Holdtime: 210
    ▽ Group 0: 234.234.234.234/32
      ▽ Join: 1
          IP address: 10.3.1.4/32 (SWR)
        Prune: 0
```

Figure 3.56 PIM Join packet sent to RP by Router 1 upon IGMP Membership Report reception (in network 10.3.1.0/24).

When the source (with IP address 10.4.1.6), located at network 10.4.1.5/24, starts to send multicast packets (in this case, ICMP packets), Router 5 sends a multicast packet encapsulated in a PIM Register message to the RP. This message is illustrated in Figure 3.57 and was captured in network 10.3.2.0/24.

The RP sends a PIM Join message to Router 5 (Figure 3.58) in order to request the native multicast traffic stream.

When the Join message reaches Router 5, this router sends a native multicast packet (an ICMP packet, in this case) to the RP, in addition to the encapsulated packet. Router R5 sends another Register message (encapsulating an ICMP multicast packet) and, finally, the RP sends a PIM Register-Stop packet to tell Router 5 to stop transmitting Register messages (Figure 3.59). Note that both the multicast session address and the address of the multicast source are identified in this message.

Box 3.58 shows the contents of the multicast routing table at Router 5. The entry starting with (*, 234.234.234.234) is a shared tree multicast entry,

No.	Time	Source	Destination	Protocol	Info
1	0.000000	10.4.1.6	234.234.234.234	PIMv2	Register
2	0.002011	10.3.2.4	224.0.0.13	PIMv2	Join/Prune
3	1.998739	10.4.1.6	234.234.234.234	ICMP	Echo (ping) request (id=0x0007, seq(be/le)=1/256, ttl=254)
4	3.998156	10.4.1.6	234.234.234.234	PIMv2	Register
5	4.000207	10.3.1.4	10.3.2.5	PIMv2	Register-stop
6	4.000282	10.4.1.6	234.234.234.234	ICMP	Echo (ping) request (id=0x0007, seq(be/le)=2/512, ttl=254)
7	51.112369	10.4.1.6	234.234.234.234	ICMP	Echo (ping) request (id=0x0007, seq(be/le)=10/2560, ttl=254)
8	111.072054	10.4.1.6	234.234.234.234	ICMP	Echo (ping) request (id=0x0007, seq(be/le)=18/4608, ttl=254)
9	111.074135	10.4.1.6	234.234.234.234	ICMP	Echo (ping) request (id=0x0007, seq(be/le)=19/4864, ttl=254)
10	111.076187	10.4.1.6	234.234.234.234	ICMP	Echo (ping) request (id=0x0007, seq(be/le)=20/5120, ttl=254)

```
▷ Frame 1: 142 bytes on wire (1136 bits), 142 bytes captured (1136 bits)
▷ Ethernet II, Src: c2:05:aa:aa:00:01 (c2:05:aa:aa:00:01), Dst: c2:04:aa:aa:00:01 (c2:04:aa:aa:00:01)
▷ Internet Protocol, Src: 10.3.2.5 (10.3.2.5), Dst: 10.3.1.4 (10.3.1.4)
▽ Protocol Independent Multicast
    0010 .... = Version: 2
    .... 0001 = Type: Register (1)
    Reserved byte(s): 00
    Checksum: 0xdeff [correct]
  ▽ PIM parameters
    ▷ Flags: 0x00000000
    ▷ Internet Protocol, Src: 10.4.1.6 (10.4.1.6), Dst: 234.234.234.234 (234.234.234.234)
    ▷ Internet Control Message Protocol
```

Figure 3.57 PIM Register packet sent to RP by Router 5 upon new multicast traffic (in network 10.3.2.0/24).

No.	Time	Source	Destination	Protocol	Info
1	0.000000	10.4.1.6	234.234.234.234	PIMv2	Register
2	0.002011	10.3.2.4	224.0.0.13	PIMv2	Join/Prune
3	1.998739	10.4.1.6	234.234.234.234	ICMP	Echo (ping) request (id=0x0007, seq(be/le)=1/256, ttl=254)
4	3.998156	10.4.1.6	234.234.234.234	PIMv2	Register
5	4.000207	10.3.1.4	10.3.2.5	PIMv2	Register-stop
6	4.000282	10.4.1.6	234.234.234.234	ICMP	Echo (ping) request (id=0x0007, seq(be/le)=2/512, ttl=254)
7	51.112369	10.4.1.6	234.234.234.234	ICMP	Echo (ping) request (id=0x0007, seq(be/le)=10/2560, ttl=254)
8	111.072054	10.4.1.6	234.234.234.234	ICMP	Echo (ping) request (id=0x0007, seq(be/le)=18/4608, ttl=254)
9	111.074135	10.4.1.6	234.234.234.234	ICMP	Echo (ping) request (id=0x0007, seq(be/le)=19/4864, ttl=254)
10	111.076187	10.4.1.6	234.234.234.234	ICMP	Echo (ping) request (id=0x0007, seq(be/le)=20/5120, ttl=254)

```
▷ Frame 2: 68 bytes on wire (544 bits), 68 bytes captured (544 bits)
▷ Ethernet II, Src: c2:04:aa:aa:00:01 (c2:04:aa:aa:00:01), Dst: IPv4mcast_00:00:0d (01:00:5e:00:00:0d)
▷ Internet Protocol, Src: 10.3.2.4 (10.3.2.4), Dst: 224.0.0.13 (224.0.0.13)
▽ Protocol Independent Multicast
    0010 .... = Version: 2
    .... 0011 = Type: Join/Prune (3)
    Reserved byte(s): 00
    Checksum: 0xe883 [correct]
  ▽ PIM parameters
      Upstream-neighbor: 10.3.2.5
      Groups: 1
      Holdtime: 210
    ▽ Group 0: 234.234.234.234/32
      ▽ Join: 1
          IP address: 10.4.1.6/32 (S)
        Prune: 0
```

Figure 3.58 PIM Join packet sent to Router 5 by RP to request native multicast traffic stream (in network 10.3.2.0/24).

corresponding to the interface in the direction of the RP and not in the direction of the multicast group source. For this shared tree, there is no multicast source and no outgoing interface to forward packets to (so, the outgoing interface list has a Null value). Note also that the RP interface is identified, and the Reverse Path Forwarding neighbor of the incoming interface is equal to 10.3.2.4, which is the IP address of the Router 4 F0/1 interface.

The Source Tree (10.4.1.6, 234.234.234.234) entry corresponds to the active multicast session: the source is identified (10.4.1.6), there is an incoming interface since multicast packets were already received (F1/0), Router 5 has

No.	Time	Source	Destination	Protocol	Info
1	0.000000	10.4.1.6	234.234.234.234	PIMv2	Register
2	0.002011	10.3.2.4	224.0.0.13	PIMv2	Join/Prune
3	1.998739	10.4.1.6	234.234.234.234	ICMP	Echo (ping) request (id=0x0007, seq(be/le)=1/256, ttl=254)
4	3.998156	10.4.1.6	234.234.234.234	PIMv2	Register
5	4.000207	10.3.1.4	10.3.2.5	PIMv2	Register-stop
6	4.000282	10.4.1.6	234.234.234.234	ICMP	Echo (ping) request (id=0x0007, seq(be/le)=2/512, ttl=254)
7	51.112369	10.4.1.6	234.234.234.234	ICMP	Echo (ping) request (id=0x0007, seq(be/le)=10/2560, ttl=254)
8	111.072054	10.4.1.6	234.234.234.234	ICMP	Echo (ping) request (id=0x0007, seq(be/le)=18/4608, ttl=254)
9	111.074135	10.4.1.6	234.234.234.234	ICMP	Echo (ping) request (id=0x0007, seq(be/le)=19/4864, ttl=254)
10	111.076187	10.4.1.6	234.234.234.234	ICMP	Echo (ping) request (id=0x0007, seq(be/le)=20/5120, ttl=254)

```
▷ Frame 5: 60 bytes on wire (480 bits), 60 bytes captured (480 bits)
▷ Ethernet II, Src: c2:04:aa:aa:00:01 (c2:04:aa:aa:00:01), Dst: c2:05:aa:aa:00:01 (c2:05:aa:aa:00:01)
▷ Internet Protocol, Src: 10.3.1.4 (10.3.1.4), Dst: 10.3.2.5 (10.3.2.5)
▽ Protocol Independent Multicast
    0010 .... = Version: 2
    .... 0010 = Type: Register-stop (2)
    Reserved byte(s): 00
    Checksum: 0xfaff [correct]
  ▽ PIM parameters
      Group: 234.234.234.234/32
      Source: 10.4.1.6
```

Figure 3.59 PIM Register-Stop packet sent to Router 5 by the RP upon arrival of native multicast traffic stream (in network 10.3.1.0/24).

```
# show ip mroute
(*, 234.234.234.234), 00:00:15/stopped, RP 10.3.1.4, flags: SPF
  Incoming interface: FastEthernet0/1, RPF nbr 10.3.2.4
  Outgoing interface list: Null

(10.4.1.6, 234.234.234.234), 00:00:15/00:02:51, flags: FT
  Incoming interface: FastEthernet1/0, RPF nbr 0.0.0.0
  Outgoing interface list:
    FastEthernet0/1, Forward/Sparse, 00:00:02/00:03:26
```

Box 3.58 IPv4 multicast routing table at Router 5 after multicast traffic source detection.

no neighbors in this interface (in fact, it is not connected to any other router through this interface) and the outgoing interface list includes interface F0/1, which is the interface used to forward multicast packets to the client of this session.

In a final experiment, the multicast client located at interface F1/0 of Router 1 is forced to leave the multicast session, simply by introducing the following command in interface configuration mode:

```
# no ip igmp join-group 234.234.234.234
```

Router 1 immediately sends a PIM Prune message to the RP: Figure 3.60 shows the contents of this packet, which was captured on network 10.3.1.0/24.

Box 3.59 shows the contents of the new multicast routing table at Router 5, after the client left the multicast session. Now, the Source Tree (10.4.1.6, 234.234.234.234) entry presents a Null outgoing interface list since there are no clients to forward these multicast packets to. Note that this entry also

No.	Time	Source	Destination	Protocol	Info
509	320.864055	10.3.1.1	224.0.0.13	PIMv2	Join/Prune

```
▷ Frame 509: 68 bytes on wire (544 bits), 68 bytes captured (544 bits)
▷ Ethernet II, Src: c2:01:aa:aa:00:01 (c2:01:aa:aa:00:01), Dst: 01:00:5e:00:00:0d (01:00:5e:00:00:0d)
▷ Internet Protocol, Src: 10.3.1.1 (10.3.1.1), Dst: 224.0.0.13 (224.0.0.13)
▽ Protocol Independent Multicast
    0010 .... = Version: 2
    .... 0011 = Type: Join/Prune (3)
    Reserved byte(s): 00
    Checksum: 0xe607 [correct]
  ▽ PIM parameters
      Upstream-neighbor: 10.3.1.4
      Groups: 1
      Holdtime: 210
    ▽ Group 0: 234.234.234.234/32
        Join: 0
      ▽ Prune: 1
          IP address: 10.3.1.4/32 (SWR)
```

Figure 3.60 PIM Prune packet sent to RP by Router 1 upon client IGMP Leave (in network 10.3.1.0/24).

```
# show ip mroute
(*, 234.234.234.234), 00:01:29/stopped, RP 10.3.1.4, flags: SPF
  Incoming interface: FastEthernet0/1, RPF nbr 10.3.2.4
  Outgoing interface list: Null

(10.4.1.6, 234.234.234.234), 00:01:29/00:02:57, flags: PFT
  Incoming interface: FastEthernet1/0, RPF nbr 0.0.0.0
  Outgoing interface list: Null
```

Box 3.59 IPv4 multicast routing table at Router 5 after client leave.

includes an additional Pruned (P) flag, since Router 5 has already received a PIM Prune message from Router 4.

3.9 Tunneling

An IP tunnel is a communications channel between two networks that is used to transport another network protocol by encapsulation of its packets. These tunnels are used to connect disjoint IP networks that don't have a native routing path to each other. Every IP packet, including its addressing information, is encapsulated within another packet format native to the transit network. Figure 3.61 illustrates this encapsulation process. IP encapsulation within IP is described in RFC 2003 [38] and is known as IP-IP tunneling. Other variants are IPv6-in-IPv4 (6in4) and IPv4-in-IPv6 (4in6).

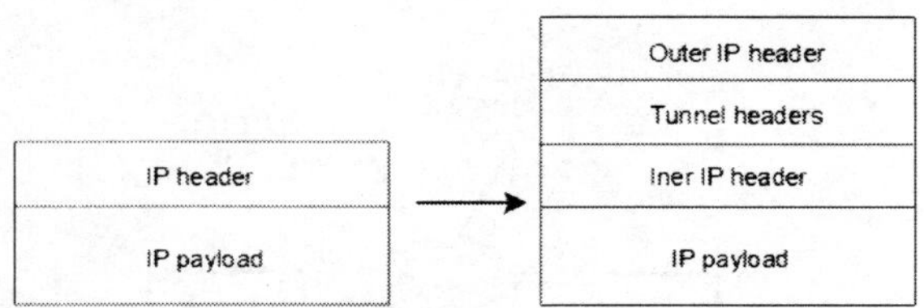

Figure 3.61 IP encapsulation process.

From an operational point of view, encapsulation requires the existence of gateways. At the borders between the source network and the transit network, as well as the transit network and the destination network, these gateways are used that establish the end-points of the IP tunnel across the transit network. Packets traversing these end-points from the transit network are stripped from their transit frame format headers and trailers used in the tunneling protocol, converted into native IP format and injected into the IP stack of the tunnel endpoints. Besides, any other protocol encapsulations that could be used during transit, such as IPsec or Transport Layer Security (TLS), are also removed.

Computer networks use a tunneling protocol when a network delivery protocol encapsulates a different payload protocol. The network layers at which the delivery and payload protocols operate can be at the same or at different levels. For example, the network layer Generic Routing Encapsulation (GRE) runs over IP and is often used to carry IP packets with private addresses over the Internet using delivery packets with public IP addresses. IP can also run over any data-link protocol, like IEEE 802.3 or the Point-to-Point Protocol (PPP).

Tunneling protocols may use data encryption to transport insecure payload protocols over a public network, thereby providing Virtual Private Network (VPN) functionality.

Exercise - IP over IP Tunnel

In the network of Figure 3.2, where OSPF has been properly configured, we want to configure an IPv4 over IPv4 tunnel between Routers 1 and 5 (using the Loopback0 virtual interfaces), as illustrated in Figure 3.62. The configuration commands are shown in Boxes 3.60, for Router 1, and 3.61 for Router 5. First of all, we have to configure the IP address of the tunnel virtual interface; then, the tunnel source and destination should be configured

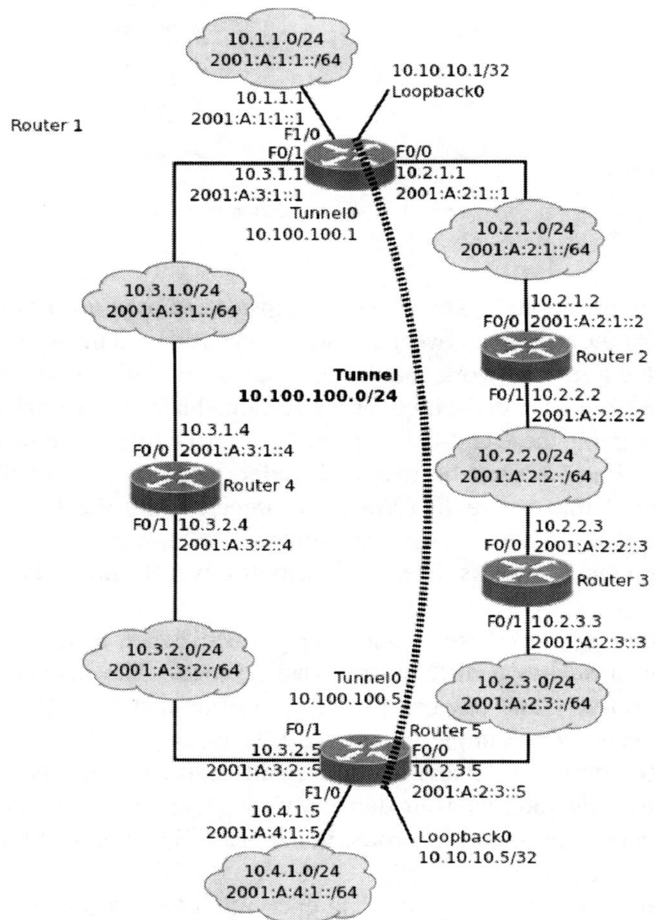

Figure 3.62 Network with an IPv4 over IPv4 tunnel between Router 1 and Router 5.

and, finally, the tunnel mode should be specified. There are several tunnel types/modes, with the simplest one being IPv4 over IPv4.

Boxes 3.62 and 3.63 show the routing tables of routers 1 and 5, respectively, after tunnel creation. We can see that there is a new entry, of the directly connected type, corresponding to network 10.100.100.0/24 and pointing to the tunnel interface.

Finally, on Router 1 we have executed the `ping 10.100.100.5` command and captured the exchanged packets: as can be seen from Figure 3.63,

```
# interface Loopback0
  # ip address 10.10.10.1 255.255.255.255
# interface Tunnel0
  # ip address 10.100.100.1 255.255.255.0
  # tunnel source Loopback0
  # tunnel destination 10.10.10.5
  # tunnel mode ipip
```

Box 3.60 Tunnel configuration at Router 1.

```
# interface Loopback0
  # ip address 10.10.10.5 255.255.255.255
# interface Tunnel0
  # ip address 10.100.100.5 255.255.255.0
  # tunnel source Loopback0
  # tunnel destination 10.10.10.1
  # tunnel mode ipip
```

Box 3.61 Tunnel configuration at Router 5.

```
# show ip route
     10.0.0.0/24 is subnetted, 8 subnets
C       10.3.1.0 is directly connected, FastEthernet0/1
C       10.2.1.0 is directly connected, FastEthernet0/0
C       10.100.100.0 is directly connected, Tunnel0
O       10.2.2.0 [110/20] via 10.2.1.2, 00:00:20, FastEthernet0/0
C       10.1.1.0 is directly connected, FastEthernet1/0
C       10.10.10.1/32 is directly connected, Loopback0
O       10.3.2.0 [110/20] via 10.3.1.4, 00:00:20, FastEthernet0/1
O       10.2.3.0 [110/30] via 10.3.1.4, 00:00:20, FastEthernet0/1
                 [110/30] via 10.2.1.2, 00:00:20, FastEthernet0/0
O       10.10.10.5/32 [110/21] via 10.3.1.4, 00:00:20, FastEthernet0/1
O       10.4.1.0 [110/21] via 10.3.1.4, 00:00:20, FastEthernet0/1
```

Box 3.62 IPv4 routing table at Router 1 after tunnel creation.

there are two IP headers, corresponding to the encapsulation of the IP datagram that includes the ICMP message into a new IP datagram whose source and destination IP addresses are the ones corresponding to the loopback router interfaces.

3.10 Use Case

3.10.1 OSPF Configuration

All interfaces that connect to lower layer routers should be configured in passive mode in order to avoid routing through lower layer equipments. So, at the core all interfaces that connect to distribution layer routers should be configured as passive; at the distribution layer, all interfaces that connect to access layer switches should also be configured as passive. In this way,

```
# show ip route
     10.0.0.0/24 is subnetted, 8 subnets
O         10.3.1.0 [110/20] via 10.3.2.4, 00:00:01, FastEthernet0/1
O         10.2.1.0 [110/30] via 10.3.2.4, 00:00:01, FastEthernet0/1
                   [110/30] via 10.2.3.3, 00:00:01, FastEthernet0/0
C         10.100.100.0 is directly connected, Tunnel0
O         10.2.2.0 [110/20] via 10.2.3.3, 00:00:01, FastEthernet0/0
O         10.1.1.0 [110/21] via 10.3.2.4, 00:00:01, FastEthernet0/1
O         10.10.10.1/32 [110/21] via 10.3.2.4, 00:00:01, FastEthernet0/1
C         10.3.2.0 is directly connected, FastEthernet0/1
C         10.2.3.0 is directly connected, FastEthernet0/0
C         10.10.10.5/32 is directly connected, Loopback0
C         10.4.1.0 is directly connected, FastEthernet1/0
```

Box 3.63 IPv4 routing table at Router 5 after tunnel creation.

No.	Time	Source	Destination	Protocol	Info
1	0.000000	10.100.100.1	10.100.100.5	ICMP	Echo (ping) request (id=0x0017, seq(be/le)=0/0, ttl=255)
2	0.010298	10.100.100.5	10.100.100.1	ICMP	Echo (ping) reply (id=0x0017, seq(be/le)=0/0, ttl=255)
3	0.014484	10.100.100.1	10.100.100.5	ICMP	Echo (ping) request (id=0x0017, seq(be/le)=1/256, ttl=255)
4	0.020540	10.100.100.5	10.100.100.1	ICMP	Echo (ping) reply (id=0x0017, seq(be/le)=1/256, ttl=255)
5	0.030705	10.100.100.1	10.100.100.5	ICMP	Echo (ping) request (id=0x0017, seq(be/le)=2/512, ttl=255)
6	0.045082	10.100.100.5	10.100.100.1	ICMP	Echo (ping) reply (id=0x0017, seq(be/le)=2/512, ttl=255)
7	0.047206	10.100.100.1	10.100.100.5	ICMP	Echo (ping) request (id=0x0017, seq(be/le)=3/768, ttl=255)
8	0.053338	10.100.100.5	10.100.100.1	ICMP	Echo (ping) reply (id=0x0017, seq(be/le)=3/768, ttl=255)
9	0.055451	10.100.100.1	10.100.100.5	ICMP	Echo (ping) request (id=0x0017, seq(be/le)=4/1024, ttl=255)
10	0.061587	10.100.100.5	10.100.100.1	ICMP	Echo (ping) reply (id=0x0017, seq(be/le)=4/1024, ttl=255)

```
▷ Frame 1: 134 bytes on wire (1072 bits), 134 bytes captured (1072 bits)
▷ Ethernet II, Src: c2:01:aa:aa:00:01 (c2:01:aa:aa:00:01), Dst: c2:04:aa:aa:00:00 (c2:04:aa:aa:00:00)
▷ Internet Protocol, Src: 10.10.10.1 (10.10.10.1), Dst: 10.10.10.5 (10.10.10.5)
▷ Internet Protocol, Src: 10.100.100.1 (10.100.100.1), Dst: 10.100.100.5 (10.100.100.5)
▷ Internet Control Message Protocol
```

Figure 3.63 ICMP packets of a ping between the tunnel end-points.

```
# router ospf 1
  # network 10.0.0.0 0.255.255.255 area 0
  # passive-interface default
  # no passive-interface Vlan101
  # no passive-interface Vlan102
```

Box 3.64 OSPFv2 configuration at Router 11.

routing at a given layer should not rely on paths passing through lower layer routers.

Box 3.64 illustrate the configuration commands used at Router 11 to configure OSPFv2 process number 1 and define the passive mode interface as the default behavior for all interfaces, with an exception for VLANs 101 and 102. The configuration of OSPFv3 at the same router is illustrated in Box 3.65. Note that only one area (area 0) was considered for the use case network, because both its physical dimension and management requirements do not require the definition of additional areas.

Default routes should be configured on both routers that connect to the Internet (routers 101 and 102). This can be done as illustrated in Box 3.66 for Router 101.

```
# interface Vlan11
  # ipv6 ospf 1 area 0
# interface Vlan12
  # ipv6 ospf 1 area 0
# interface Vlan13
  # ipv6 ospf 1 area 0
# interface Vlan101
  # ipv6 ospf 1 area 0
# interface Vlan102
  # ipv6 ospf 1 area 0
!
# ipv6 router ospf 1
  # passive-interface default
  # no passive-interface Vlan101
  # no passive-interface Vlan102
```

Box 3.65 OSPFv3 configuration at Router 11.

```
# router ospf 1
  # passive-interface FastEthernet2/1
  # network 10.0.0.0 0.255.255.255 area 0
  # default-information originate always metric 10 metric-type 1
# ipv6 router ospf 1
  # default-information originate always metric 10 metric-type 1
  # passive-interface FastEthernet2/1
```

Box 3.66 OSPFv2 and OSPFv3 configuration at Router 101.

After completing the OSPF configuration, the routing tables of the different routers should include the paths that are needed to provide full connectivity. Box 3.67 show the IPv4 routing table of Router 11, where the last entry corresponds to the default route that is announced by the two routers connecting to the Internet.

Box 3.68 show the IPv6 routing table of Router 11, where we have suppressed the link local entries in order to improve readability. In this case, the first entry is the one that corresponds to the default route announced by the two routers connecting to the Internet.

Obviously, similar routing tables can be seen at the other routers.

3.10.2 NAT and SNAT Configuration

NAT ans SNAT were configured on the routers that connect to outside, routers 101 and 102. Router 101 is the primary server for SNAT, while Router 102 is the backup server. The address range 192.1.1.1-192.1.1.31 is used for address translations. Box 3.69 show the configuration commands used at Router 101, while Box 3.70 illustrate the configuration commands used at Router 102.

```
     10.0.0.0/8 is variably subnetted, 13 subnets, 3 masks
O       10.0.8.0/24 [110/3] via 10.0.2.2, 00:38:58, Vlan102
                    [110/3] via 10.0.1.1, 00:38:58, Vlan101
C       10.2.0.0/24 is directly connected, Vlan12
C       10.0.2.0/24 is directly connected, Vlan102
C       10.3.0.0/24 is directly connected, Vlan13
O       10.0.3.0/24 [110/2] via 10.0.2.2, 00:38:58, Vlan102
                    [110/2] via 10.0.1.1, 00:38:58, Vlan101
C       10.1.0.0/24 is directly connected, Vlan11
C       10.0.1.0/24 is directly connected, Vlan101
O       10.0.6.0/24 [110/2] via 10.0.2.2, 00:38:58, Vlan102
O       10.0.7.0/24 [110/2] via 10.0.2.2, 00:38:58, Vlan102
O       10.0.4.0/24 [110/2] via 10.0.1.1, 00:38:58, Vlan101
O       10.0.5.0/24 [110/2] via 10.0.1.1, 00:38:58, Vlan101
O       10.0.64.0/18 [110/3] via 10.0.2.2, 00:38:58, Vlan102
                     [110/3] via 10.0.1.1, 00:38:58, Vlan101
O       10.0.128.0/17 [110/2] via 10.0.2.2, 00:38:58, Vlan102
                      [110/2] via 10.0.1.1, 00:38:58, Vlan101
O*E1 0.0.0.0/0 [110/12] via 10.0.2.2, 00:38:58, Vlan102
               [110/12] via 10.0.1.1, 00:38:58, Vlan101
```

Box 3.67 Sample IPv4 routing table at Router 11.

No.	Time	Source	Destination	Protocol	Info
1	0.000000	10.0.0.101	10.0.0.102	TCP	11015 > cisco-snat [PSH, ACK] Seq=2490432978 Ack=3273835674
2	0.219239	10.0.0.102	10.0.0.101	TCP	cisco-snat > 11015 [ACK] Seq=3273835674 Ack=2490433090
3	2.552494	10.0.0.102	10.0.0.101	TCP	cisco-snat > 11015 [PSH, ACK] Seq=3273835674 Ack=2490433090
4	2.768538	10.0.0.101	10.0.0.102	TCP	11015 > cisco-snat [ACK] Seq=2490433090 Ack=3273835786

Figure 3.64 SNAT synchronization example.

Figure 3.64 shows the TCP packets that were exchanged between routers 101 and 102 to synchronize their NAPT tables, according to the SNAT service.

In order to test the NAPT mechanism, we executed a ping from the private IP address 10.1.0.1 to the public IP address 192.5.5.1. By accessing the NAT translation tables of both routers 101 (see Box 3.71) and 102 (see Box 3.72), we can conclude that SNAT synchronization is working because both translation tables have exactly the same information.

```
OE1   ::/0 [110/12], tag 1
       via FE80::C004:4FFF:FE6F:0, Vlan101
       via FE80::C005:4FFF:FE6F:0, Vlan102
C     2001:A:A:1::/64 [0/0]
       via ::, Vlan101
C     2001:A:A:2::/64 [0/0]
       via ::, Vlan102
O     2001:A:A:3::/64 [110/2]
       via FE80::C004:4FFF:FE6F:0, Vlan101
       via FE80::C005:4FFF:FE6F:0, Vlan102
O     2001:A:A:4::/64 [110/3]
       via FE80::C005:4FFF:FE6F:0, Vlan102
O     2001:A:A:5::/64 [110/2]
       via FE80::C004:4FFF:FE6F:0, Vlan101
O     2001:A:A:6::/64 [110/2]
       via FE80::C005:4FFF:FE6F:0, Vlan102
O     2001:A:A:7::/64 [110/2]
       via FE80::C005:4FFF:FE6F:0, Vlan102
O     2001:A:A:8::/64 [110/3]
       via FE80::C005:4FFF:FE6F:0, Vlan102
       via FE80::C004:4FFF:FE6F:0, Vlan101
O     2001:A:A:40::/64 [110/3]
       via FE80::C004:4FFF:FE6F:0, Vlan101
       via FE80::C005:4FFF:FE6F:0, Vlan102
O     2001:A:A:80::/64 [110/2]
       via FE80::C005:4FFF:FE6F:0, Vlan102
       via FE80::C004:4FFF:FE6F:0, Vlan101
C     2001:A:A:100::/64 [0/0]
       via ::, Vlan11
C     2001:A:A:200::/64 [0/0]
       via ::, Vlan12
C     2001:A:A:300::/64 [0/0]
       via ::, Vlan13
```

Box 3.68 Sample IPv6 routing table at Router 11 (Local entries were suppressed).

```
# interface loopback 0
  # ip address 10.0.0.101 255.255.255.0
# ip nat pool NATPOOL 192.1.1.1 192.1.1.31 netmask 255.255.255.192
# access-list 1 permit 10.0.0.0 0.255.255.255
# ip nat Stateful id 101
  # primary 10.0.0.101
    # peer 10.0.0.102
    # mapping-id 10
# ip nat inside source list 1 pool NATPOOL mapping-id 10 overload
# interface FastEthernet1/0
  # ip nat outside
# interface FastEthernet0/0
  # ip nat inside
# interface FastEthernet0/1
  # ip nat inside
# interface FastEthernet2/0
  # ip nat inside
# interface FastEthernet2/1
  # ip nat inside
```

Box 3.69 NAT configuration at Router 101.

```
# interface loopback 0
  # ip address 10.0.0.102 255.255.255.0
# ip nat pool NATPOOL 192.1.1.1 192.1.1.31 netmask 255.255.255.192
# access-list 1 permit 10.0.0.0 0.255.255.255
# ip nat Stateful id 102
  # backup 10.0.0.102
    # peer 10.0.0.101
    # mapping-id 10
# ip nat inside source list 1 pool NATPOOL mapping-id 10 overload
# interface FastEthernet1/0
  # ip nat outside
# interface FastEthernet0/0
  # ip nat inside
# interface FastEthernet0/1
  # ip nat inside
# interface FastEthernet2/0
  # ip nat inside
# interface FastEthernet2/1
  # ip nat inside
```

Box 3.70 NAT configuration at Router 102.

```
# sh ip nat translations
  Pro Inside global Inside local  Outside local Outside global
  icmp 192.1.1.1:6   10.1.0.1:6    192.5.5.1:6    192.5.5.1:6
```

Box 3.71 NAT translation table at Router 101.

```
# sh ip nat translations
  Pro Inside global Inside local  Outside local Outside global
  icmp 192.1.1.1:6   10.1.0.1:6    192.5.5.1:6    192.5.5.1:6
```

Box 3.72 NAT translation table at Router 102.

4

Quality of Service

4.1 Introduction

Some network applications, such as Voice and Video, are particularly sensitive to network delay, while others, like File Transfer, have a large tolerance for network delay or bandwidth limitation: from a user point of view, FTP simply takes longer to download a file to the target system which, besides being annoying to the user, does not normally prevent the operation of the application. On the other hand, if voice packets take too long to reach their destination, the resulting speech sounds distorted.

The main causes of degradation of the service level provided by an IP network can be summarized as:

- Lack of bandwidth - Multiple flows contest for a limited amount of bandwidth;
- Too much delay - Packets have to traverse many network devices and links that add up to the overall delay;
- Variable delay - Sometimes there is a lot of other traffic, which results in more delay;
- Drops - Packets have to be dropped when a link is congested.

Quality of Service (QoS) mechanisms should be deployed in the network to solve different problems that can affect specific services or the network as a whole: give priority to certain critical applications, maximize the use of the network infrastructure, give a better performance to delay sensitive applications or adapt to changes in network traffic flows.

Frequently, the simplest method to achieve better performance on a network is to increase available bandwidth. Today, higher capacities are easily available at a low price, but more bandwidth does not always guarantee the desired level of performance. It may happen that protocols are the first cause of congestion, consuming the additional bandwidth and leading to the same congestion issues experienced before the bandwidth upgrade. So, more

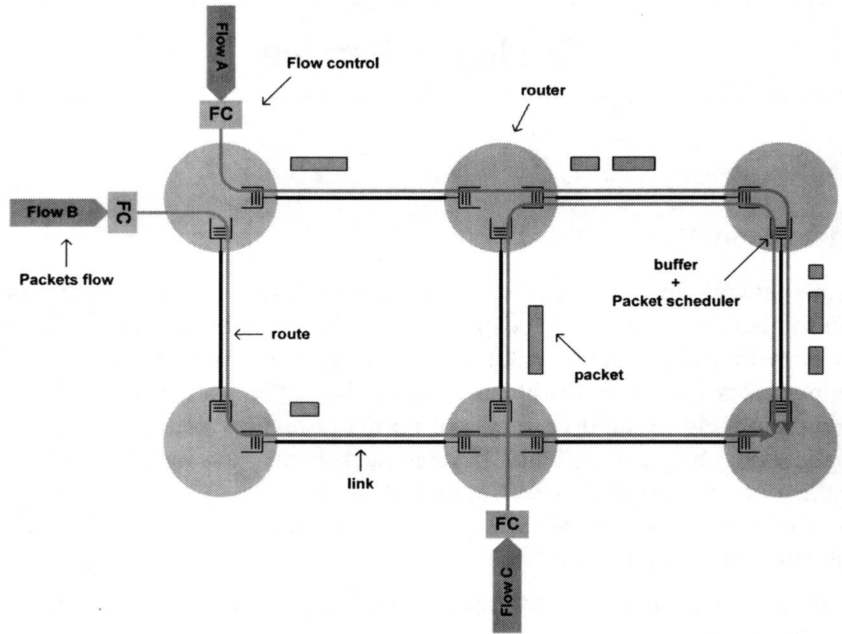

Figure 4.1 Quality of service support in the IP network.

intelligent approaches should rely on analyzing the traffic flowing through the bottlenecks, determining the importance of each protocol and application and design strategies to prioritize the access to the available bandwidth. QoS mechanisms allow the network administrator to have control over bandwidth, latency (which is the delay of a packet traversing the network), jitter (the change of latency over a given period of time) and minimize packet loss by prioritizing flows of some protocols/applications.

By knowing the characteristics of the traffic flow, the network entry points should control its admission into the network, after guaranteeing that the network is able to support the required QoS level. Then, each node of the network (router) should implement the appropriate congestion management mechanism (including a buffer management mechanism and a packet scheduler discipline) and route the different IP packets that compose the flow. This set of procedures is illustrated in Figure 4.1.

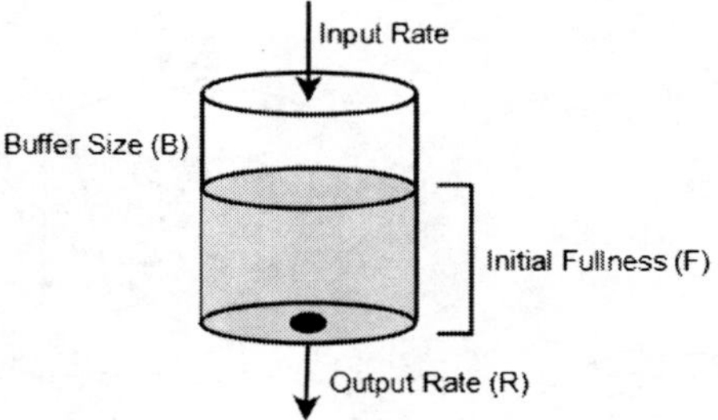

Figure 4.2 The Leaky Bucket model.

4.2 Preliminary Concepts

4.2.1 Leaky Bucket and Token Bucket

The very simple leaky bucket concept is usually used in queuing theory and can be easily explained. A network queue can be compared to a bucket into which network packets are poured (Figure 4.2). The bucket has a hole at the bottom that lets packets drip out at a constant rate. Three parameters define the bucket: the capacity (B), the rate at which packets flow out of the bucket (R) and the initial fullness of the bucket (F). If packets drop in the bucket faster than the hole can let them drip out, the bucket slowly fills up. If too many packets drop in the bucket, the bucket may eventually overflow. Those packets are lost since they do not drip out of the bucket. However, the input rate can vary around R without overflowing the bucket, as long as the average input rate does not exceed the capacity of the bucket. The larger the capacity, the more the input rate can vary within a given time window.

The token bucket mechanism is a related concept frequently used in QoS engineering. It represents a pool of resources that can be used by a service whenever it needs it. As time passes, tokens are added to the bucket by the network. When an application needs to send something out to the network, it must remove an amount of tokens equal to the amount of data it needs to transmit. If there are not enough tokens in the bucket, the application must wait until the network adds more tokens to the bucket. If the application does not make use of its tokens, the token bucket may eventually spill over. The spilled tokens are then lost and the application cannot make use of them. This means that each token bucket has a clearly defined maximum token capacity. Figure 4.3 illustrates the token bucket concept.

The token bucket model is characterized by the following parameters: r, the token filling rate (bytes/s), b, the bucket size (bytes), p, the peak transmis-

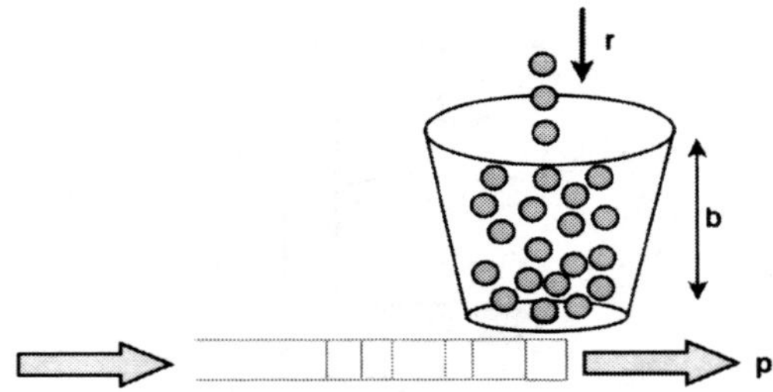

Figure 4.3 The Token Bucket model.

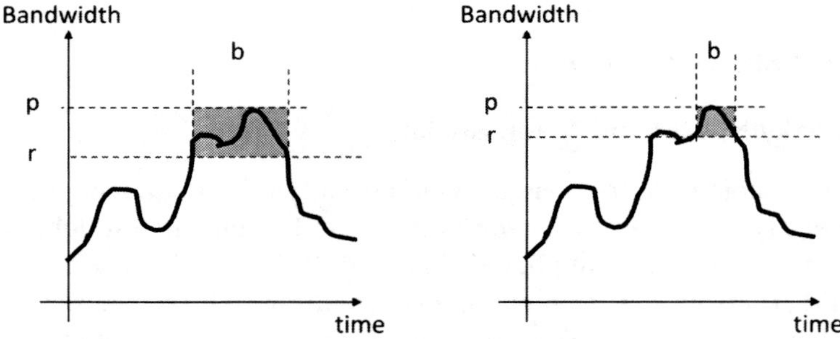

Figure 4.4 Number of admissible packets.

sion rate (bytes/s), *M*, the maximum packet size (bytes) and m, the minimum policed unit size (bytes). In a time interval with length Δt the number of packets that can be admitted by the network must be less than or equal to $r * \Delta t + b$, as shown in Figure 4.4.

This model is frequently used in traffic shaping, a computer network traffic management technique that delays some or all datagrams to bring them into compliance with a desired traffic profile. When performing traffic shaping according to the token bucket model, the following variables are defined:

- Mean Rate or Committed Information Rate (CIR) - Specifies how much data can be sent on average;
- Burst Size or Committed Burst Size - Specifies how much data can be sent over a single time interval without causing scheduling problems;
- Time Interval - The time for a single burst.

The burst size is the amount of data that can be sent with the token bucket over a single time interval. The mean rate is the burst size divided by the time interval. Therefore, when a token bucket is regulating an output interface, its rate over an interval of time cannot exceed the mean rate. However, within that interval, the bit rate may be arbitrarily fast. In this way, large data flows are regulated down to what the network can actually handle, and momentary bursts are smoothed out by buffering, rather than being dropped.

4.2.2 Tail Dropping

Network queues correspond to memory space, so they are finite and can only hold a predetermined amount of information. When a queue fills up, packets are placed in the queue in the order that they were received. When the amount of packets that enter the queue exceed the queue's capacity to hold them, the queue experiences a tail drop. These tail drops represent packets that never entered the queue. They are instead simply discarded by the router.

Tail drops can obviously impact user response. Dropped packets mean requests for retransmissions. With more and more applications running over the TCP/IP protocol, tail drops can also introduce another phenomenon known as global synchronization, which results from the interaction of the TCP/IP sliding window mechanism. In a TPC/IP communication, if a block is successfully sent without errors, the window size increases, allowing the sender to transmit more packets per interval. If an error occurs in the transmission, the window size slides down to a lower value and starts increasing again. When many TCP/IP conversations occur simultaneously, each conversation increases its window size as packets are successfully transmitted. Eventually, these conversations use up all the available bandwidth, which causes the interface's queue to drop packets. These dropped packets are interpreted as transmission errors for all conversations, which simultaneously reduces their window sizes to send fewer packets per interval. This global synchronization causes a fluctuating network usage.

Congestion avoidance techniques attempt to eliminate this global synchronization by dropping packets from TCP conversations at random as the link approaches the congestive state. A well known early drop method is Early Random Drop (ERD): when a router detects that congestion is about to happen, it starts dropping packets before the buffer overflows, signaling the sources that congestion is imminent and they should reduce their transmission rate. The algorithm is invoked every time a packet arrives and works by dropping packets, with a fixed drop probability, when the queue length exceeds the Drop Level. The Random Early Detection (RED) mechanism, an improvement on ERD, randomly drops queued packets based on configurable thresholds. By dropping only some of the traffic before the queue is saturated, instead of all newly arriving traffic (tail drop strategy), RED limits the impact of TCP global synchronization. Besides, as the link approaches saturation, RED can increase the rate at which it drops packets. Another mechanism, designated by Weighted Random Early Detection (WRED), uses a calculated weight to make selective packet dropping decisions. IP precedence is used to calculate the packet weight, which allows network administrators to impact the WRED algorithm to provide preferential service to certain applications.

In a network controlled by RED or WRED, the congestion avoidance mechanism starts discarding packets randomly as the link approaches a pre-configured threshold value. As conversations throttle back their transmission rates, the overall bandwidth usage of the link can be kept near an optimal value. This results in a better utilization of network resources and increased total throughput over time.

4.2.3 Traffic Policing and Traffic Shaping

Traffic policing techniques use the token bucket model to strictly limit the transmission to a predetermined rate. Conversations that exceed the capacity of their token bucket see their traffic being dropped by the policing agent or have their ToS (Type of Service) field rewritten to a lesser precedence value.

Traffic shaping is a mechanism that restricts traffic going out an interface to a particular speed, at the same time attempting to buffer bursts in excess of this maximum speed. Traffic shaping thereby acts to smooth out or shape traffic into a stream that conforms to downstream requirements.

Generic Traffic Shaping (GTS) is the most relevant variant of traffic shaping and acts to limit packet rates sent out an interface to a mean rate, while allowing for buffering of momentary bursts. With GTS parameters configured to match the network architecture, downstream congestion can be avoided, eliminating bottlenecks in topologies with data rate mismatches.

In the GTS variant, when packets arrive at the router an interrupt occurs. If the queue is empty, GTS consults the credit manager (token bucket) to see if there is enough credit to send the packet. If there is not, the packet is sent to the configured queue. If there is credit available, the packet is sent to the output interface, and the associated credit value is deducted from the token bucket. Queued packets are serviced at regular time intervals. The credit manager is checked each time interval to determine if there is enough credit to transmit the next packet waiting in queue. If there is, the packet is sent to the output interface, and the appropriate number of credits is charged.

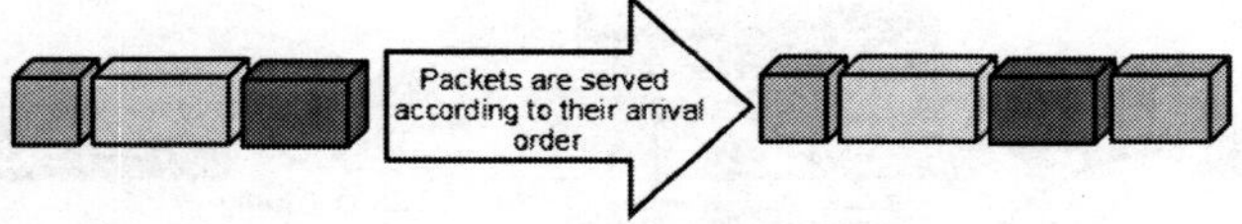

Figure 4.5 The FIFO queuing discipline.

GTS can be applied on a per-interface basis and make use of extended access lists to classify network traffic into different traffic shaping policies.

4.3 Congestion Management

Congestion management is a general term that encompasses different types of queuing strategies used to manage situations where the bandwidth demands of network applications exceed the total bandwidth that can be provided by the network. Congestion management controls the injection of traffic into the network so that certain network flows have priority over others. We will examine the following congestion management techniques:

- First in First Out Queuing
- Priority Queuing
- Custom Queuing
- Weighted Fair Queuing

4.3.1 First in First Out Queuing

First in First Out (FIFO) queuing is the simplest queuing management mechanism: it simply states that the first packet entering the interface will be the first packet to exit the interface. No special packet classification is made. Figure 4.5 illustrates the FIFO queuing mechanism.

FIFO can be the default queuing mechanism on some IOS versions, while others use Weighted Fair Queuing (WFQ). If this is the case, the interface configuration command *no fair-queue* can be used to disable WFQ, activating the FIFO mechanism. In order to verify the operation of the FIFO queuing process on an interface, the *show queueing* command can be used.

The FIFO queuing mechanism is clearly inappropriate for real-time applications, especially during times of congestion. FIFO will never discriminate or give preference to higher-priority packets. Thus, applications such as VoIP can be starved out during periods of congestion. However, FIFO can be efficient in certain circumstances: suppose, for example, that a 10 Mbps Ethernet

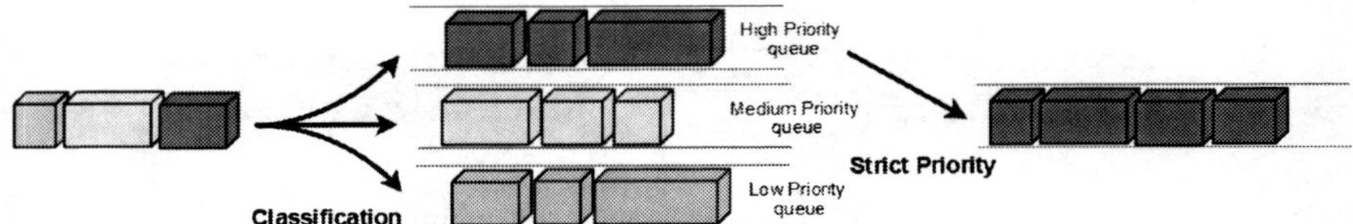

Figure 4.6 The Priority Queuing discipline.

segment is connected to router that in turn connects to a network through a 45 Mbps segment. Since the Ethernet traffic can not overwhelm the outbound connection, the use of a simple queuing mechanism like FIFO reduces the delay experienced by packets as the router processes them.

4.3.2 Priority Queuing

Priority Queuing (PQ) is a powerful form of congestion management that allows the network administrator to define up to four queues for network traffic: High, Medium, Normal, and Low priority queues. The router processes the queues strictly based on their priority. If there are packets in the high priority queue, this queue will be processed until it is empty, independently of the state of the other queues. Once the high priority queue is empty, the router moves to the medium queue and dispatches a single packet. Immediately the router checks the high queue to ensure it is still empty. If it is, it will go to the medium queue, then the normal, then the low. All three, high, medium, and normal, must be completely empty before a single packet is dispatched out of the low queue. Every time a packet is dispatched the router checks the high queue. Figure 4.6 shows PQ in operation.

Traffic can be classified and assigned to the different queues based on the protocol or sub-protocol type, the source interface, the transport layer port number, the packet size or any other parameter identifiable through a standard or extended access list.

With this queuing mechanism low priority traffic can be denied the chance to transmit at all. When the traffic of a lower priority queue is not serviced because there is too much traffic in higher priority queues, a condition called queue starvation is said to have happened. Queue starvation is a serious problem of Priority Queuing, and the ability to completely starve lower priority traffic is something that we should carefully consider before designing a PQ strategy. Typically, PQ is used when delay-sensitive applications encounter problems on the network.

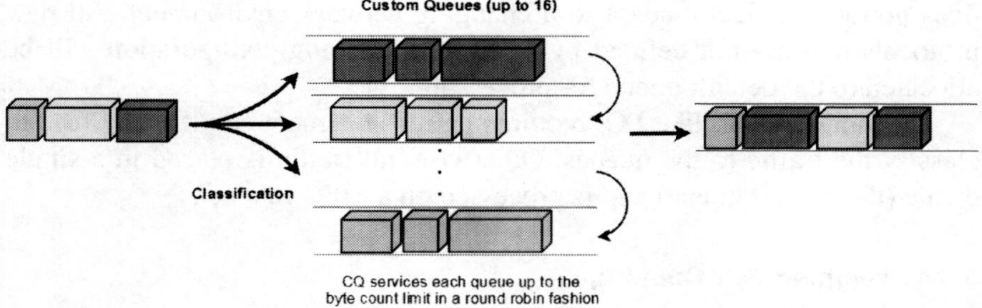
Custom Queues (up to 16)
Classification
CQ services each queue up to the
byte count limit in a round robin fashion

Figure 4.7 The Custom Queuing discipline.

4.3.3 Custom Queuing

Custom queuing (CQ) shifts the service of queues from an absolute mechanism based on priority to a round-robin approach, servicing each queue sequentially. CQ allows the creation of up to 16 user queues, each queue being serviced in succession by the CQ process. Each of the user configurable queues represents an individual "leaky bucket", which is also susceptible to tail drops. Unlike priority queuing, however, custom queuing ensures that each queue gets serviced, thereby avoiding the potential situation in which a certain queue never gets processed. The amount of bytes for each queue can be adjusted in order for the CQ process to spend more time on certain queues. CQ can therefore offer a more refined queuing mechanism, but it cannot ensure absolute priority like PQ.

Custom queuing operates by servicing the user-configured queues individually and sequentially for a specific amount of bytes. The default byte count for each queue is 1500 bytes, so without any customization, CQ would process 1500 bytes from queue 1, then 1500 bytes from queue 2, then 1500 bytes from queue 3, and so on. Traffic can be classified and assigned to any queue through the same methods as priority queuing, namely, protocol or subprotocol types, source interface, source or destination IP addresses, source or destination transport layer port numbers, packet size, fragments, or any parameter identifiable through a standard or extended access list. Figure 4.7 shows CQ in action.

Custom queuing is an excellent mechanism to perform bandwidth allocation on a high traffic link. It allows network administrators to control the flow of packets and provide assured throughput to preferred services. Unlike priority queuing, the custom queuing mechanism ensures that each queue is serviced sequentially. However, as with priority queuing, custom queuing does not automatically adapt to a changing network environment. All new protocols that are not defined in the custom queuing configuration will be allocated to the default queue for processing.

Custom queuing, like PQ, requires policy statements on the interface to classify the traffic to the queues. Otherwise, all traffic is placed in a single queue (the default queue) and is processed on a FIFO manner.

4.3.4 Weighted Fair Queuing

Weighted Fair Queuing (WFQ) dynamically classifies network traffic into individual flows and assigns each flow a fair share of the total bandwidth. Each flow is classified as a high bandwidth or low bandwidth flow. Low bandwidth flows, such as Telnet, get priority over high bandwidth flows, such as FTP. If multiple high bandwidth flows occur simultaneously, they share the remaining bandwidth evenly after low bandwidth flows have been serviced. Each of these flows is placed into an individual queue that follows the leaky bucket analogy. If packets from a specific flow exceed the capacity of the queue to which it is allocated, that queue is subject to tail drops like all other queues.

Once classified, flows are placed in a fair queue. The default number of dynamic queues is 256. Each queue is serviced in a round-robin fashion, like custom queuing, with service priority being given to low bandwidth queues. Each queue is configured with a default congestive discard threshold that limits the number of messages in each queue.The default congestive discard value for each queue is 64 packets. For high bandwidth flows, messages attempting to enter the queue once the discard threshold is reached are discarded. Low bandwidth messages, however, can still enter the queue even though the congestive discard threshold is exceeded for that queue. The limits for dynamic queues and the congestive discard threshold can both be adjusted up to a value of 4096 packets.

So far, the process described shows an equal treatment of all the conversations occurring on an outbound interface. This process would thus be better referred to as "fair queuing" (Figure 4.8). The "weighted" factor is related to the ToS or IP precedence field. WFQ takes into account IP precedence and gives preferential treatment to higher precedence flows by adjusting their weight. If all packets have the same default precedence value, then the weighted factor does not affect the WFQ process. The weight of flows when different ToS values are present is calculated by adding 1 to the packet precedence. The total weight of all flows represents the total bandwidth to

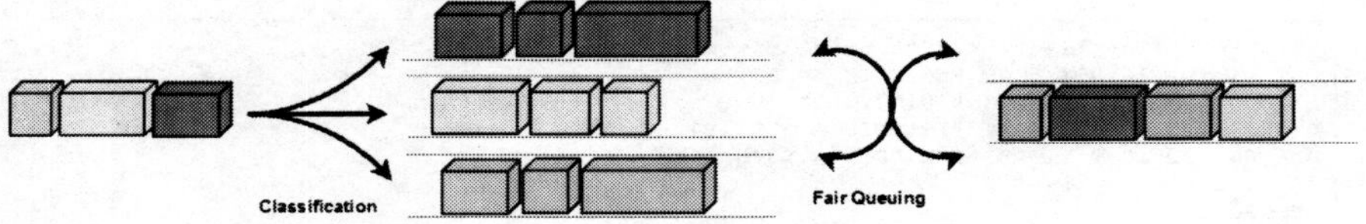

Figure 4.8 The Fair Queuing discipline.

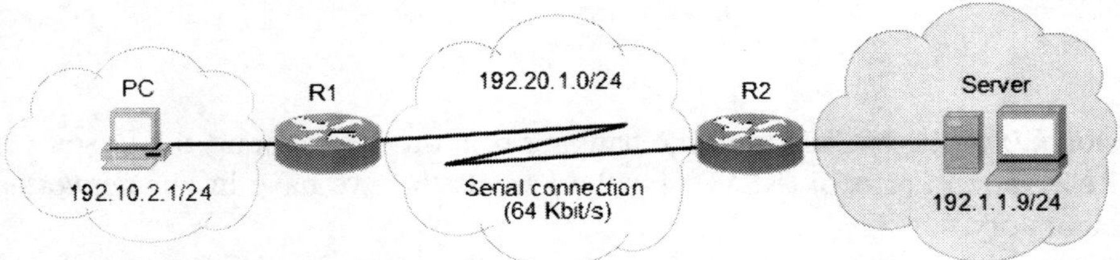

Figure 4.9 Network setup used to illustrate the congestion management mechanisms.

be divided amongst the individual flows. For example, if three flows all use the default IP precedence of 0, they are each given a weight of 1 (0 + 1). The weight of the total bandwidth is 3 (1 + 1 + 1), and each flow is given one-third of the total bandwidth. On the other hand, if two flows have an IP precedence of 0, and a third flow has a precedence of 5, the total weight is 8 (1 + 1 + 6).The first two flows are each given one-eighth of the bandwidth, whereas the third flow receives six-eighths of the total bandwidth.

Exercise - Congestion Management Mechanisms

FIFO

In the network of Figure 4.9, let us configure the FIFO mechanism in both serial interfaces. If the default queuing management mechanism in each interface is already FIFO, we don't have to take any action; otherwise, if the configured mechanism is WFQ, we have to deactivate it using the command

```
no fair-queue
```

in serial interface configuration mode.

In order to generate traffic, in this chapter we use the Multi-Generator (MGEN) open source software, developed by the Naval Research Laboratory (NRL) PROTocol Engineering Advanced Networking (PROTEAN) Research Group and available for download at [65]. The MGEN toolset generates real-time traffic patterns to load the network according to different profiles. The generated traffic can be received and logged for analysis. We used ver-

```
//<time> <flow_id> ON <addr:port> <pattern> <rate> <size>
0 1 ON 192.10.2.1:5000 PERIODIC 50 36
0 2 ON 192.10.2.1:5001 PERIODIC 25 100
0 3 ON 192.10.2.1:5002 PERIODIC 9.375 484
0 4 ON 192.10.2.1:5003 PERIODIC 4.6875 996
//<time> <flow_id> OFF
120000 1 OFF
120000 2 OFF
120000 3 OFF
120000 4 OFF
```

Box 4.1 MGEN commands.

sion 5.0 of the MGEN toolset, changing it in order to support the Resource Reservation Protocol (RSVP) for the kernels that we have in our networks laboratory.

In this example, we want to generate at the server four UDP packet flows with the following characteristics:

- Flow 1 - Packet length of 64 bytes, data rate of 25.6 kbps, destination address equal to the PC address; destination port equal to 5000;
- Flow 2 - Packet length of 128 bytes, data rate of 25.6 kbps, destination address equal to the PC address; destination port equal to 5001;
- Flow 3 - Packet length of 512 bytes, data rate of 38.4 kbps, destination address equal to the PC address; destination port equal to 5002;
- Flow 4 - Packet length of 1024 bytes, data rate of 38.4 kbps, destination address equal to the PC address; destination port equal to 5003.

The commands used at the server are shown in the script of Box 4.1: *time* is the flow starting and is expressed in milliseconds; *flow_id* is the identifier of each flow; *ON* is a reserved word of the command that initiates the flow; the packet flow is then identified by its destination IP address and destination port, separated by :; the supported *patterns* are PERIODIC (for deterministic traffic) and POISSON; *rate* is the flow rate is expressed in packets/s and *size* is the fixed packet size, that should not include the size of the IP (20 bytes) and UDP (8 bytes) headers. All four flows should have a 2 minutes duration, so the *time* parameter on the OFF command should be equal to 120000 milliseconds.

At the PC, the Dynamic-Receiver (DREC) utility receives and logs the traffic generated by the MGEN program. In this case, only one command is needed (Box 4.2) in order to specify the port numbers that are used to receive the incoming traffic flows (4 different port numbers, in this case).

Regarding the results obtained in this example, the received bit rates for flows 1 to 4 were equal to 14.132 kbps, 9.537 kbps, 20.547 kbps and 17.062

```
//PORT <port range>
PORT 5000-5003
```

Box 4.2 DREC commands.

kbps, respectively. So, we can conclude that using the FIFO mechanism higher bit rate flows achieve a higher service bandwidth because they are able to put more packets in the queue. On the other hand, for similar bit rates flows with smaller packets are able to achieve higher service rates because when a small packet arrives at the queue it has a higher probability of finding enough available space. In conclusion, the performance achieved by the different flows results from a compromise between the bit rate and the packet size values.

Weighted Fair Queuing

Let us now configure the WFQ discipline in the serial interface of Router 2 using the command:

```
fair-queue cdt dq rq
```

in interface configuration mode, where *cdt* is the number of messages allowed in each queue, *dq* is the number of dynamic queues used for best effort flows and *rq* is the number of queues that can be reserved for the RSVP protocol. In this case, we set a limit of 64 messages per queue, 16 queues and 0 queues that can be reserved for RSVP (since no reservations will be allowed).

Using exactly the same flows of the previous experiment, the received bit rates for flows 1 to 4 are equal to 14.658 kbps, 15.112 kbps, 15.488 kbps and 15.574 kbps, respectively. From these results, we can conclude that the WFQ mechanism equitably divides the available bandwidth by the different flows, independently from their individual bit rates or packet lengths.

Priority Queuing

Finally, let us change the queuing management mechanism of the Router 2 serial interface in order to consider a higher strict priority for all UDP traffic destined to port 5003 and a lower strict priority for all UDP traffic destined to port 5000. In order to configure an interface with strict priority, first we have to create a priority list in global configuration mode by using the *priority-list* command, as shown in Box 4.3. The high, medium, normal and low priority levels are available for configuration. Then, the sizes of the queues associated with the different priorities should be defined. In this case, the queue sizes

```
# priority-list 1 protocol ip low udp 5000
# priority-list 1 protocol ip high udp 5003
  # priority-group 1
# priority-list 1 queue-limit 2 4 6 8
```

Box 4.3 Configuration of the priority queuing mechanism.

for the high, medium, normal and low priorities are set to 2, 4, 6 and 8 packets, respectively. Finally, we have to activate priority list 1 in interface configuration mode using the *priority-group 1* command.

Using the same UDP flows of the previous experiments, the received bit rates for flows 1 to 4 are equal to 0.036 kbps, 9.78 kbps, 14.66 kbps and 38.402 kbps, respectively. The high priority flow (flow 4, destined to UDP port 5003) achieves its full bandwidth. Flows 2 and 3 were both considered as normal priority flows, since no explicit configuration was made. So, these flows share the same queue according to a FIFO queuing discipline, which justifies the difference in the obtained bandwidths. Finally, flow 1 (the low priority flow) takes the remaining available bandwidth (almost zero), justifying the insignificant bandwidth it was able to obtain.

4.4 QoS Services

There are three service models designed to provide a specific level of service to network traffic from one end of the network to the other: best-effort service, integrated service, and differentiated service.

4.4.1 Best-Effort Service

The Best-Effort Service occurs when the network make its best to deliver packet to its destination. With best-effort service there are no guarantees that the packet will ever reach its intended destination. An application can send data in any amount, whenever it needs to, without requesting permission or notifying the network. Certain applications, such as FTP and HTTP, can operate under this model. This is, however, not an optimal service model for applications that are sensitive to network delays, bandwidth fluctuations, and other changing network conditions. Network telephony applications, for example, may require a more consistent amount of bandwidth in order to function properly.

4.4.2 Integrated Service

The Integrated Service (Intserv) model provides applications with a guaranteed level of service by negotiating network parameters end-to-end [6]. Applications request the level of service necessary for them to operate properly and rely on the QoS mechanism to reserve the necessary network resources prior to the application beginning its transmission. The application will not send traffic until it receives a signal from the network stating that it can provide the requested QoS end-to-end.

To accomplish this, the network uses a process called Admission Control. Admission control is the mechanism that prevents the network from being overloaded. The network will not send a signal to the application to start transmitting the data if the requested QoS cannot be delivered. Once the application begins the transmission of data, the network resources reserved for the application are maintained end-to-end until the application finishes or until the bandwidth reservation exceeds what is allowable for this application. The network will perform its tasks of maintaining the per-flow state, classification, policing, and intelligent queuing per packet to meet the required QoS. The Resource Reservation Protocol (RSVP) is used to signal the network of the QoS requirements of an application. RSVP works in conjunction with routing protocols to determine the best path through the network that provides the required QoS. RSVP enabled routers actually create dynamic access lists to provide the requested QoS and ensure that packets are delivered at the prescribed minimum quality parameters.

The Intserv architecture is based on five key points: definition of the QoS parameters that establish the level of service, admission requirements, packet classification, scheduling and resource reservation using RSVP.

The Intserv model includes a lot of QoS parameters, where the following are some of the most relevant:

- Available_Path_Bandwidth - Bandwidth available for the data flow.
- Minimum_Path_Latency - The latency associated with the current node. This value is very important for real-time applications, such as voice and video. In fact, knowing the upper and lower limits of this value allow the receiving node to properly adjust its QoS reservation and buffer requirements to yield an acceptable service level.
- Non_IS_Hop - Provides information about any node on the data flow that does not provide QoS services. The presence of such nodes can have a severe impact on the end-to-end functioning of Intserv.

- Number_Of_IS_Hops - This is simply a counter of the number of Intserv-aware hops that a packet takes in his path from source to destination.
- Path_MTU - This value informs the end point of the maximum packet size that can transverse the network.
- Token_Bucket_Tspec - Describes the exact traffic parameters according to the token bucket model.

An Intserv router must perform a set of important tasks. First of all, it is necessary to determine which flows are to be delivered as standard IP best-effort and which are to be delivered as dedicated Intserv flows with a corresponding QoS requirement. Priority queuing mechanisms can be used to segregate Intserv traffic from the normal best-effort traffic. The admission requirements determine if a specific data flow can be admitted without disrupting the current data streams in progress.

Then, each packet must be mapped to a corresponding data flow and the accompanying class of service. The packet classifier then sets each class to be acted upon as an individual data flow subject to the negotiated QoS for that flow. Adequate queuing mechanisms must be enabled to ensure correct sequential delivery of packets in a minimally disruptive fashion.

Figure 4.10 shows the internal architecture of an Intserv router: the RSVP module has to interact with the routing, admission control and policy control modules to determine the characteristics of the different reservations that should be made, sending the corresponding parameters' values to the packet classifier and packet scheduler blocks. So, the architecture is quite complex and involves a lot of inter-dependencies.

4.4.2.1 Resource Reservation Protocol

Intserv delivers quality of service via a reservation process that allocates a fixed bandwidth and a maximum delay value to a data flow. This reservation is performed on an end-to-end basis using the Resource Reservation Protocol (RSVP) [8, 7, 107]. RSVP is independent of Intserv: whereas Intserv specifies the set of service classes and the parameters to deliver QoS, RSVP requests this service level from network nodes and, in some cases, carries the Intserv parameters.

RSVP is a signaling protocol that negotiates reservations of resources for client applications in order to guarantee a certain QoS level. It is an out-of-band signaling protocol. RSVP makes reservations of resources for data flows across a network. These reserved flows are usually referred to as sessions. A session is defined as packets having the same destination address (uni-

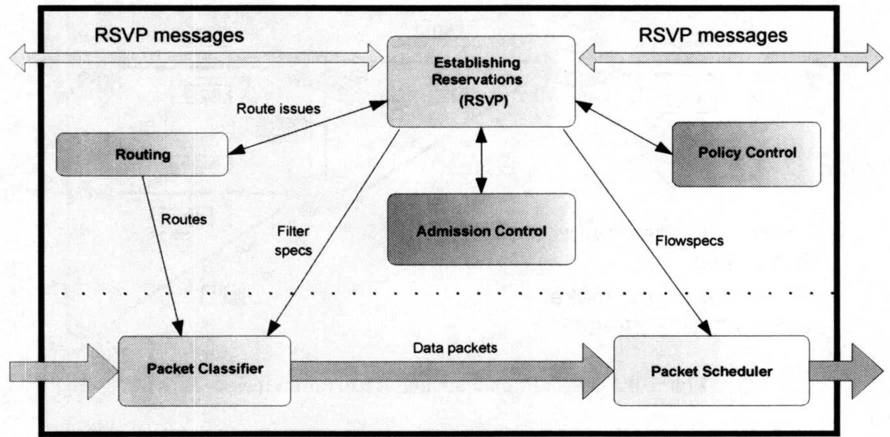

Figure 4.10 The internal architecture of an Intserv router.

cast or multicast), IP protocol identification and destination port. Clients use RSVP to request a guarantee of QoS across the network. Routers participate in RSVP by allocating resources to particular flows, or denying resources if there are none available, and by forwarding RSVP information to other routers.

RSVP is often set up between two clients in a point-to-point situation, or between multiple senders and multiple recipients (multicast). It is even possible for RSVP to negotiate a multipoint to single-point transmission. In any case, the RSVP session startup process reserves resources in a single direction only. To have full-duplex QoS guarantees, it is necessary for the session startup process to be performed twice, once in each direction.

In the network scenario of Figure 4.11 we have one sender and three receivers. The first step is for the sender to transmit a RSVP Path message to the receivers. This Path message travels across the network according to the underlying routing protocol. At each hop through the network, the Path message is modified to include the current hop. In this way, a history of the route taken across the network is built and passed to the receivers. Now that the receivers have the complete route from the Path message, a Reservation (Resv) message is constructed and sent to the sender along the exact reverse path. At each hop, the router determines if the reservation can be made, based on available bandwidth, CPU cycles and other parameters. If the reservation is possible, resources in the router are allocated and the Resv packet is for-

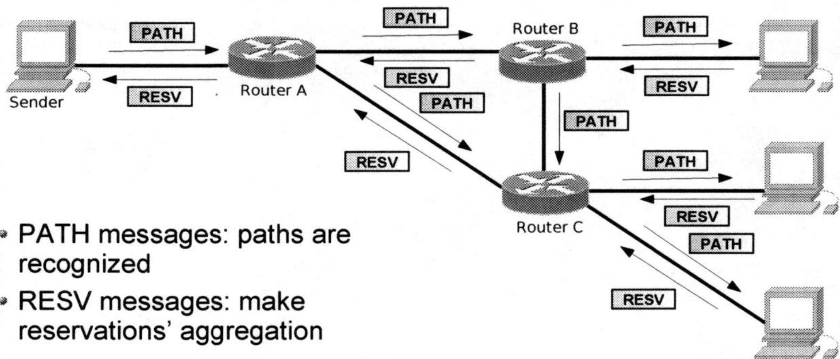

Figure 4.11 RSVP: the Path and Reservation messages.

warded upstream to the previous hop, based on the information contained in the Resv packet. If the reservation is declined, an error message is sent to the receivers and the Resv packet is not forwarded. Only when the sender receives a Resv packet does it know that it can start sending data and guarantee a particular QoS to the downstream receivers. Routers B and C in Figure 4.11 have the capacity to aggregate reservations, in order to ask for a reservation that can accommodate the QoS needs of its downstream routers.

Figure 4.12 shows the general format of the RSVP messages. *Version* is the RSVP protocol version; *Flags* are not defined yet; the Message Type can be 1 for a Path message, 2 for a Resv message, 3 for a PathErr message, 4 for a ResvErr message, 5 for a PathTear message, 6 for a ResvTear message, 7 for a ResvConf message, 10 for a ResvTearConf message or 20 for an Hello message; the *RSVP Checksum* is the checksum of the RSVP message; *Send TTL* is the TTL value on the IP packet this message was sent with; *Reserved* is not used and, finally, *RSVP Length* is the length of the RSVP message in bytes, including the common header. RSVP *Length*, therefore, is always at least 8.

The format of each RSVP object includes the following fields: *Object Length* is the length of the RSVP object, including the object header (so, it is always at least 4 and must be a multiple of 4); *Class-Num* is the object's class; *C-Type* is the object's class type and *Object Contents* is the object itself.

The Path message, whose format is shown in Figure 4.13, includes three mandatory RSVP objects:

- Session - Identifies the session by the destination IP address, destination port and protocol ID;

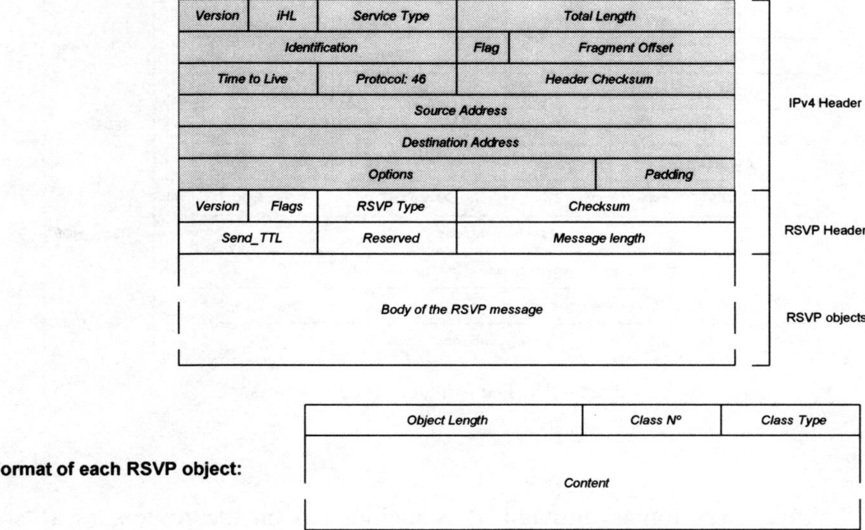

Figure 4.12 The format of the RSVP messages.

- Flowspec (Flow Specification) - Includes TSpec, the Flow Traffic Specification containing the parameters that describe the traffic source based on the Token Bucket model;
- RSVP_Hop - Indicates to the next router the sending IP address and port;
- Time_Values - Indicates the time period between successive sending of Path messages.

The Resv message, whose format is shown in Figure 4.14, includes the following important objects:

- Style - Identifies the style of the reservation;
- Flowspec (Flow Specification) - Includes TSpec, the Flow Traffic Specification containing the parameters that describe the traffic source based on the Token Bucket model, and RSpec (Flow Reservation Specification), containing the parameters describing the reservation that the receiver wants to become supported;
- Filter_Spec (Filter Specification) - Contains the flow descriptor that enables routers to identify packets belonging to this reservation (source address, destination address, protocol type, source port number, destination port number, or any combination of these parameters).

1	0	RSVP Type: 1	Checksum		PATH header
Send_TTL		0	Message length: 40		
SESSION object length: 12			Class Nº : 1	Class Type: 1	SESSION
Destination Address					
Protocol ID		Flags	Destination port		
RSVP_HOP object length: 12			Class Nº : 3	Class Type: 1	RSVP_HOP
Last Hop Address					
Logical Interface Handle of the last node (LIH)					
TIME_VALUES object length: 8			Class Nº : 5	Class Type: 1	TIME_VALUES
Update period (ms)					

Figure 4.13 The format of the Path message.

After a session is initiated, it is maintained on the routers as a "soft state". This soft state session must be refreshed by periodic Path and Resv messages; otherwise, it will be terminated after a timeout interval (that is a multiple of the period of the Path and Resv messages). There is no real difference between the process of initiating a new reservation and refreshing an old one: in both cases, the Path message is built with the previous hop and next hop information, and the Resv statement is adjusted with any new QoS requirements.

Tear-down messages are originated when the reservation is no longer needed, instead of waiting for the cleanup timeout to remove the session. There are two types of tear-down messages: PathTear and ResvTear. PathTear messages, like Path messages, flow in the downstream direction, whereas ResvTear messages, like Resv messages, flow upstream. In addition to clients issuing immediate requests for tear-downs, routers detecting a session time-out or a loss of resources will send their own tear-down messages to upstream (ResvTear) and downstream (PathTear) neighbors.

Besides these, there are other RSVP messages: PathErr, which are sent by routers in error situations; ResvErr, sent by routers when a reservation cannot be supported and ResvConfirmation (Type = 0x07), sent by routers to confirm the establishment of a reservation. PathErr and ResvErr messages are simply sent upstream to the sender that created the error, but they do not modify the path state in any of the nodes that they pass through. Error messages that can be included are: admission failure, ambiguous path, bandwidth unavailable, bad flow specification, service not supported. Confirmation messages can be

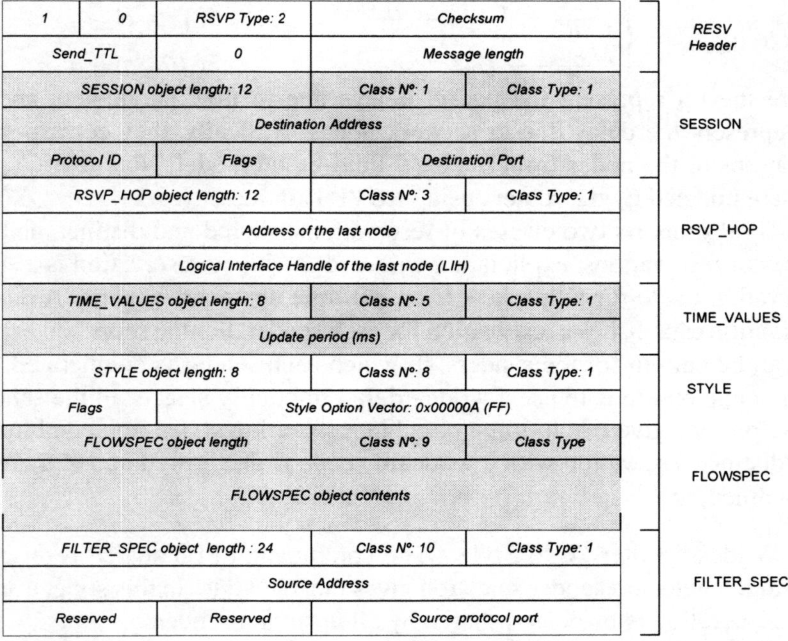

Figure 4.14 The format of the Resv message.

sent by each node in a data flow path if a RSVP reservation from the receiving node is received containing an optional reservation confirmation object.

RSVP allows two types of reservations: controlled-load [108] and guaranteed-rate [90]. According to the Intserv definition, controlled-load gives applications service as if they were traversing an unloaded network. Applications requesting controlled load can expect low latency and a low number of packet drops. These applications are usually considered tolerant real-time applications. An example could be the playback of a recorded conference call. Guaranteed-rate delivers assured bandwidth with constant delay. Applications that require this service to function well are usually considered intolerant real-time applications, like Voice over IP (VoIP). Both services demand that the sender condition the packet sending process according to a token bucket model.

The guaranteed-rate service is able to assure a maximum delay, given by the following equation:

$$Delay_{\max} = \begin{cases} \left[\frac{b-M}{R} \times \frac{p-R}{p-r}\right] + \frac{M+C_{\text{tot}}}{R} + D_{\text{tot}} & \text{if } p > R \geq r \\ \frac{M+C_{\text{tot}}}{R} + D_{\text{tot}} & \text{if } R \geq p \geq r \end{cases}$$

where the Cs represent the packet delays due to flow parameters and the Ds represent the delay due to network nodes. Basically, they correspond to deviations of the nodes from the ideal fluid-based model. Note that C_{tot} and D_{tot} are updated by the routers using RSVP Path messages.

RSVP supports two classes of reservations, shared and distinct, and two scopes of reservations, explicit and wildcard. A shared reservation is a single reservation made for all packets from multiple upstream senders. A distinct reservation establishes a reservation for each sender. For the scope, an explicit list can be chosen for the senders, in which each sender is enumerated. The other scope option is to use a wildcard that implicitly selects all the senders. These options give rise to three possible reservation styles (the combination of a distinct reservation with a wildcard scope is disallowed and is therefore not defined):

- Wildcard-Filter (WF) Style - The combination of a shared reservation and a wildcard sender selection gives the WF style. In this style, a single reservation is made and shared by all upstream senders.
- Shared-Explicit (SE) Style - The combination of a shared reservation and an explicit sender list gives rise to the SE style. This style creates a single reservation shared by a list of upstream senders. Both the WF and SE reservation styles are appropriate for data flows that are known not to interfere with each other. An example of this would be an audio conference where it could be assumed that the multiple senders would not typically talk at the same time.
- Fixed-Filter (FF) Style - The combination of a distinct reservation and an explicit sender list gives rise to the FF style. In this style, a distinct reservation is created for data packets from a particular sender. This reservation is not shared with any other sender. However, if another receiver is added for that sender, the reservation is not doubled, but merged. This kind of style would be appropriate for video signals where the data from each sender are different.

The reservation style is declared by the Style object of the RSVP messages, while senders are declared on the Filter_Spec object.

An RSVP flow descriptor is the combination of a flowspec and a filter-spec. A flowspec is the QoS requested, and the filterspec is the set of packets that will receive this QoS. When new flows are added to the group of reservations in a node, they will often need to be merged into a common reservation. In the case of multicast traffic where the same data is going to multiple recipients, the recipients will still make a Resv request. It is up to RSVP to join this request with the active reservations. When this is done, the flows are referred to as "merged". The RSVP rules do not allow the merging of distinct and shared reservations, nor the merging of explicit sender selection and wildcard sender selection. So, all three styles are mutually incompatible.

So, we can summarize the main advantages of using RSVP in the following way:

- Admissions Control - RSVP not only provides QoS but also helps other applications by not transmitting when the network is busy.
- Network Independence - RSVP is not dependent on a particular networking architecture.
- Interoperability - RSVP works inside existing protocols and with other QoS mechanisms.
- Distributed - RSVP is a distributed service and therefore has no central point of failure.
- Transparency - RSVP can tunnel across an RSVP-unaware network.

What happens when QoS reservations have to cross non-RSVP network regions? Figure 4.15 shows such a scenario. The RSVP_Hop object sent on Path messages allows Resv messages to be correctly routed. The Logical Interface Handler (LIH) of the RSVP_Hop object allows to solve the problem when a Resv message arrives to a router through an interface different from the one it has used to send the Path message.

A RSVP router routes a Resv message without processing it if it did not receive any Path message, as illustrated in Figure 4.16.

However, RSVP has also some disadvantages that dissuade its utilization:

- Scaling and Performance Issues - Multi-field (source and destination addresses, protocol, source and destination port) classification and statefulness of reservations consume router memory and CPU resources.
- Route Selection and Stability - The shortest path may not have available resources, and the active path may go down.
- Setup Time - An application cannot start transmitting until the reservation has been completed, and reservation establishment times are high.
- Low flexibility - Only two service classes are supported.

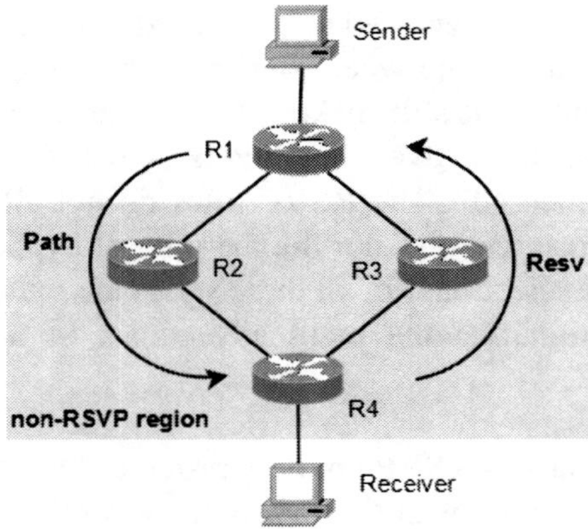

Figure 4.15 How RSVP works on non-RSVP network regions.

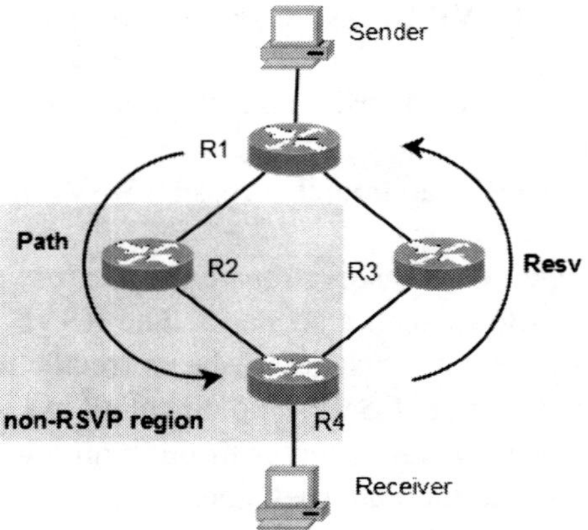

Figure 4.16 Routing a Resv message.

Exercise - RSVP Operation

Using the same network illustrated in Figure 4.9, let us now configure the WFQ mechanism with a limit of 64 messages per queue, 32 queues and 8 queues available for reservation through RSVP.

In order to activate the RSVP protocol, the following command should be used in interface configuration mode:

```
ip rsvp bandwidth rb lrf
```

where *rb* is the maximum bandwidth that can be reserved through RSVP (this value must not exceed 75% of the link bandwidth) and *lrf* is the maximum

```
# fair-queue 64 32 8
# ip rsvp bandwidth 48 48
```

Box 4.4 Configuration of the RSVP protocol.

```
//<time> <flow_id> ON <addr:port> <pattern> <rate> <size>
//[RSVP <Tspec>]
0 1 ON 192.10.2.1:5000 PERIODIC 50 36
0 2 ON 192.10.2.1:5001 PERIODIC 25 100
0 3 ON 192.10.2.1:5002 PERIODIC 9.375 484 RSVP [t 4800 512 4800 512 512]
0 4 ON 192.10.2.1:5003 PERIODIC 4.6875 996
120000 1 OFF
120000 2 OFF
120000 3 OFF
120000 4 OFF
```

Box 4.5 MGEN script considering a RSVP reservation for flow 3.

bandwidth that can be reserved by each RSVP reservation (this value must not exceed the *rb* value). In this example we used 48 Kbps for both *rb* and *lrf*. The configuration commands are illustrated in Box 4.4.

Controlled Load Reservation

Let us suppose that we want to configure a controlled load RSVP reservation for the traffic destined to port 5002 (flow 3). The MGEN script will look like Box 4.5. Tspec has the following format:

```
[t <r> <b> <p> <m> <M>]
```

where r is the data Rate (in bytes/sec), b is the token Bucket size (in bytes), p is the peak data rate (in bytes/sec), m is the minimum policed unit size (bytes) and M is the maximum policed unit size (bytes). So, the token bucket parameters will be set to $r = 4800$, $b = 512$, $p = 4800$, $m = 512$ and $M = 512$.

If the reservation style is of the WildCard Filter type, that is, the receiver specifies a unique reservation value for all possible senders, the DREC script could be the one shown in Box 4.6. The reservation style can be one of the following: Wildcard (WF), Fixed-Filter (FF), or Shared-Explicit (SE). For the WF reservation style, the reservation parameters should include the Flowspec, the specification of the flow according to the following syntax:

```
[cl <r> <b> <p> <m> <M>]
```

where *cl* stands for a controlled-load service.

During the execution of the experiment (the 2 minutes duration of the flows), the reservation establishment can be confirmed using the `show ip rsvp reservation` command, as shown in Box 4.7 for Router 2 (a similar

```
PORT 5000-5003
//<time> RESV <session> <style> <reservation parameters>
2000 RESV 192.10.2.1:5002 WF [cl 4800 512 4800 512 512]
```

Box 4.6 DREC script considering a controlled load reservation for flow 3.

```
# show ip rsvp reservation
To             From         Pro      DPort     SPort
192.10.2.1     0.0.0.0      UDP      5002      0

Next Hop      I/F        Fi      Serv      BPS
192.10.2.2    Se0/0/0    WF      LOAD      38400
```

Box 4.7 State of the controlled load reservation at Router 2.

```
PORT 5000-5003
2000 RESV 192.10.2.1:5002 WF [gx 4800 0 4800 512 4800 512 512]
```

Box 4.8 DREC script considering a guaranteed service reservation for flow 3.

reservation was made by Router 1). Flow 3 was able to achieve its full bandwidth of 38.4 kbps. The RSVP Reservation packet is shown in Figure 4.17, where both the reservation style and the flow specification parameters can be confirmed.

Guaranteed Rate Reservation

In a second experiment, let us change the reservation type for the traffic destined to port 5002 (flow 3) to guaranteed-rate service. The DREC script have to changed to the one shown in Box 4.8 but the MGEN script remained unchanged. The reservation style is also WF but its type should be changed to guaranteed-rate. Now, the Flowspec should be specified according to the following syntax:

```
[gx <R> <S> <r> <b> <p> <m> <M>]
```

where *gx* stands for a guaranteed-rate service, *R* is the Reservation rate and *S* is the Slack term (or admissible delay). The reservation rate will be *4800* packets/s, while the admissible slack term will be defined as *0*. The remaining parameters of the Flowspec are the same of the controlled-load service.

Once again, the reservation establishment can be confirmed using the `show ip rsvp reservation` command, as shown in Box 4.9. Flow 3 was also able to achieve its full bandwidth of 38.4 kbps.

Let us execute another experiment: we want to make a guaranteed service reservation for flow 3, with a reservation rate of 64 kbits/s. The DREC script should be the one shown in Box 4.10. Obviously, in this case the reservation

No.	Time	Source	Destination	Protocol	Info
1	0.000000	192.1.1.9	192.10.2.1	RSVP	PATH Message. SE
2	1.6[illegible]1491	192.10.2.1	192.10.2.2	RSVP	RESV Message. SE
3	1.721785	192.10.2.2	192.10.2.1	RSVP	CONFIRM Message.

```
+ Frame 2 (134 bytes on wire, 134 bytes captured)
+ Ethernet II, Src: Shuttle_3e:d4:13 (00:30:1b:3e:d4:13), Dst: Cisco_22:97:f8 (00:22:55::
+ Internet Protocol, Src: 192.10.2.1 (192.10.2.1), Dst: 192.10.2.2 (192.10.2.2)
- Resource ReserVation Protocol (RSVP): RESV Message. SESSION: IPv4, Destination 192.10.:
  + RSVP Header. RESV Message.
  + SESSION: IPv4, Destination 192.10.2.1, Protocol 17, Port 5002.
  + HOP: IPv4, 192.10.2.1
  + TIME VALUES: 30000 ms
  + CONFIRM: Receiver 192.10.2.1
  + SCOPE
  - STYLE: Wildcard Filter (17)
      Length: 8
      Object class: STYLE object (8)
      C-type: 1
      Flags: 0x00
      Style: 0x000011 - Wildcard Filter
  - FLOWSPEC: Controlled Load: Token Bucket, 4800 bytes/sec.
      Length: 36
      Object class: FLOWSPEC object (9)
      C-type: 2
      Message format version: 0
      Data length: 7 words, not including header
      Service header: 5 - Controlled Load
      Length of service 5 data: 6 words, not including header
    + Token Bucket: Rate=4800 Burst=512 Peak=4800 m=512 M=512
```

Figure 4.17 Reservation packet.

```
# show ip rsvp reservation
To              From        Pro    DPort    SPort
192.10.2.1      0.0.0.0     UDP    5002     0

Next Hop      I/F       Fi     Serv    BPS
192.10.2.2    Se0/0/0   WF     RATE    38400
```

Box 4.9 State of the guaranteed service reservation at Router 2.

can not be established because there is not enough bandwidth in the serial link connecting routers 1 and 2 to assure the intended reservation. Exchanged packets were captured and are illustrated in Figure 4.18: as can be seen, the Reservation Error message clearly states that an admission control failure has happened due to the unavailability of the requested bandwidth.

Let us now execute a final experiment. Let us configure MGEN to generate three packet flows with the following characteristics (different starting and duration times):

- Flow 1 - Best effort flow, with a bit rate of 64 kbits/s and a fixed packet length of 128 bytes, starting at instant 0 seconds and having a duration of 3 minutes;

```
PORT 5000-5003
//<time> RESV <session> <style> <reservation parameters>
2000 RESV 192.10.2.1:5002 WF [gx 8000 0 4800 512 4800 512 512]
```

Box 4.10 DREC script for a 64 Kbit/s guaranteed service reservation.

No.	Time	Source	Destination	Protocol	Info
1	0.000000	192.1.1.9	192.10.2.1	RSVP	PATH Message. SESSI
2	1.489297	192.10.2.1	192.10.2.2	RSVP	RESV Message. SESSI
3	1.493070	192.10.2.2	192.10.2.1	RSVP	RESV ERROR Message.
4	31.513949	192.10.2.1	192.10.2.2	RSVP	RESV Message. SESSI
5	31.517729	192.10.2.2	192.10.2.1	RSVP	RESV ERROR Message.
6	44.362624	192.1.1.9	192.10.2.1	RSVP	PATH Message. SESSI
7	55.399401	192.10.2.1	192.10.2.2	RSVP	RESV Message. SESSI
8	55.403362	192.10.2.2	192.10.2.1	RSVP	RESV ERROR Message.

```
Frame 3 (142 bytes on wire, 142 bytes captured)
Ethernet II, Src: Cisco_22:97:f8 (00:22:55:22:97:f8), Dst: Shuttle_3e:d4:13 (00:30:1b:3e:(
Internet Protocol, Src: 192.10.2.2 (192.10.2.2), Dst: 192.10.2.1 (192.10.2.1)
Resource Reservation Protocol (RSVP): RESV ERROR Message. SESSION: IPv4, Destination 192.1
  RSVP Header. RESV ERROR Message.
  SESSION: IPv4, Destination 192.10.2.1, Protocol 17, Port 5002.
  HOP: IPv4, 192.10.2.2
  ERROR: IPv4, Error code: Admission Control Failure , Value: 2, Error Node: 192.10.2.2
    Length: 12
    Object class: ERROR object (6)
    C-type: 1 - IPv4
    Error node: 192.10.2.2
    Flags: 0x00
    Error code: 1 - Admission Control Failure
    Error value: 2 - Requested bandwidth unavailable
  SCOPE
  STYLE: Wildcard Filter (17)
  FLOWSPEC: Guaranteed Rate: Token Bucket, 4800 bytes/sec. RSpec, 8000 bytes/sec.
```

Figure 4.18 Exchanged packets during the frustrated reservation attempt.

- Flow 2 - Controlled load flow, with a bit rate of 32 kbits/s and a fixed packet length of 256 bytes, starting at instant 60 seconds and having a duration of 2 minutes;
- Flow 3 - Guaranteed service flow, with a bit rate of 32 kbits/s and a fixed packet length of 512 bytes, starting at instant 120 seconds and a duration of 1 minute.

The MGEN script is illustrated in Box 4.11. Note that flows 1 to 3 are destined to IP address 192.10.2.1 and port numbers 5000, 5001 and 5002, respectively.

The DREC script corresponding to this experiment is show in Box 4.12: flow 2 will undergo a controlled load reservation, while flow 3 will be subject to a guaranteed rate reservation.

Using the `show rsvp reservation` command, we could verify that the reservation for flow 2 was successful because there was enough bandwidth available (Box 4.13). However, reservation for flow 3 did not succeed because

```
0 1 ON 192.10.2.1:5000 PERIODIC 62.5 100
60000 2 ON 192.10.2.1:5001 PERIODIC 15.625 228
              RSVP [t 4000 256 4000 256 256]
120000 3 ON 192.10.2.1:5002 PERIODIC 7.8125 484
              RSVP [t 4000 512 4000 512 512]
180000 1 OFF
180000 2 OFF
180000 3 OFF
```

Box 4.11 MGEN script to generate three different flows.

```
PORT 5000-5002
0 RESV 192.10.2.1:5001 WF [cl 4000 256 4000 256 256]
0 RESV 192.10.2.1:5002 WF [gx 4000 0 4000 512 4000 512 512]
```

Box 4.12 DREC script considering three flows and two reservations.

```
# show ip rsvp reservation
To            From        Pro    DPort    Sport
192.10.2.1    0.0.0.0     UDP    5001     0

Next Hop      I/F         Fi    Serv      BPS
192.10.2.2     Se0/0/0    WF    LOAD
```

Box 4.13 State of the reservations at Router 2.

the available bandwidth after the establishment of the reservation for flow 2 is not enough. So, the remaining bandwidth of approximately 32 Kbps will be shared between flows 1 and 3 using a WFQ mechanism.

4.4.3 Differentiated Service

The disadvantages of the Integrated Services model lead to the proposal of the Differentiated Service (DiffServ) model [68, 5]. DiffServ includes a set of classification tools and queuing mechanisms to provide certain protocols or applications with a certain priority over other network traffic. Differentiated services rely on the edge routers to perform the classification of the different types of packets traversing the network. Network traffic can be classified by network address, protocols and ports, ingress interfaces or any other classification that can be accomplished through the use of an access list (or an equivalent feature).

With DiffServ, a parsimonious number of service classes are defined and individual data flows are grouped together within each individual service class and treated identically. Each service class is entitled to certain queuing and priority mechanisms within the entire network. Marking, classifying and

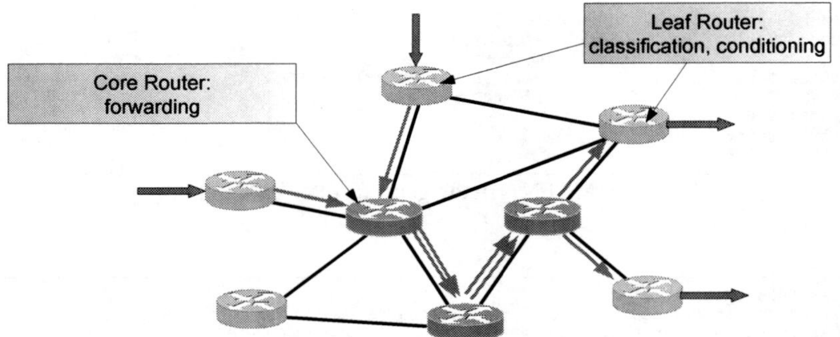

Figure 4.19 The basic idea of the DiffServ model.

admittance to a Diffserv network occur only at the network edge or points of ingress. So, Edge Routers (i) classify packets, marking each packet on the Type of Service field of the IPv4 header or Traffic Class field of the IPv6 header; (ii) condition traffic, using the Token Bucket model to verify if the incoming traffic is conforming to the contracted traffic and, if not, delay or drop excess traffic. Interior routers (Core Routers) are only concerned with Per Hop Behaviors (PHB) as marked in the packet header. Figure 4.19 illustrates the basic idea behind the DiffServ architecture.

Figure 4.20 represents the main DiffServ actions a router has to perform in order to implement QoS policies:

- Classification - Each class-oriented QoS mechanism has to support some type of classification (access lists, route maps, class maps, etc.).
- Metering - Some mechanisms measure the rate of traffic to enforce a certain policy (e.g. rate limiting, shaping, scheduling, etc.).
- Marking - Some mechanisms have the capability to mark packets based on classification and/or metering (e.g. Committed Access Rate (CAR), QoS Policy Propagation through BGP (QPPB), Policy-based Routing (PBR), class-based marking, etc).
- Shaping - Some mechanisms are used to enforce a rate limit based on the metering (excess traffic is delayed), like for example Generic Traffic Shaping (GTS), Frame Relay Traffic Shaping (FRTS) or Class-based Shaping.
- Dropping - Some mechanisms are used to drop packets (e.g. Random Early Detection (RED) and Weighted Random Early Detection (WRED)).

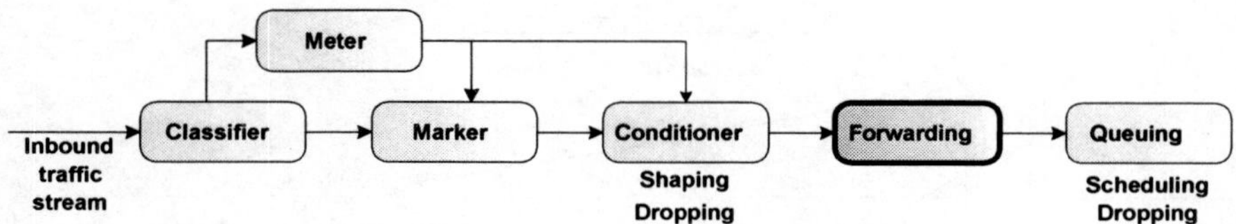

Figure 4.20 The main DiffServ mechanisms.

- Forwarding - There are several forwarding mechanisms available (fast switching, CEF switching, routing, etc).
- Queuing - Each interface has a queuing mechanism, ranging from traditional queuing mechanisms (FIFO, Priority Queuing (PQ), Custom Queuing (CQ), Weighted Fair Queuing (WFQ), etc) to advanced queuing mechanisms like Class-based WFQ or Class-based Low-Latency Queuing (LLQ).

4.4.3.1 The DiffServ Code Point

DiffServ uses, as its packet marker, the Differentiated Services Code Point (DSCP), defined in RFC 2474 [73] and RFC 2475 [5]. The DSCP is found within the Differentiated Services (DS) field, which is a replacement for the ToS field of RFC 791 [75] (Figure 4.21): the 8-bit ToS field is repartitioned into a 6-bit DSCP field and a 2-bit unused portion. The DS field is incompatible with the older ToS field. The structure is identical for both IP precedence and DSCP but the meanings of the bit structures varies. So, an IP precedence aware router will interpret the first 3 bits based on IP precedence definitions. The meaning inferred as to how the packets are handled may be considerably different to what is intended by DSCP definitions. Figure 4.21 illustrates the format and setting of the DSCP field.

The DSCP field maps to a provisioned Per Hop Behavior (PHB) that is not necessarily one to one or consistent across service providers or networks. Remember that the DS field should only be marked at the ingress point of a network, by the network ingress device for best performance. However, it may be marked, as needed, anywhere within the network with a corresponding efficiency penalty. This is significantly different from both the old ToS field and from the Intserv model in which the end host marked the packet. This marker was carried unaltered throughout the network. In Diffserv, the DS field may be remarked every time a packet crosses a network boundary to represent the current settings of that service provider or network. No fixed

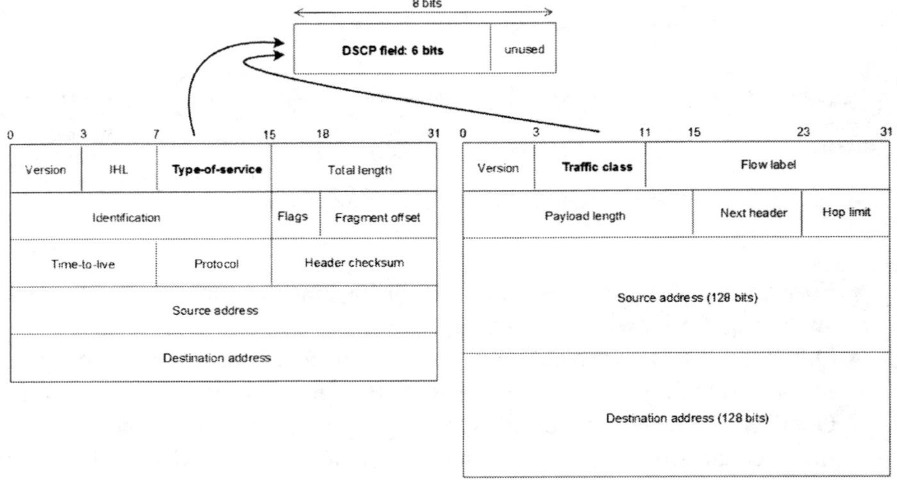

Figure 4.21 The DSCP field.

meanings are attributed to the DS field. Interpretation and application are reserved for the network administrator or service provider to determine based on negotiated service level agreements (SLA) with the customers or other criteria.

4.4.3.2 Per Hop Behavior

RFC 2744 [106] defines four suggested code points and their recommended corresponding behaviors. The Diffserv specification does attempt to maintain some of the semantics of the old ToS field and, as such, specifies that a packet header that contains the bit structure of xxx000 is to be defined as a reserved DSCP value.

The default PHB corresponds to a value of 000000 and states that the packet shall receive the traditional best-effort based delivery with no special characteristics or behaviors. This is the default packet behavior: the queuing mechanism is FIFO and the buffer discarding policy is tail-drop.

The Class Selector Code Points utilize the old ToS field and as such are defined as being up to 8 values of corresponding PHBs. There is no defined value to these code points; however, RFC 2474 [73] states that packets with a higher class selector code point and PHB must be treated in a priority manner over ones with a lower value. It also states that every network that makes use of this field must map at least two distinct classes of PHBs. These values are only locally significant within the particular network.

The Express or Expedited Forwarding (EF) PHB is the highest level of service possible in a Diffserv network. It utilizes RSVP to assure low loss, low jitter, a minimum departure rate and guaranteed bandwidth priority connections with prioritized forwarding through a Diffserv enabled network (RFC 2598 [58]). This traffic is defined as having minimal, if any, queuing delays throughout the network. The bandwidth used by this class is policed, so the class is not allowed to exceed the guaranteed amount (excess traffic is dropped). Note that this is analogous to a data flow (or micro flow in Diffserv) in the Intserv architecture and care must be taken to provide sufficient resources for this class. Extreme importance is assigned to admission controls for this class of service. Even a priority queue will give poor service if it is allowed to become saturated. Essentially, this level of service emulates a dedicated circuit within a Diffserv network. In terms of traffic that would utilize such a channel, Voice traffic would be one of the prime data types. The EF PHB uses a DSCP value of 101110, which looks like IP precedence 5 to non-DiffSserv compliant devices.

Assured forwarding (AF) PHB is the most commonly used on the Diffserv architecture. Within this PHB there are 4 AF groups (called class 1, 2, 3 and 4 by Cisco) that are further divided into three Olympics groups, gold, silver, and bronze, to represent differing packet drop allowances. Tables 4.1 and 4.2 illustrate the encoding strategy of the different AF classes. According to this strategy, the bit structure and the corresponding precedence and class values are the ones illustrated in Table 4.3.

RFC 2597 [49] states that each packet will be delivered in each service class as long as the traffic conforms to a specific traffic profile. Any excess traffic will be accepted by the network, but will have a higher probability of being dropped based on its service class and group. Differentiated Random Early Discard is used within each class to prevent congestion within the class. The Diffserv specification does not lay out drop rates but states that class 4 is preferentially treated to class 3 and that within each AF, each class gets preferential treatment over other classes. The individual values are left to the network administrator to assign as desired based on their existing service level agreements and network requirements.

To support the assured forwarding functionality, each node in a Diffserv network must implement some form of bandwidth allocation to each AF class and some form of priority queuing mechanism to allow for policing of this categorization. Excess resources may be allocated between non-idle classes

Table 4.1 Assured Forwarding encoding

Class	Value
AF1	001dd0
AF2	010dd0
AF3	011dd0
AF4	100dd0

Table 4.2 Drop probability values

Drop Probability (dd)	Value
Low	01
Medium	10
High	11

Table 4.3 Assured Forwarding Bit Drop Precedence Values

AF Codepoints	Class 1	Class 2	Class 3	Class 4
Low Drop Precedence	001010	010010	011010	100010
Medium Drop Precedence	001100	010100	011100	100100
High Drop Precedence	001110	0101110	011110	100110

in a specified way. Reordering of IP packets of the same flow is not allowed if they belong to the same AF class.

4.4.3.3 Diffserv Functionality

For traffic to flow in a Diffserv enabled network, several steps must sequentially occur. First, the edge device classifies the traffic. In this process, the individual data flows are marked according to their precedence in a manner predetermined by the network administrator. This classification can be either on the value present in the DSCP value if available, or a more general set of conditions including, but not limited to, IP address, port numbers, destination address, or even the ingress port. Once traffic has been classified within the Diffserv service provider definitions, it is then passed to an admission filter that shapes or conditions the traffic to meet existing network traffic streams and behavioral aggregates. This includes measuring each stream against relative token bucket burst rates and buffer capacities to determine if ingress will be allowed or if packets should be differentially dropped or delayed. If the packet is admitted into the Diffserv network, its DSCP field is either written or rewritten (if it already existed) and passed as an aggregate stream through the network. The DSCP field then triggers a pre-configured PHB at each node along the Diffserv network.

Exercise - Diffserv Operation

In order to illustrate the operation of the Diffserv mechanism, we will consider the network illustrated in Figure 4.22, where several UDP flows, emulating different Internet services, are sent from the server to the PC through a corporate network. For simplicity reasons, traffic is considered as unidirectional (from the Server to the PC), although it is usually bidirectional in operating scenarios. Let us consider 5 traffic classes with the following characteristics and requirements:

- Voice is considered the **premium** class and its traffic will be marked with a DSCP value of 46.
- The **gold** class consists of Video sessions and its traffic is marked with a DSCP value of 10.
- The **silver** traffic class consists of FTP sessions, with a DSCP value of 22.
- The **bronze** traffic class consists of Web traffic and traffic is marked with a DSCP value of 26.
- Everything else is considered as belonging to the **best-effort** traffic class.
- The premium class should be forwarded with the lowest delay possible up to a maximum of 40% of the link bandwidth during periods of congestion.
- The gold class should be treated preferentially over the silver class, which in turn should be treated preferentially over the bronze class.
- The gold, silver, and bronze classes should have 25%, 20%, and 10%, respectively, of the interface bandwidth as the minimum bandwidth guarantees.
- Best-effort class should be policed to 16 kbps.

In order to configure Diffserv, the Cisco Express Forwarding capability should be enabled by issuing at all routers the command:

```
ip cef
```

in global configuration mode. The different services will be emulated using the MGEN utility at the Server to generate traffic to a specific port number. Traffic classification can be implemented using Class Maps.

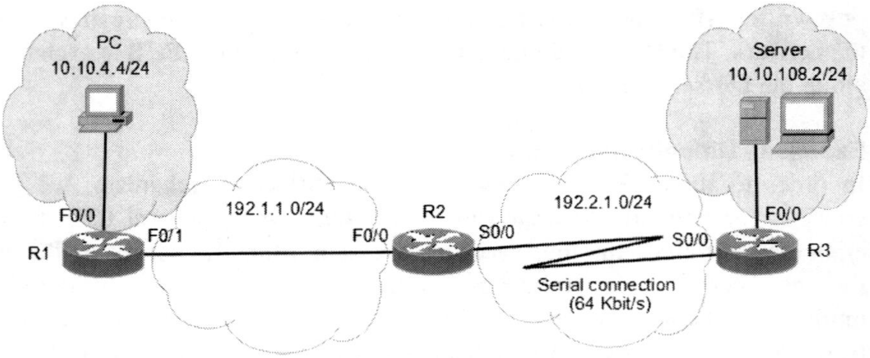

Figure 4.22 Setup used to illustrate the Diffserv mechanism.

Cisco Class Maps allow us to classify network traffic based on:

- Layer 3 and layer 4 traffic flow information, such as source or destination IP address, source or destination port, virtual IP address, IP protocol and port, management protocol;
- Layer 7 protocol information, such as HTTP cookie, HTTP URL, HTTP header, HTTP content, FTP request commands, among other.

The traffic classification process comprises three main steps:

- Creating a class map using the *class-map* command and the associated *match* commands, which comprise a set of match criteria related to layer 3 and layer 4 traffic classifications or layer 7 protocol classifications;
- Creating a policy map using the *policy-map* command, which refers to the class maps and identifies a series of actions to perform based on the traffic match criteria.
- Activating the policy map and attaching it to a specific VLAN interface or globally to all VLAN interfaces associated with a context using the *service-policy* command.

So, in order to classify traffic in the different classes, five extended access lists (where the port numbers for the different services vary from 5001 to 5005) must be defined at Router 3, as well as the class maps that will correspond to the different Diffserv services. Extended ACLs will be covered in detail in chapter 5: for now, let us say that an extended ACL allows to control traffic based on its source and destination addresses, protocol information, port number(s) for TCP and UDP or message types for ICMP. The necessary configuration is shown in Box 4.14.

```
# access-list 101 permit udp any any eq 5001
# access-list 102 permit udp any any eq 5002
# access-list 103 permit udp any any eq 5003
# access-list 104 permit udp any any eq 5004
# access-list 105 permit udp any any eq 5005
# class-map match-all EF
  # match access-group 101
# class-map match-all AF1
  # match access-group 102
# class-map match-all AF2
  # match access-group 103
# class-map match-all AF3
  # match access-group 104
```

Box 4.14 Configuration of the access lists and class maps at Router 3.

```
# policy-map SETDSCP
  # class EF
    # set ip dscp 46
  # class AF1
    # set ip dscp 10
  # class AF2
    # set ip dscp 22
  # class AF3
    # set ip dscp 26
```

Box 4.15 Policy map configuration at Router 3.

In order to mark packets, a policy map (named SETDSCP) should be configured at Router 3. The code is presented in Box 4.15 and should be applied in the input direction of interface F0/0 by using the command:

```
service-policy input SETDSCP
```

In order to configure the different behavior aggregate requirements for the traffic classes, the traffic classes have to be defined and a policy map (named EDGE) should be created at Router 3, as shown in Box 4.16.

The policy map is applied in the output direction of interface S0/1 of Router 3 by using the command:

```
service-policy output EDGE
```

At the core router, the traffic classes correspond to the service classes that were defined in the edge router (Box 4.17).

At Router 2, a new policy map (named CORE) should be created in order to assure prioritization and bandwidth guarantees for each class. The code is illustrated in Box 4.18.

This policy map should be applied in the output direction of interface F0/0 of Router 2.

In order to test the Diffserv configuration, the MGEN tool was used to generate five IP packet flows that emulate the different services, with dis-

```
# class-map match-all premium
  # match ip dscp 46
# class-map match-all gold
  # match ip dscp 10
# class-map match-all silver
  # match ip dscp 22
# class-map match-all bronze
  # match ip dscp 26
# class-map best-effort
  # match access-group 105

#policy-map EDGE
  # class premium
  # priority percent 40
  # class gold
  # bandwidth percent 25
  # class silver
  # bandwidth percent 20
  # class bronze
  # bandwidth percent 10
  # class best-effort
  # police 16000 2000 2000 conform-action set-dscp-transmit 0
//In this case the rate limit is 16Kbps; the normal burst
// has 2000 bytes and the maximum burst 2000 bytes
```

Box 4.16 Configuration of different behavior aggregate requirements at Router 3.

```
# class-map match-all premium
  # match ip dscp 46
# class-map match-all gold
  # match ip dscp 10
# class-map match-all silver
  # match ip dscp 22
# class-map match-all bronze
  # match ip dscp 26
# class-map match-all best-effort
  # match ip dscp 0
```

Box 4.17 Traffic classes at Router 2.

tinct QoS requirements. All flows are composed by packets with 64 bytes of length, have a bit rate equal to 200 Kbit/s and are generated by the server to different port numbers at the PC (from 5001 to 5005). The MGEN script is shown in Box 4.19. The DREC script is quite simple and is shown in Box 4.19.

With these configurations, the premium class source was able to transmit at almost its full rate, 204.10 Kbit/s; the gold class could transmit at a rate of 127.68 Kbit/s (around 25% of the total available bandwidth); the silver class transmitted at a rate of 101.85 Kbit/s (around 20% of the total available bandwidth); the bronze class transmitted at a rate of 51.23 Kbit/s (around 10% of the total available bandwidth) and, finally, the best effort class was able to achieve a transmission rate of 18.23 Kbit/s. Note that the best effort traffic

```
# policy-map CORE
  # class premium
  # priority percent 40
  # class gold
  # bandwidth percent 25
  # random-detect dscp-based
  # random-detect dscp 10 1 4000
//(10 is the DSCP value; 1 is the minimum threshold;
//4000 is the maximum threshold)
  # class silver
  # bandwidth percent 20
  # random-detect dscp-based
  # random-detect dscp 22 1 4000
  # class bronze
  # bandwidth percent 10
  # random-detect dscp-based
  # random-detect dscp 26 1 4000
```

Box 4.18 Policy map definition at Router 2.

```
0 1 ON 10.10.4.4:5001 PERIODIC 390.625 36
0 2 ON 10.10.4.4:5002 PERIODIC 390.625 36
0 3 ON 10.10.4.4:5003 PERIODIC 390.625 36
0 4 ON 10.10.4.4:5004 PERIODIC 390.625 36
0 5 ON 10.10.4.4:5005 PERIODIC 390.625 36
180000 1 OFF
180000 2 OFF
180000 3 OFF
180000 4 OFF
180000 5 OFF
```

Box 4.19 MGEN script for the Diffserv experiment.

```
PORT 5001-5005
```

Box 4.20 DREC script for the Diffserv experiment.

was not limited to 16 Kbits/s, as expected, because the available bandwidth was slightly higher than this value.

Box 4.21 shows the result of executing the `show policy-map` command at Router R3, confirming that it is configured as expected.

If we capture the UDP packets in the serial interface of Router 3, we should expect to see the DSCP mark corresponding to each class. This can be confirmed by looking at the example shown in Figure 4.23 for the premium class.

```
# show policy-map EDGE
 POLICY-MAP VOIP
  Class premium
   Bandwidth 40 (%) Max Threshold 64 (packets)
  Class gold
   Bandwidth 25 (%) Max Threshold 64 (packets)
  Class silver
   Bandwidth 20 (%) Max Threshold 64 (packets)
  Class bronze
   Bandwidth 10 (%) Max Threshold 64 (packets)
  Class best-effort
   police cir 16000 bc 1750 be 1750
   conform-action set-dscp-transmit default
   exceed-action drop
```

Box 4.21 Configured EDGE policy-map.

```
No.  Time          Source       Destination    Protocol  Info
  81 20.451593     10.10.4.4    10.10.108.2    UDP       Source port: 58461
  82 20.453849     10.10.4.4    10.10.108.2    UDP       Source port: 57992
  83 20.455859     10.10.4.4    10.10.108.2    UDP       Source port: 57184
  84 20.457942     10.10.4.4    10.10.108.2    UDP       Source port: 36508
  85 20.460012     10.10.4.4    10.10.108.2    UDP       Source port: 34373
  86 20.462338     10.10.4.4    10.10.108.2    UDP       Source port: 42715

▶ Frame 81 (78 bytes on wire, 78 bytes captured)
▶ Ethernet II, Src: cc:15:6c:bf:f1:00 (cc:15:6c:bf:f1:00), Dst: ca:13:6c:bf:00:1c (ca:13:6c:bf:00:1c)
▼ Internet Protocol, Src: 10.10.4.4 (10.10.4.4), Dst: 10.10.108.2 (10.10.108.2)
   Version: 4
   Header length: 20 bytes
 ▶ Differentiated Services Field: 0xb8 (DSCP 0x2e: Expedited Forwarding; ECN: 0x00)
   Total Length: 64
   Identification: 0x0000 (0)
 ▶ Flags: 0x02 (Don't Fragment)
   Fragment offset: 0
   Time to live: 63
   Protocol: UDP (0x11)
 ▶ Header checksum: 0xb6db [correct]
   Source: 10.10.4.4 (10.10.4.4)
   Destination: 10.10.108.2 (10.10.108.2)
▼ User Datagram Protocol, Src Port: 58461 (58461), Dst Port: commplex-link (5001)
   Source port: 58461 (58461)
   Destination port: commplex-link (5001)
   Length: 44
 ▶ Checksum: 0x8536 [validation disabled]
```

Figure 4.23 UDP packet marked with a DSCP value of 0x2e (46 in decimal).

4.4.4 How IntServ and DiffServ Determine Network Design

Integrated services, by their nature, require that the end nodes in a data flow mark the packets with the required QoS characteristics. The internetwork then provides the required data flow up to the limit of available resources.

In the Diffserv architecture, services mark and classify packets only at the network ingress points. The core network then simply imposes the Per Hop Behaviors as defined by the service provider in response to information contained within the DSCP. This simplified operating mode is responsible for the scalability of this architecture.

Figure 4.24 IntServ and DiffServ integration.

Tipically, the core of a network consists of a large number of high-speed connections and data flows. For every data flow that must be monitored by a node, significant resources are allocated. The amount of resources required to control data flows at the core backbone speeds would require significant investments and configurations and impose undue switching latency. If a node in the core must make decisions based on ingress and queuing characteristics, the whole network model is compromised.

The best architectures classify traffic at the ingress edges of the network. In this way, each router has to deal only with a small volume of manageable traffic. This allows for maximum response, utilization, and minimal latency from all network components concerned. As the number of data streams decreases, the amount of classification and queuing is less at the edge, which means that delay and jitter are minimized: thus, the chances of exhausting resources is reduced and over all costs are lowered as less expensive hardware can be utilized.

4.4.5 Integration of the IntServ and DiffServ QoS Models

The IntServ architecture is appropriate for small/access networks, while the DiffServ architecture is appropriate for large networks, mainly transit networks [4]. Both architectures can coexist and inter operate, as shown in Figure 4.24, providing IntServ services on large networks and explicit admission control instead of SLAs on DiffServ networks. In this case, border routers of both network types should classify RSVP requests on the appropriate DiffServ service classes and, if the are no sufficient resources, should refuse the RSVP reservation requests.

When mapping IntServ services to the DiffServ domain, some basic requirements should be satisfied:

- PHBs in the DiffServ domain must be appropriately selected for each requested service in the IntServ domain.

- The required policing, shaping and marking must be done at the edge routers of the DiffServ domain.
- Taking into account the resource availability in the DiffServ domain, admission control must be implemented for the requested traffic in the IntServ domain.

So, in practice a mapping function should be implemented at the boundary between both QoS domains to assign an appropriate DSCP to a flow specified by Tspec parameters in the IntServ domain, in such a way that the same QoS could be achieved for IntServ when running over DiffServ domain.

4.5 Use Case

4.5.1 DiffServ Configuration

We want to differentiate VoIP and Video Conference traffic, giving a higher quality of service (QoS) level to these services (in this case, the QoS level corresponds to reserving a percentage of the link bandwidth equal to 30%). So, we will configure DiffServ, creating a class-map named *realtime* that will correspond to the Expedited Forwarding (EF) service level of DiffServ and will be used to provide the required QoS level to the VoIP and Video Conference services.

In order to implement such a strategy, packets from these services should be marked at the distribution layer edge routers, differentiating the traffic that will go the core part of the infrastructure. Box 4.22 illustrate all the commands that should be used in the distribution layer edge routers (in this case, the commands were inserted in Router 11).

Box 4.23 illustrate all the commands that should be used in the core layer routers (in this case, the commands were inserted in Router 1). Core routers should differentiate traffic in all interfaces.

Obviously, Internet access routers (routers 101 and 102) should do the symmetric operation, that is, they should mark traffic coming from the outside and destined to VLANs 12 and 13 (the ones that correspond to the VoIP and Video Conference services), thus differentiating traffic that goes to the core layer of the network infrastructure.

In order to test this configuration, we have executed the *ping* command from a host belonging to subnet 10.2.0.0/24 and captured the generated packets. As expected, this traffic was marked with a DSCP decimal value of 46, the one that corresponds to the EF service class, as illustrated in Figure 4.25.

```
# access-list 20 permit 10.2.0.0 0.0.0.255
# access-list 20 permit 10.2.1.0 0.0.0.255
# access-list 20 permit 10.3.0.0 0.0.0.255
# access-list 20 permit 10.3.1.0 0.0.0.255
!
# class-map match-all EF
  # match access-group 20
# policy-map SETDSCP
  # class EF
    # set ip dscp ef
!
# interface vlan 12
  # service-policy input SETDSCP
# interface vlan 13
  # service-policy input SETDSCP
!
# class-map match-all realtime
  # match ip dscp ef
# policy-map QoS
  # class realtime
    # priority percent 30
!
# interface vlan 101
  # service-policy output QoS
# interface vlan 102
  # service-policy output QoS
```

Box 4.22 DiffServ configuration in an edge router (at Router 11).

```
# class-map match-all realtime
  # match ip dscp ef
# policy-map QoS
  # class realtime
    # priority percent 30
!
# interface FastEthernet 0/0
  # service-policy output QoS
# interface FastEthernet 0/1
  # service-policy output QoS
(... repeat in all interfaces ...)
```

Box 4.23 DiffServ configuration in a core router (at Router 1).

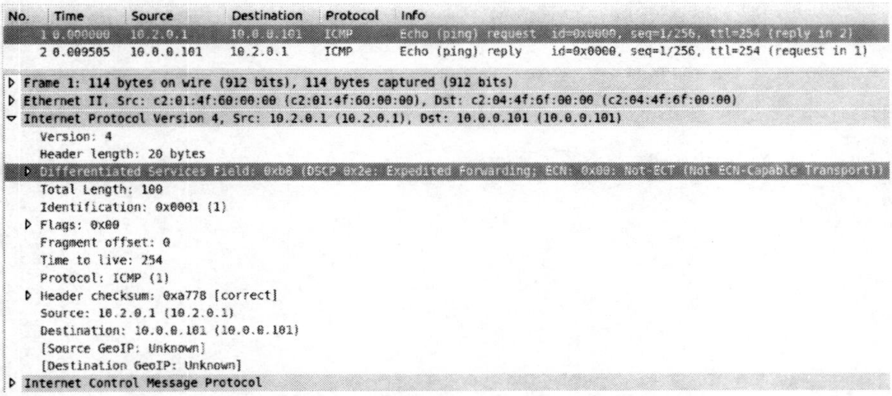

No.	Time	Source	Destination	Protocol	Info
1	0.000000	10.2.0.1	10.0.0.101	ICMP	Echo (ping) request id=0x0000, seq=1/256, ttl=254 (reply in 2)
2	0.009505	10.0.0.101	10.2.0.1	ICMP	Echo (ping) reply id=0x0000, seq=1/256, ttl=254 (request in 1)

```
▷ Frame 1: 114 bytes on wire (912 bits), 114 bytes captured (912 bits)
▷ Ethernet II, Src: c2:01:4f:60:00:00 (c2:01:4f:60:00:00), Dst: c2:04:4f:6f:00:00 (c2:04:4f:6f:00:00)
▽ Internet Protocol Version 4, Src: 10.2.0.1 (10.2.0.1), Dst: 10.0.0.101 (10.0.0.101)
    Version: 4
    Header length: 20 bytes
  ▷ Differentiated Services Field: 0xb8 (DSCP 0x2e: Expedited Forwarding; ECN: 0x00: Not-ECT (Not ECN-Capable Transport))
    Total Length: 100
    Identification: 0x0001 (1)
  ▷ Flags: 0x00
    Fragment offset: 0
    Time to live: 254
    Protocol: ICMP (1)
  ▷ Header checksum: 0xa778 [correct]
    Source: 10.2.0.1 (10.2.0.1)
    Destination: 10.0.0.101 (10.0.0.101)
    [Source GeoIP: Unknown]
    [Destination GeoIP: Unknown]
▷ Internet Control Message Protocol
```

Figure 4.25 DiffServ EF marked packet.

5

Access Control and Secure Communications

5.1 Introduction

Network security design should be a concern for clients and organizations that are designing, upgrading, moving or re-designing their networks and Internet access points and want to incorporate layered security mechanisms and controls into the network. Security features should be incorporated into the network from the beginning, reducing or even eliminating the need for future downtime and expensive installations/upgrades.

Today, networks and network users are subject to a huge variety of security threats. The most common include:

- Viruses, worms, and Trojan horses;
- Spyware and adware;
- Zero-day or zero-hour attacks;
- Hacker attacks;
- Denial of service attacks;
- Data interception and theft;
- Identity theft.

There is not a single solution to protect a network or a user from such a variety of threats. Multiple layers of security are needed and, if one fails, others still stand. Network security can be accomplished using hardware and software based solutions/components.

A network security system usually consists of many components which, ideally, should work together in order to minimize maintenance and improve security. Network security components often include:

- Anti-virus and anti-spyware.
- Firewalls to block unauthorized access to the network.
- Intrusion Detection and Prevention Systems (IDS and IPS) to identify fast-spreading threats, such as zero-day or zero-hour attacks.
- Virtual Private Networks (VPNs), to provide secure remote access.

In this chapter we will not cover all these components, being essentially focused on access control mechanisms and methods to implement secure communications.

5.2 Security Policy

According to RFC 2196 [28], a security policy is a "Formal statement of the rules by which people who are given access to an organization's technology and information assets must abide." The policy should address physical access, authentication, accountability, authorization and data encryption.

Physical security refers to limiting access to key network resources by keeping them protected from inadvertent misuses, hackers, competitors and terrorists. The network equipment should also be protected from natural disasters such as floods, fires, storms, and earthquakes.

Authentication is intended to identify who is requesting network services (users, devices or software processes) and is traditionally based on a combination of the following proofs:

- Something the user knows - A unique secret that is shared by the authenticating parties (in the case of a user it can be a classic password, a Personal Identification Number (PIN), or a private cryptographic key).
- Something the user has - Something that is unique to the user, like for example password token cards, security cards and hardware keys.
- Something the user is - A unique physical characteristic of the user, such as a fingerprint, retina pattern, voice or face.

Accountability defines the responsibilities of users, operations staff and management.

Authorization defines what users can do after they have accessed the resources. Authorization grants privileges to processes and users. The authorization mechanism should give a user only the minimum access permissions that are necessary.

Finally, data encryption is a process that scrambles data to protect it from being read by anyone but the intended receiver. An encryption device encrypts data before placing it on a network. A decryption device decrypts the data before passing it to an application. A router, server, end system, or dedicated device can act as an encryption or decryption device. Data that is encrypted is called ciphered data (or simply encrypted data). Data that is not encrypted is called plain text or clear text. Encryption is a useful security feature for providing data confidentiality and can also be used to identify the sender of data. Although authentication and authorization should also protect the confidentiality of data and identify senders, encryption is a good security feature to implement in case other types of security fail.

5.3 Network Access Control

One of the most common security problems that network managers face is to control access to certain network segments, services or hosts. Several mechanisms are available to implement access control policies.

5.3.1 Network Firewall

A network firewall provides a single point of defense between networks, protecting one network from the others [56]. It can be considered as a system or group of systems that enforces a control policy between two or more networks (access control, flow control and content control). It minimizes local vulnerabilities by evaluating each packet against the policies of network security. These systems can monitor all the network traffic and alert to any attempts to bypass security or to any patterns of inappropriate use. They can be hardware or software based and can provide gateway services: Network Address Translation, proxing and application gateway and security perimeter extension (tunneling).

Network firewalls can be based on the following technologies:

- Stateless packet filtering - Route packets between internal and external hosts, but do it selectively. A static stateless packet-filter firewall looks at individual packets and is optimized for speed and configuration simplicity. However, this approach presents some problems: undesirable packets can be fitted to a packet rule criteria and, therefore, pass through the filter; packets can pass through the filter by being fragmented; complex rule sets are difficult to implement and maintain correctly.
- Stateful/dynamic packet filtering - Combines the best of packet filtering and proxy services technologies by maintaining the complete connection states. A stateful firewall can track communication sessions and more intelligently allow or deny traffic. For example, a stateful firewall can remember that a protected client initiated a request to download data from an Internet server and allow data back in for that connection. A stateful firewall can also work with protocols, such as active (port-mode) FTP, that require the server to also open a connection to the client.

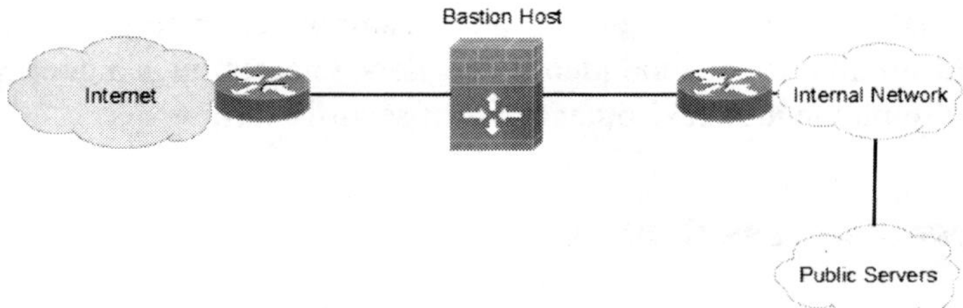

Figure 5.1 Dual-homed network firewall.

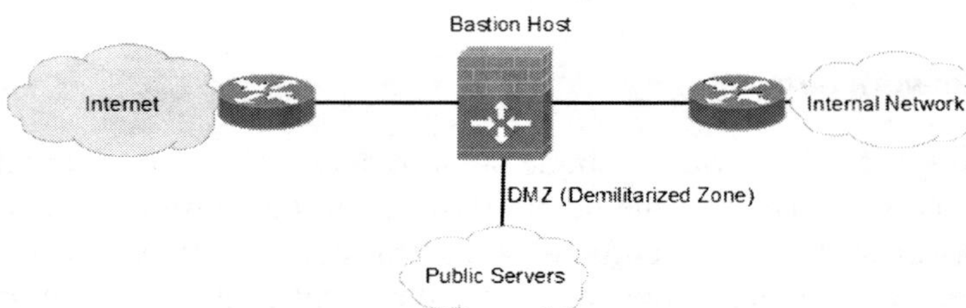

Figure 5.2 Multi-homed network firewall.

- Proxy services - Specialized application or server programs that take users' requests for Internet services and forward them to the actual services. The proxies provide replacement connections and act as gateways to the services. Proxy firewalls are the most advanced type of firewall but also the least common. They examine packets and support stateful tracking of sessions. These types of firewalls can block malicious traffic and content that is deemed unacceptable. However, this approach also presents some problems: they usually represent a single point of failure, it is difficult to add new services to them, are CPU intensive and often perform slower under stress.

The most common network firewall architectures are single-homed and multi-homed. A single-homed or standalone firewall is typically a machine with a single network connection that runs its own access control rules. A dual-homed firewall is usually placed between a trusted and an untrusted network (Figure 5.1), while multi-homed firewalls usually include a Demilitarized Zone (DMZ) (Figure 5.2). A DMZ is a perimeter network outside the protected internal network that is used to place public servers/services. It is a "semi-protected" zone, so it must be assumed that any machine placed on the DMZ is at risk.

A bastion host is a computer that is fully exposed to attacks, is located on the public side of the DMZ and functions as an application-level gateway, redirecting traffic to servers on the DMZ. Bastion hosts must be designed and configured to minimize the chances of penetration, so all unnecessary services, protocols, programs, and network ports should be disabled or removed. Some bastions are deliberately exposed to potential hackers to both delay and facilitate tracking of attempted break-ins.

5.3.2 Intrusion Detection and Prevention Systems

An Intrusion Detection System (IDS) detects malicious events and notifies an administrator, using email, paging, or logging of the occurrence. An IDS can also perform statistical and anomaly analysis. Some IDS devices can report to a central database that correlates information from multiple sensors to give an administrator an overall view of the real-time security of a network. An Intrusion Prevention System (IPS) can dynamically block traffic by adding rules to a firewall or by being configured to inspect (and deny or allow) traffic as it enters a firewall. So, an IPS is an IDS that can detect and prevent attacks.

There are two types of IDS devices:

- Host IDS - Resides on an individual host and monitors that host.
- Network IDS - Monitors all network traffic that it can see, watching for predefined signatures of malicious events. A network IDS is often placed on a subnet that is directly connected to a firewall so that it can monitor the traffic that has been allowed and look for suspicious activity.

A false alarm occurs when an IDS or IPS reports a network event as a serious problem when it actually is not a problem. This false alarm problem has been mitigated by sophisticated software and services existing on modern IDS/IPS devices.

5.3.3 Cisco Access Control Lists

Cisco implements security policies in its packet filters, called Access Control Lists (ACLs). An ACL is a sequential collection of `permit` and `deny` conditions. Each packet is tested against the conditions in the ACL, one by one. The first match determines whether the software accepts or rejects the packet: since the software stops testing conditions after the first match, the order of the conditions is critical. If no conditions match, the software rejects the packet because there is an implicit `deny any` at the end of all access lists.

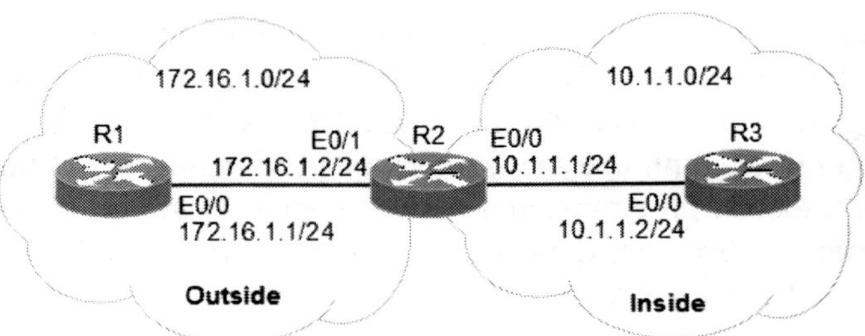

Figure 5.3 Network scenario used to illustrate ACLs.

ACLs can be applied to inbound or outbound traffic (the inbound or outbound interface should be referenced as if looking at the port from inside the router or switch) and only one ACL should be applied per protocol and per direction.

ACLs can be used to prevent unwanted traffic in the network in order to avoid hackers from penetrating the network or just to prevent employees from using systems they should not be using. IP access lists can also be used to filter routing updates and to match packets for prioritization, VPN tunneling and implementing quality of service features.

There are several types of ACLs, with different complexity levels. We will present the most common ones.

5.3.3.1 Standard ACL

Standard ACLs control traffic by comparing the source address of the IP packets with the addresses configured in the ACL. The command syntax format of a standard ACL is the following:

```
# access-list <access-list-number> {permit | deny}
        {<source-ip> | <source-net> <wildcard-mask> | any}
```

The `<access-list-number>` can be anything from 1 to 99. The `<wildcard-mask>` is the binary inverse of the subnet mask. A `<source-net> <wildcard-mask>` setting of 0.0.0.0 255.255.255.255 can be specified as `any`. The wildcard can be omitted if it is all zeros.

After defining the ACL, it must be applied to the interface (inbound or outbound) using the following command in interface configuration mode:

```
# ip access-group <access-list-number> {in | out}
```

Referring to Figure 5.3, the example shown in Box 5.1 corresponds to a standard ACL that blocks all traffic except that from source 10.1.1.x (any host located in the inside network).

```
# interface Ethernet0/0
  # ip address 10.1.1.1 255.255.255.0
  # ip access-group 1 in
# access-list 1 permit 10.1.1.0 0.0.0.255
```

Box 5.1 Standard ACL example

5.3.3.2 Extended ACL

Extended ACLs control traffic by the comparison of the source and destination addresses of the IP packets to the addresses configured in the ACL. The command syntax format of an extended ACL includes the following fields, although it can also include many more optional fields:

```
# access-list <access-list-number> {permit | deny} <protocol>
    {<source-ip> | <source-net> <wildcard-mask> | any} [{eq|lt|gt} <source-port>]
    {<dest-ip> | <dest-net> <wildcard-mask> | any} [{eq|lt|gt} <dest-port>]
```

In this case, the <access-list-number> can an integer between 101 to 199. The <protocol> parameter indicates the type of IP packet that should be filtered: it is possible to specify a well-known name for any protocol whose number is less than 255, while for other protocols their number should be typed. Operators eq|lt|gt specify a comparison operator for the TCP or UDP source and destination port numbers and <source-port>/<dest-port> specifies the TCP or UDP port number or well-known name. A lot of optional fields can be included in the extended ACL, like for example: icmp-type | icmp-num, which specifies the ICMP protocol type; precedence name | num, which specifies the IP precedence; several other options related to Quality of Service control, like priority, priority-force, priority-mapping, dscp-mapping, and dscp-marking.

After defining the ACL, it must be applied to the interface (inbound or outbound direction), using the following command in interface mode:

```
# ip access-group <access-list-number> {in | out}
```

The extended ACL shown in Box 5.2 is used to permit traffic on the 10.1.1.x network (inside) and to receive ping responses from the outside, while it prevents unsolicited pings from people located outside, permitting all other traffic.

> **Placement of standard and extended ACLs**
>
> Standard access lists should be applied closest to the destination, while extended access lists should be applied closest to the source.

```
# interface Ethernet0/1
  # ip address 172.16.1.2 255.255.255.0
  # ip access-group 101 in

# access-list 101 deny icmp any 10.1.1.0 0.0.0.255 echo
# access-list 101 permit ip any 10.1.1.0 0.0.0.255
```

Box 5.2 Extended ACL example

```
# interface Ethernet0/0
  # ip address 10.1.1.1 255.255.255.0
  # access-group intoout in

# ip access-list extended intoout
# permit tcp host 10.1.1.2 host 172.16.1.1 eq telnet
```

Box 5.3 IP named ACL example

```
# ip access-group <access-list-number | name> {in | out}
# ip access-list extended name
# permit protocol any any reflect name [timeoutseconds]
# ip access-list extended name
# evaluate name
```

Box 5.4 Reflexive ACL commands

5.3.3.3 IP Named ACL

IP named ACLs allow standard and extended ACLs to be given names instead of numbers. The command syntax format for IP named ACLs is:

```
ip access-list extended | standard name
```

Box 5.3 shows an example of the use of a named ACL (intoout) in order to block all traffic except the Telnet connection from host 10.1.1.2 to host 172.16.1.1.

5.3.3.4 Reflexive ACLs

Reflexive ACLs allow IP packets to be filtered based on upper-layer session information. They are generally used to allow outbound traffic and to limit inbound traffic in response to sessions that originate inside the router.

Reflexive ACLs can be defined only with extended named IP ACLs. They cannot be defined with numbered or standard named IP ACLs, or with other protocol ACLs. Reflexive ACLs can be used in conjunction with other standard and static extended ACLs.

Box 5.4 shows the syntax of the most relevant commands used in reflexive ACLs but it is easier to look at a concrete example.

```
# ip reflexive-list timeout 120
# interface Ethernet0/1
  # ip address 172.16.1.2 255.255.255.0
# ip access-group inboundfilter in
# ip access-group outboundfilter out

# ip access-list extended inboundfilter
# permit icmp 172.16.1.0 0.0.0.255 10.1.1.0 0.0.0.255
# evaluate tcptraffic

# ip access-list extended outboundfilter
# permit icmp 10.1.1.0 0.0.0.255 172.16.1.0 0.0.0.255
# permit tcp 10.1.1.0 0.0.0.255 172.16.1.0 0.0.0.255
        reflect tcptraffic
```

Box 5.5 Reflexive ACL example

Box 5.5 corresponds to a reflexive ACL that permits ICMP outbound and inbound traffic, while only permitting TCP traffic that has initiated from inside; other traffic is denied.

The `outboundfilter` ACL evaluates all outbound traffic on interface E0/1. The first command of this ACL permits all ICMP traffic from network 10.1.1.0/24 (inside) to network 172.16.1.0/24 (outside); the second command defines the reflexive access list `tcptraffic`. This entry permits all outbound TCP traffic and creates a new access list named `tcptraffic`. Also, when an outbound TCP packet is the first in a new session, a corresponding temporary entry will be automatically created in the reflexive access list `tcptraffic`.

The `inboundfilter` access list evaluates all inbound traffic on interface E0/1. The first command allows all ICMP traffic from the outside network to the inside one, while the last entry points to the reflexive access list `tcptraffic`. If a packet does not match the first entry, the packet will be evaluated against all the entries in this reflexive access list.

5.3.3.5 Context-Based Access Control

Context-based access control (CBAC) inspects traffic that travels through the firewall in order to discover and manage state information for TCP and UDP sessions. This state information is used in order to create temporary openings in the access lists of the firewall.

CBAC is a type of extended access list and is similar in approach to a reflexive access list. Its main advantages over reflexive ACLs are that it will allow monitoring of packet traffic beyond the transport layer, having the ability to monitor traffic that may originate on one port but has continuity on another port (such as File Transfer Protocol traffic). Its initial operation is similar to reflexive access lists, since it opens temporary "holes" in defined

```
# ip inspect name mycbac ftp timeout 3600
# ip inspect name mycbac http timeout 3600
# ip inspect name mycbac tcp timeout 3600
# ip inspect name mycbac udp timeout 3600
# interface Ethernet0/1
  # ip address 172.16.1.2 255.255.255.0
  # ip access-group 111 in
  # ip inspect mycbac out
# access-list 111 deny icmp any 10.1.1.0 0.0.0.255 echo
# access-list 111 permit icmp any 10.1.1.0 0.0.0.255
```

Box 5.6 CBAC ACL example

access lists to allow return traffic either into or out of the network. It uses a type of stateful inspection table that is stored locally (until the exchange is completed) to determine whether or not to permit network access through the router.

Box 5.6 shows an example of the use of CBAC in order to inspect outbound traffic. Extended ACL 111 normally blocks the return traffic other than ICMP without CBAC opening holes for the return traffic corresponding to FTP, HTTP, TCP and UDP.

The syntax of the CBAC `ip inspect` command is the following:

```
ip inspect name <inspection-name> <protocol> [timeout <seconds>]
```

Filtering logic

Access list entries should filter from specific to general.

Whenever a packet is blocked by an ACL, the router that implements it sends an ICMP `host unreachable` message to the sender of the rejected packet and discards the packet.

5.3.4 Linux IPtables

IPtables are a Linux product that can be used to perform:

- Stateful packet inspection - The firewall keeps track of each connection passing through it and in certain cases will view the contents of data flows in an attempt to anticipate the next action of certain protocols.
- Filter packets based on a MAC address and the values of the flags in the TCP header - This helps preventing attacks using malformed packets and in restricting access from locally attached servers to other networks in spite of their IP addresses.

- System logging, with the possibility of adjusting the level of detail of the reporting.
- Network Address Translation.
- Rate limiting, being able to block some types of denial of service (DoS) attacks.

All packets inspected by IPtables pass through a sequence of built-in tables (queues) for processing. Each of these queues is dedicated to a particular type of packet activity and is controlled by an associated packet transformation/filtering chain. There are three tables:

- Mangle table - Responsible for changing the quality of service bits in the TCP header. Packets are subject to the following chains: Pre-routing, Post-routing, Output, Input and Forward.
- Filter queue - Responsible for packet filtering. It has three built-in chains where we can place the firewall policy rules: Forward chain, which filters packets to servers protected by the firewall; Input chain, which filters packets destined for the firewall; Output chain, which filters packets originating from the firewall.
- NAT queue - Responsible for network address translation. It has two built-in chains: Pre-routing, which NATs packets when the destination address needs to be changed; Post-routing, which NATs packets when the source address of the packet needs to be changed.

For each firewall rule that is created, we have to specify the table and chain, with an exception: since most rules are related to filtering, IPtables assume that the filter table is the default. Figure 5.4 illustrates how packets are handled by IPtables. The packet is first examined by rules located in the mangle table's PREROUTING chain, if any. It is then inspected by the rules in the NAT table's PREROUTING chain to see whether the packet requires DNAT. It is then routed.

If the packet is destined for a protected network, then it is filtered by the rules in the FORWARD chain of the filter table and, if necessary, the packet undergoes SNAT in the POSTROUTING chain before arriving at Network B. When the destination server decides to reply, the packet undergoes the same sequence of steps. Both the FORWARD and POSTROUTING chains may be configured to implement quality of service (QoS) features in their mangle tables.

If the packet is destined for the firewall itself, then it passes through the mangle table of the INPUT chain, if configured, before being filtered by the

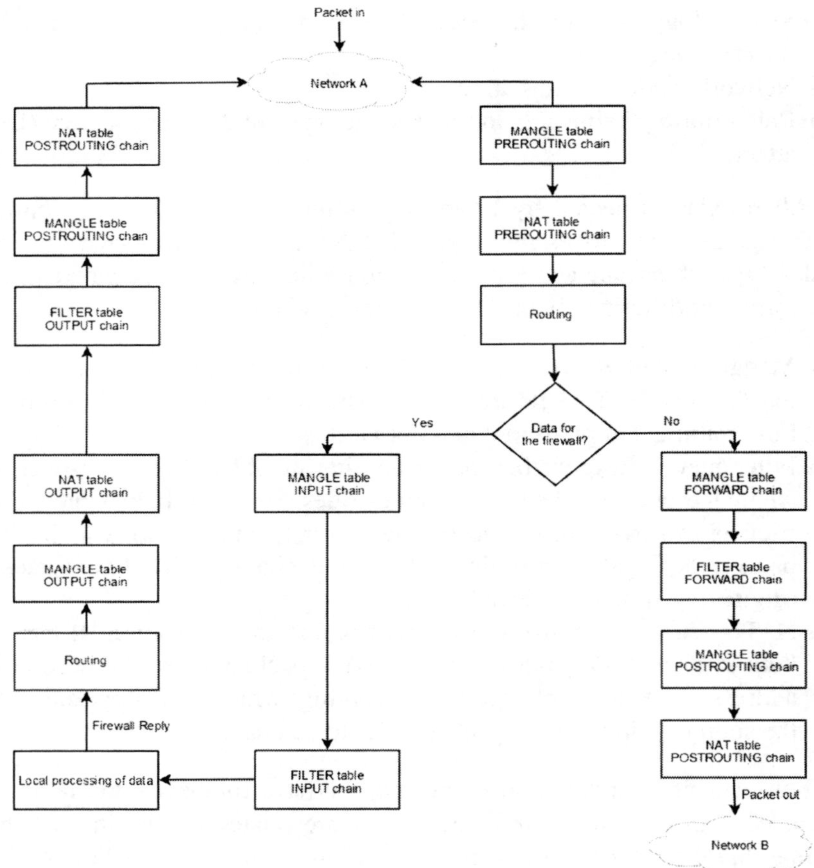

Figure 5.4 IPtables packet flow diagram.

rules in the INPUT chain of the filter table before. If it successfully passes these tests then it is processed by the intended application on the firewall.

At some point, the firewall needs to reply. This reply is routed and inspected by the rules in the OUTPUT chain of the mangle table, if any. Next, the rules in the OUTPUT chain of the nat table determine whether DNAT is required and the rules in the OUTPUT chain of the filter table are then inspected to help restrict unauthorized packets. Finally, before the packet is sent back to the Internet, SNAT and QoS mangling is done by the POSTROUTING chain

In addition to the built-in chains, the user can create any number of user-defined chains within each table, which allows them to group rules logically. Each chain contains a list of rules: if a given rule does not match, then processing continues with the next rule; if, however, the rule does match the packet, then the rule's target instructions are followed (and further processing of the chain is usually aborted).

Each firewall rule inspects each IP packet and then tries to identify it as the target of some sort of operation. Once a target is identified, the packet needs to jump over to it for further processing. Targets can be thought of as subroutines. The built-in targets that IPtables use are:

- Accept - The packet is handed over to the end application or the operating system for processing;
- Drop - The packet is blocked;
- Log - The packet information is sent to the syslog daemon for logging;
- Reject - Works like the DROP target, but will also return an error message to the host sending the packet that the packet was blocked;
- DNAT - Used to do Destination NAT;
- SNAT - Used to do Source NAT;
- Masquerade - This is basically the same as the SNAT target, but it can be used with dynamically assigned IP connections (like dial-up or DHCP connections, which get dynamic IP addresses).

Some targets can only be used in certain chains, and/or certain tables: for example, the SNAT and the Masquerade targets can only be used in the POSTROUTING chain of the NAT table. The target of a rule can also be the name of a user-defined chain.

5.4 Secure Network Communications

This section introduces some concepts and technologies that are currently used to provide secure network communication, like cryptography, hash functions and the IPSec protocol.

5.4.1 Cryptography Basics

Cryptography can be used to provide message confidentiality and integrity and sender verification. The basic functions of cryptography are encryption, decryption and cryptographic hashing. In order to encrypt and decrypt messages, the sender and recipient need to share a secret. Typically this is a key

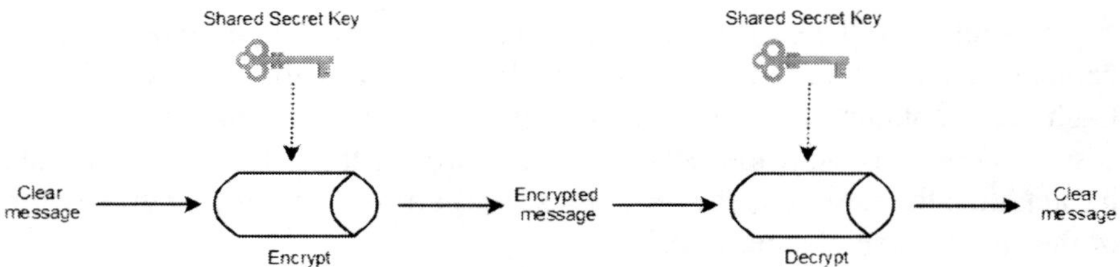

Figure 5.5 Symmetric key cryptography.

that is used by the cryptographic algorithm. The key is used by the sender to encrypt the message and by the recipient to decrypt the message. This process can be done on a fixed message or a communications stream. Cryptographic hashing is the process of generating a fixed-length string from a message of arbitrary length. If the sender provides a cryptographic hash with the message, the recipient can verify its integrity.

The three basic types of cryptography are symmetric key, asymmetric (public) key systems and cryptographic hash functions. Typically, the strength of a cryptography system is directly related to the length of the key.

Symmetric key cryptography, which is also referred to as shared key or shared secret encryption, uses the same key to encrypt and decrypt data, as illustrated in Figure 5.5. Some common symmetric key algorithms are the Data Encryption Standard (DES) [14], Triple DES (DES applied three times, using different keys) [105], Blowfish [89] and the Advanced Encryption Standard (AES) [76]. DES is ineffective because it uses a 64-bit key and has been broken.

The main advantage of symmetric key cryptography is speed, while its main problems are key distribution and scalability: keys need to be distributed securely, and each secure channel needs a separate key. Symmetric key systems provide confidentiality but do not provide authenticity of the message, and the sender can deny having sent the message.

Asymmetric (public) key cryptography uses a pair of mathematically related keys. Each key can be used to encrypt or decrypt. However, a key can only decrypt a message that has been encrypted by the related key. The public key, which may be known by anybody, can be used to encrypt messages and verify signatures. The private key is known only to the recipient and is used to decrypt messages and sign (create) signatures. The key pair is called the public/private key pair, as illustrated in Figure 5.6. Common public key systems are the Rivest-Shamir-Adelman (RSA) [88] and the Diffie-Hellman [85].

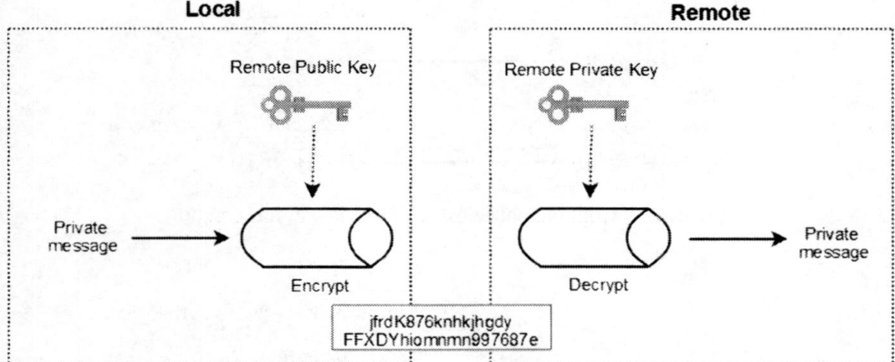

Figure 5.6 Public key cryptography.

Asymmetric key systems solve the key distribution and scalability problems associated with symmetric systems. Asymmetric key systems provide a greater range of security services than symmetric systems. They provide for confidentiality, authenticity and nonrepudiation. The main problem with these systems is speed. It takes significantly more computer resources to encrypt and decrypt with asymmetric systems than with symmetric ones.

If we want to use the public key mechanism to ensure confidentiality between two hosts A and B (Figure 5.7), the procedure to send an encrypted message from A to B should be:

- Host A encrypts data with Host B public key (PU_B);
- Host A sends encrypted data to Host B;
- Host B decrypts data with Host B private key (PR_B).

while the procedure to send an encrypted message from B to A should be:

- Host B encrypts data with Host A public key (PU_A);
- Host B sends encrypted data to Host A;
- Host A decrypts data with Host A private key (PR_A).

In order to send an authenticated message from A to B, the procedure should be (Figure 5.8):

- Host A creates a signature by encrypting data with Host A private key (PR_A);
- Host A sends data and signature to host B;
- Host B verifies data by decrypting signature with Host A public key (PU_A) and compares with received message.

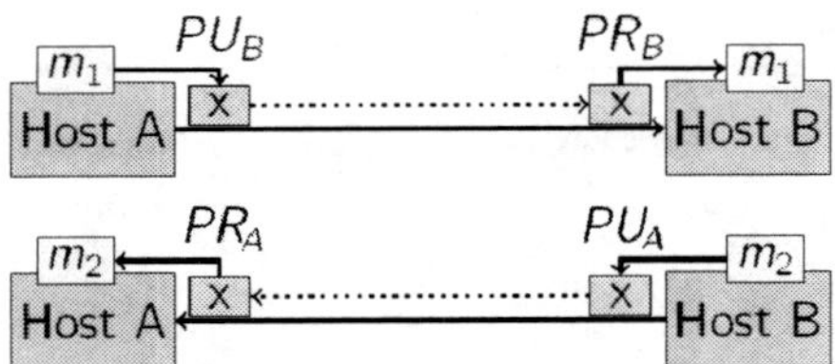

Figure 5.7 Confidentiality using public key cryptography.

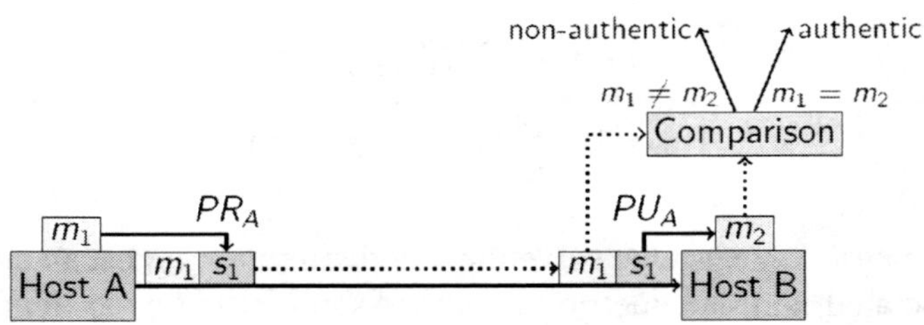

Figure 5.8 Authentication using public key cryptography.

A cryptographic hash function is a hash function, that is, an algorithm that takes an arbitrary block of data and returns a fixed-size bit string, the (cryptographic) hash value, such that an (accidental or intentional) change to the data will (with very high probability) change the hash value. The data to be encoded is often called the "message," and the hash value is sometimes called the message digest or simply digest. The main properties of these functions are that it is difficult to find different files that produce the same digest and that the function is one-way. Therefore, it is not computationally feasible to recover a message given its digest.

Two common examples of hash functions are the Secure Hash Algorithm (SHA), commonly SHA-1 [23] and the Message-Digest algorithm 5 (MD5) [87]. SHA-1 is used in many common security applications including SSL, TLS and IPSec. MD5 is generally used to create a digital fingerprint for verifying file integrity.

5.4.2 IPSec

Nowadays, the most important security challenges faced by a communications network are:

- Authentication - Make sure that the data received is indeed coming from the expected partner (avoiding unauthorized sources to transmit data to a given station);
- Integrity - Make sure that the data received is indeed what was transmitted by the source (avoiding modifications to the data by unknown parties, executed while the data traveled from source to destination);
- Rejection of replayed data - Make sure that receivers identify and discard packets that have been already received;
- Confidentiality - Make sure that nobody "listens and understands" the data on its way from source to destination.

IPsec addresses all these issues and defines the necessary tools for their provision. The basic idea of IPsec is to "mark" packets before being injected into the communications network, and use this mark at the receiving side in order to decide whether the packet arrived from the correct source (authentication), whether the packet content is exactly the one generated by the source, without any modifications (integrity), and whether the packet is not a replay of one of the previous packets, already received (rejection of replayed data). In addition, IPsec also defines a framework for data encryption ensuring that potential "listeners" in the network would not be able to understand the information carried in the packet (confidentiality). The marking process results in new fields being added to the packet to be protected.

Packets can be marked by the user computer (client) or by the ingress edge router. When the user computer is the one marking the transmitted packets, it is said that IPsec is used in "transport mode". When the ingress router is doing the job on behalf of the user (acting as a proxy IPsec entity), it is said that IPsec is used in "tunnel mode". Figure 5.9 illustrates both operation modes.

Authentication and integrity check processes are based on the addition of an Integrity Check Value (ICV) field to the packet that will be protected. ICV is the result of two consecutive processes: a hash value is calculated for the packet to protect and the hash value is encrypted using a secret key, generating the ICV value to be added to the packet.

The first step relies on a hash function, which is (as already said) a function that takes variable length input data and produces fixed length output data that can be regarded as the fingerprint of the input data. Hash functions should be collision resistant, i.e., it should be hard to find two different inputs generating the same hash value.

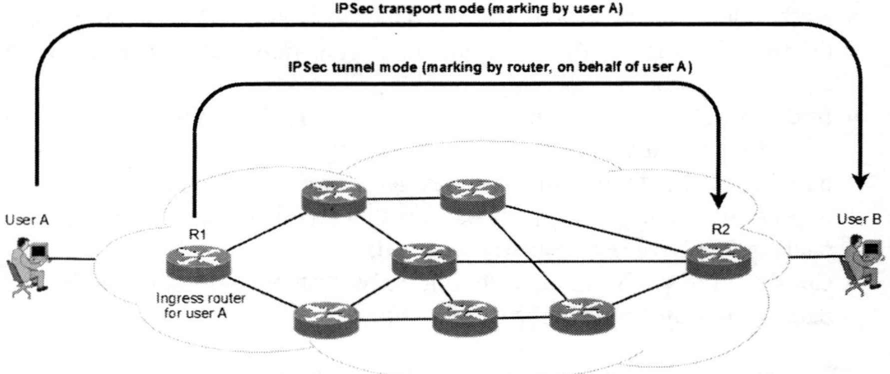

Figure 5.9 IPSec tunnel and transport modes.

In the second step, several encryption functions can be used, such as DES, Triple DES, International Data Encryption Algorithm (IDEA) [17], Triple IDEA, etc.

The receiving party calculates the hash value of the received packet. If the packet has not been tampered with, the hash value calculated at the reception should be the same as the hash value calculated by the transmitter. The hash value as calculated by the transmitter is not available in the packet, but its encrypted value is there (the ICV). Applying the decryption key (that has been agreed between the parties as being the one to be used for decoding messages arriving from that specific user) to the ICV should result in the same hash value as the one calculated at the transmitter. If the values are indeed equal, the decoding process is considered to be successful, and the packet is accepted. Packets for which the two values are different are considered as being corrupted, and are therefore rejected.

Confidentiality is obtained by encrypting the packet to be protected. The receiving party, and only it, should be able to decrypt the packet. Some identifiers should however remain unencrypted, allowing the receiving party to identify the packet and to decode it. Figure 5.10 illustrates this process.

Depending on the specific IPSec protocol used and the selected mode (transport or tunnel), IPSec marking appears in different locations in the protected packet. When authentication is used, the marking protects the whole packet. When confidentiality marking is used, the transport and tunnel mode provide different types of services. In transport mode the encryption process is executed by the end station, and routers in the network are expected to route

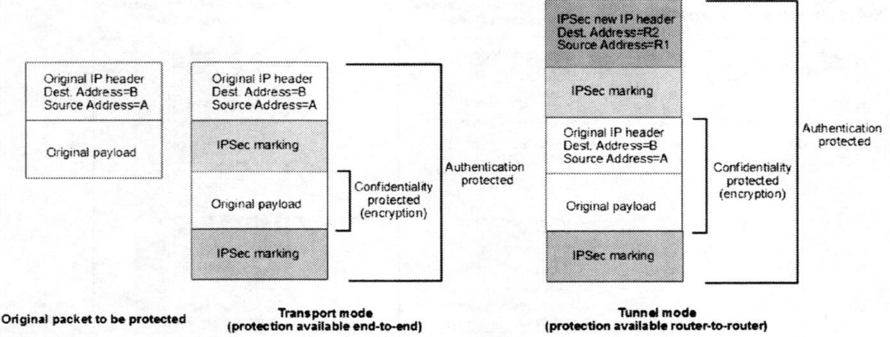

Figure 5.10 IPSec marking and protection.

the encrypted packet towards its destination. As a result, the encrypting station can not encrypt the IP header, as it includes the vital information needed by routers for correct operation. The only part that can be protected in this case is the payload itself (the upper layers). In tunnel mode the encryption process is executed by the ingress router, and the resulting packet is sent via the network to another router (the other end of the tunnel) as indicated in a new IP header, added by the ingress router. As a result, the ingress router can encrypt the whole original packet.

Obviously, the receiving party should be aware of the marking executed by the transmitting party. The two parties, have to enter into a logical relationship, known as a Security Association (SA), where the actual parameters regarding the algorithms and keys to be used are agreed. Security Associations are unidirectional. In a bi-directional link in which both directions have to be protected by IPsec, there will be two SAs: A to B and B to A. Security Associations are uniquely identified by fields present in the protected packet, allowing the receiving party to associate a received packet to a specific SA and (as a result) to activate the correct algorithms for its processing.

So, SA represent a policy contract between two peers or hosts and contains the following security parameters: authentication/encryption algorithm; key length and other encryption parameters; session keys for authentication and encryption, which can be entered manually or negotiated automatically; a specification of network traffic to which the SA will be applied; IPSec AH or ESP encapsulation protocol and tunnel or transport mode.

IPSec consists of three major parts:

- AH - Authentication Header protocol (defined in RFC 2402 [59]), which provides marking for authentication, integrity and replay protection;

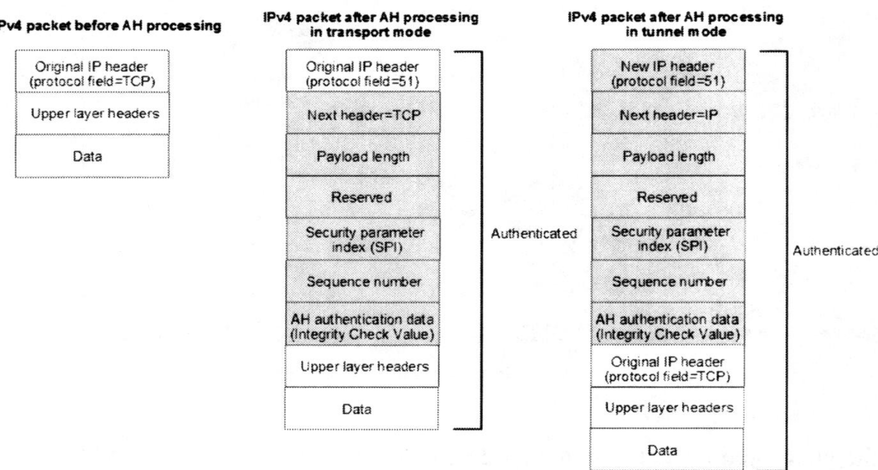

Figure 5.11 AH packet format.

- ESP - Encapsulating Security Payload (defined in RFC 2406 [60]), which provides marking for authentication, integrity and replay protection, as well as for confidentiality.
- IKE - Internet Key Exchange (defined in RFC 2409 [48]), which establishes and maintains Security Associations.

The format of the AH packet is illustrated in Figure 5.11. The Next Header field represents the type of the next payload and the SPI field is an arbitrary value that, together with the IP destination address of the packet, identifies the actual SA to which the packet belongs. Stations identify the SA to be used for the processing of an incoming packet based on the IP destination address, the SPI and the security service (as indicated by the "protocol" value in the field preceding the security header). The Sequence Number field avoids replay of the packet, while the Authentication Data field includes the ICV used to authenticate the packet.

⚠ AH code 51

AH packets are identified by code 51, which appears in the IP header preceding the AH header. The code appears in the (i) `protocol` field when the preceding header is IPv4; (ii) `next header` field when the preceding header is IPv6.

Figure 5.12 illustrates the format of the ESP packet.

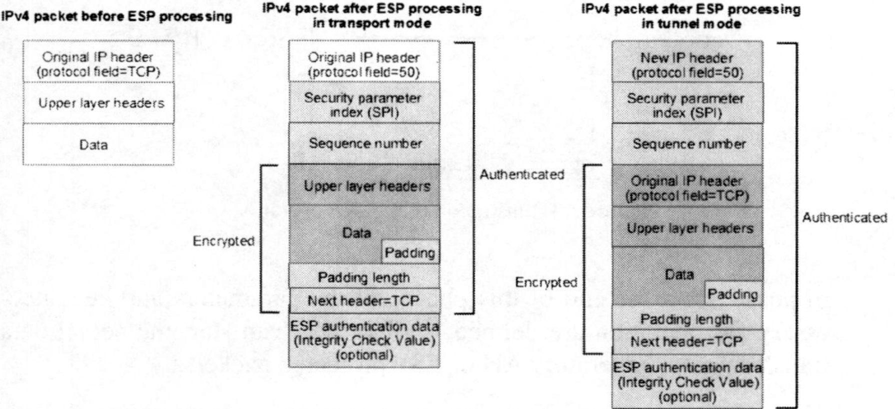

Figure 5.12 ESP packet format.

> **ESP code 50**
>
> ESP packets are identified by code 50, which appears in the IP header preceding the AH header. The code appears in the (i) `protocol` field when the preceding header is IPv4; (ii) `next header` field when the preceding header is IPv6.

AH and ESP use symmetric secret key algorithms, although public key algorithms are also feasible.

The purpose of the IKE protocol is to negotiate the protection parameters to be used by the partners. IKE is a protocol built as a combination of ISAKMP (Internet Security Association and Key Management Protocol), defined in RFC 2408 [69] for the negotiation of the SA parameters, and the OAKLEY protocol, defined in RFC 2412 [78] for the selection of the encryption keys. For an effective protection of the data, the parameters negotiation should be itself protected.

IKE defines therefore two different phases:

- Phase 1 - Creates a protected environment (an SA) between the partners in order to protect the negotiation of the authentication and encryption parameters that will be used in the data transfer phase. At the end of this phase, the two partners have been authenticated, and keys have been generated for the protection of the messages to be exchanged.
- Phase 2 - Partners negotiate the protection parameters to be used in the data transfer phase. The negotiation is protected by the elements defined

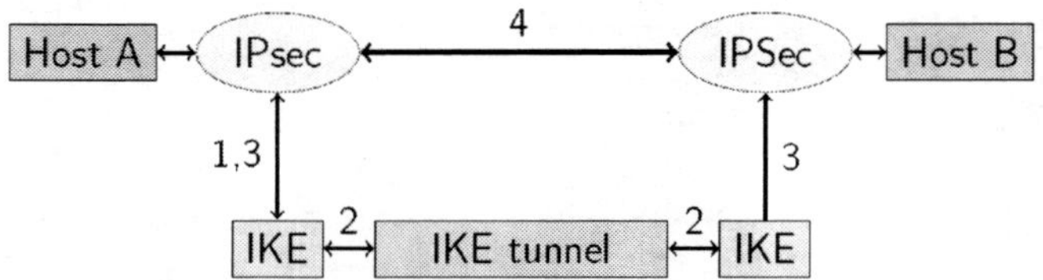

Figure 5.13 Illustration of the IKE operation.

in phase 1. At the end of this phase, all the parameters and keys necessary for operation are defined, and partners can start the actual data transfer phase, generating AH or ESP protected packets.

IKE modes control the efficiency versus security tradeoff during the initial IKE key exchange: the Main Mode requires six packets back and forth and provides complete security during the establishment of an IPsec connection; the Aggressive mode uses half the exchanges and provides less security because some information is transmitted in clear text.

The main advantages of IKE can be summarized as follows:

- Eliminates the need to manually specify IPSec security parameters at both peers;
- Allows administrators to specify a lifetime for the IPSec SA;
- Allows encryption keys to change during IPSec sessions;
- Allows IPSec to provide anti-replay services;
- Permits Certification Authority (CA) support for a manageable and scalable IPSec implementation;
- Allows dynamic authentication of peers.

Exercise - IPSec Tunnel Mode Operation

Let us reconfigure the IPv4 over IPv4 tunnel configured in Chapter 4, and depicted in Figure 3.62, in order to become an IPsec over IPv4 tunnel with an Authentication Header (AH).

The configuration that is needed at the tunnel endpoints (routers 1 and 5) are shown in Boxes 5.7 and 5.8. First, we establish an ISAKMP protection policy with priority 10 (the highest priority is 1 and the lowest priority 10,000). This is the beginning of IKE negotiation process for IPSec. Then, we specify that a pre-shared key must be used to apply the security policy and configure a pre-shared authentication key, used in global configuration mode. Then, we define the IPSec security associations by establishing the

```
# crypto isakmp policy 10
  # authentication pre-share
# crypto isakmp key 0 12345 address 10.10.10.5
# crypto ipsec transform-set Tauth ah-sha-hmac
# crypto ipsec profile Ptunnel
  # set transform-set Tauth

# interface Tunnel0
  # ip address 10.100.100.1 255.255.255.0
  # tunnel source FastEthernet1/0
  # tunnel destination 10.10.10.5
  # tunnel mode ipsec ipv4
  # tunnel protection ipsec profile Ptunnel
```

Box 5.7 IPsec Tunnel with AH configuration at Router 1.

```
# crypto isakmp policy 10
  # authentication pre-share
# crypto isakmp key 0 12345 address 10.10.10.1
# crypto ipsec transform-set Tauth ah-sha-hmac
# crypto ipsec profile Ptunnel
  # set transform-set Tauth

# interface Tunnel0
  # ip address 10.100.100.5 255.255.255.0
  # tunnel source FastEthernet1/0
  # tunnel destination 10.10.10.1
  # tunnel mode ipsec ipv4
  # tunnel protection ipsec profile Ptunnel
```

Box 5.8 IPsec Tunnel with AH configuration at Router 5.

transform sets, which include the tunnel type and the algorithm that is used. It is possible to define more than one transform set. Note that the transform set is an acceptable combination of security protocols and algorithms, which has to be matched on the peer router. The IPSec profile will define the preference order of the established transform sets. Finally, the IPSec profile is applied to the tunnel. Note that the tunnel mode was changed from IPv4 over IPv4 (command `tunnel mode ipip`) to IPSec over IPv4 (command `tunnel mode ipsec ipv4`).

The tunnel establishment and ISAKMP negotiation is depicted in Figure 5.14, showing the exchanged ISAKMP messages.

On Router 1 we executed the `ping 10.100.100.5` command and captured the exchanged packets at network 10.3.1.0/24: as can be seen from Figure 5.15, there are two IP headers, corresponding to the encapsulation of the IP datagram that includes the ICMP message into a new IP datagram (source and destination IP addresses are the ones corresponding to the loopback router interfaces) with an IPSec AH header.

No.	Time	Source	Destination	Protocol	Info
1	0.000000	10.10.10.5	10.10.10.1	ISAKMP	Identity Protection (Main Mode)
2	0.043447	10.10.10.1	10.10.10.5	ISAKMP	Identity Protection (Main Mode)
3	0.072234	10.10.10.5	10.10.10.1	ISAKMP	Identity Protection (Main Mode)
4	0.104173	10.10.10.1	10.10.10.5	ISAKMP	Identity Protection (Main Mode)
5	0.129056	10.10.10.5	10.10.10.1	ISAKMP	Identity Protection (Main Mode)
6	0.141612	10.10.10.1	10.10.10.5	ISAKMP	Identity Protection (Main Mode)
7	0.153972	10.10.10.5	10.10.10.1	ISAKMP	Quick Mode
8	0.162329	10.10.10.1	10.10.10.5	ISAKMP	Quick Mode
9	0.174706	10.10.10.5	10.10.10.1	ISAKMP	Quick Mode

Figure 5.14 IPsec Tunnel establishment (ISAKMP initial negotiation).

No.	Time	Source	Destination	Protocol	Info
1	0.000000	10.100.100.1	10.100.100.5	ICMP	Echo (ping) request
2	0.008244	10.100.100.5	10.100.100.1	ICMP	Echo (ping) reply
3	0.012449	10.100.100.1	10.100.100.5	ICMP	Echo (ping) request
4	0.022746	10.100.100.5	10.100.100.1	ICMP	Echo (ping) reply

```
▷ Frame 1: 158 bytes on wire (1264 bits), 158 bytes captured (1264 bits)
▷ Ethernet II, Src: c2:01:aa:aa:00:01 (c2:01:aa:aa:00:01), Dst: c2:04:aa:aa:00:00 (c2:04:aa:aa:00:00)
▷ Internet Protocol, Src: 10.10.10.1 (10.10.10.1), Dst: 10.10.10.5 (10.10.10.5)
▽ Authentication Header
    Next Header: IPIP (0x04)
    Length: 24
    AH SPI: 0x891353f6
    AH Sequence: 19
    AH ICV: 296f0a3a3d4d5d159207234c
▷ Internet Protocol, Src: 10.100.100.1 (10.100.100.1), Dst: 10.100.100.5 (10.100.100.5)
▷ Internet Control Message Protocol
```

Figure 5.15 ICMP packets transported over an IPsec AH tunnel.

```
# crypto ipsec transform-set Tcipher esp-des
# crypto ipsec profile Ptunnel
  # set transform-set Tcipher
```

Box 5.9 IPsec Tunnel with ESP (re)configuration at Router 1.

In a second experiment, we changed the tunnel to become an IPSec over IPv4 tunnel with an Encapsulating Security Payload (ESP) Header. The necessary configurations are shown in Box 5.9 for Router 1 and Box 5.10 for Router 5.

On Router 1, we executed again the `ping 10.100.100.5` command and captured the exchanged packets at network 10.3.1.0/24: as can be seen from Figure 5.16, there is an IP header whose source and destination IP addresses are the ones corresponding to the loopback router interfaces, an IPSec ESP header and the ciphered payload.

```
# crypto ipsec transform-set Tcipher esp-des
# crypto ipsec profile Ptunnel
  # set transform-set Tcipher
```

Box 5.10 IPsec Tunnel with ESP (re)configuration at Router 5.

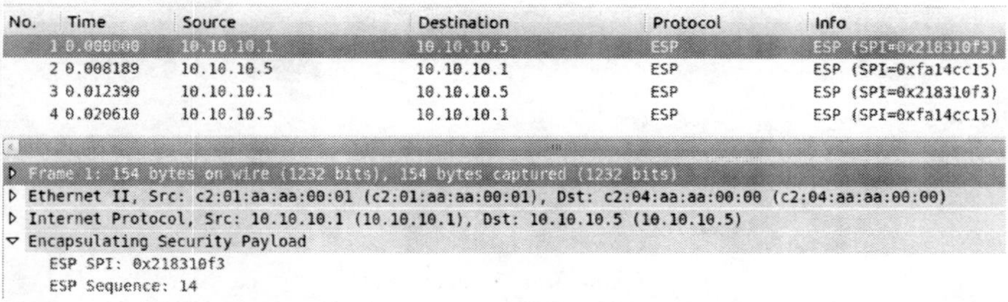

No.	Time	Source	Destination	Protocol	Info
1	0.000000	10.10.10.1	10.10.10.5	ESP	ESP (SPI=0x218310f3)
2	0.008189	10.10.10.5	10.10.10.1	ESP	ESP (SPI=0xfa14cc15)
3	0.012390	10.10.10.1	10.10.10.5	ESP	ESP (SPI=0x218310f3)
4	0.020610	10.10.10.5	10.10.10.1	ESP	ESP (SPI=0xfa14cc15)

```
▷ Frame 1: 154 bytes on wire (1232 bits), 154 bytes captured (1232 bits)
▷ Ethernet II, Src: c2:01:aa:aa:00:01 (c2:01:aa:aa:00:01), Dst: c2:04:aa:aa:00:00 (c2:04:aa:aa:00:00)
▷ Internet Protocol, Src: 10.10.10.1 (10.10.10.1), Dst: 10.10.10.5 (10.10.10.5)
▽ Encapsulating Security Payload
    ESP SPI: 0x218310f3
    ESP Sequence: 14
```

Figure 5.16 Ciphered ICMP packets transported over an IPsec ESP tunnel.

5.4.3 Virtual Private Networks

A Virtual Private Network (VPN) is a private network that uses a public network (usually the Internet) to connect remote sites or users together. The VPN uses "virtual" connections routed through the Internet from the business's private network to the remote site or employee. By using a VPN, businesses ensure security because anyone intercepting the encrypted data can't read it.

Companies have replaced leased lines with new technologies that use Internet connections without sacrificing performance and security. Businesses started by establishing intranets, which are private internal networks designed for use only by company employees. Intranets enabled distant colleagues to work together through technologies such as desktop sharing. By adding a VPN, a business can extend all its intranet's resources to employees working from remote offices or their homes.

There are two types of VPNs: remote-access and site-to-site. A remote-access VPN (Figure 5.17) allows individual users to establish secure connections with a remote computer network. Those users can access the secure resources on that network as if they were directly plugged in to the network's servers.

A site-to-site VPN (Figure 5.18) allows offices in multiple fixed locations to establish secure connections with each other over a public network such as the Internet. Site-to-site VPN extends the company's network, making computer resources from one location available to employees at other locations. There are two types of site-to-site VPNs:

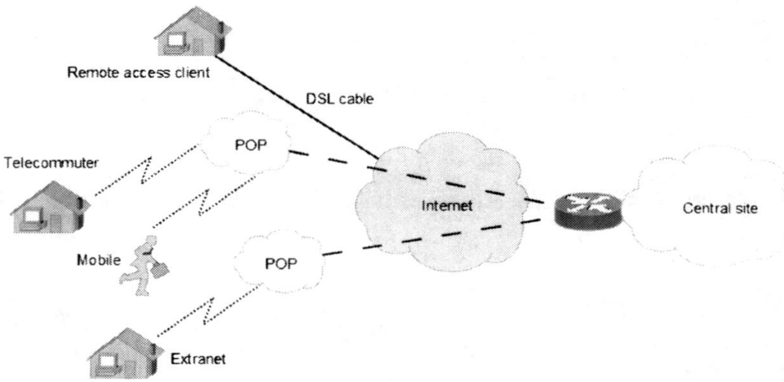

Figure 5.17 Remote VPN.

- Intranet-based - If a company has one or more remote locations that they wish to join in a single private network, they can create an intranet VPN to connect each separate LAN to a single WAN.
- Extranet-based - When a company has a close relationship with another company (such as a partner, supplier or customer), it can build an extranet VPN that connects those companies' LANs. This extranet VPN allows the companies to work together in a secure, shared network environment while preventing access to their separate intranets.

The most relevant remote-access VPNs are the following:

- PPTP - Based on PPTP, which packages data within PPP packets and encapsulates the PPP packets within IP packets; uses a form of General Routing Encapsulation (GRE) to get data to and from its final destination; supports authentication based on several protocols (PAP, EAP, CHAP, etc); creates a TCP control connection between the VPN client and VPN server to establish a tunnel; can support only one tunnel at a time for each user.
- L2TP/IPsec - Authentication is performed with Digital Certificates (RSA) or with the same PPP authentication mechanisms as PPTP; provides data integrity, authentication of origin and replay protection; encryption is provided by IPSec; can support multiple, simultaneous tunnels for each user; is slower than PPTP.
- SSL/TLS VPN - SSL/TLS protocol handles the VPN tunnel creation; RSA handshake (or DH) is used exactly as IKE in IPSec.
- SSH VPN - VPN over a SSH connection; uses SSH tunneling.

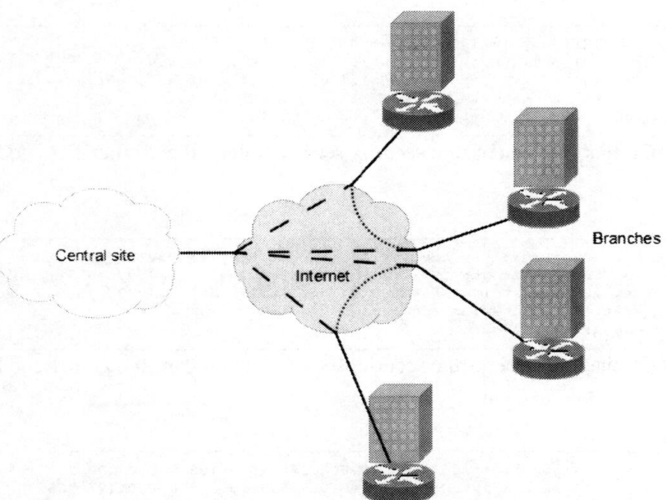

Figure 5.18 Site-to-site VPN.

- Open VPN - Implements a SSL/TLS VPN; allows PSK, certicate, and login/password based authentication; encryption is provided by OpenSSL; is compatible with dynamic and NAT addresses.

The most relevant site-to-site VPNs are:

- IPsec tunnels with static configuration - Requires the knowledge of all peers (IP addresses and security parameters) and has a high configuration overhead.
- IPsec tunnels with dynamic configuration (at the headend/hub) - Provide a hub and spoke configuration.
- IPsec + GRE tunnels - Generic Routing Encapsulation (GRE) allows the protection of multicast traffic over IPsec; Dynamic Multipoint VPN (DMVPN) provides full meshed connectivity with the simple configuration of hub and spoke.

5.5 Use Case

5.5.1 ACLs Configuration

Let us configure some ACLs that will implement a simple security policy in our use case scenario.

```
# access-list 101 deny ip 192.1.1.0 0.0.0.255 any
# access-list 101 permit tcp any 192.1.1.32 0.0.0.31 eq www
# access-list 101 permit tcp any 192.1.1.0 0.0.0.255 gt 1024 established
# interface FastEthernet0/1
  # ip access-group 101 in
```

Box 5.11 ACL that implements the security set 1 (configured at Router 101 and Router 102).

```
# access-list 102 permit tcp any 192.1.1.32 0.0.0.31 eq www
# access-list 102 permit tcp 192.1.1.0 0.0.0.255 192.1.1.32 0.0.0.31 eq 143
# access-list 102 permit tcp 10.0.0.0 0.255.255.255 192.1.1.32 0.0.0.31 eq 143
# interface FastEthernet2/1
  # ip access-group 102 out
```

Box 5.12 ACL that implements the security set 2 (configured at Router 101 and Router 102).

```
# access-list 103 permit ip 10.1.0.0 0.0.0.255 10.0.128.0 0.0.127.255
# access-list 103 permit ip 10.1.1.0 0.0.0.255 10.0.128.0 0.0.127.255
# interface FastEthernet1/15
  # ip access-group 103 out
```

Box 5.13 ACL that implements the security set 3 (configured at Router 1 and Router 2).

Security Rules Set 1

First of all, we want to configure an ACL that allows access, from the outside world, to the Web servers that are located in the DMZ area. Besides, TCP traffic from outside should only be allowed if it belongs to previously established (from the corporate network) sessions that use ports higher than 1024. Finally, all traffic from the outside with source IP address equal to any address of the corporate network should also be blocked.

The ACL that is illustrated in Box 5.11 is able to implement this set of rules and should be applied in both routers 101 and 102, specifically in the incoming direction of their F0/1 interfaces.

Security Rules Set 2

Now, we want to restrict DMZ access only to TCP traffic destined to ports 80 and 143 and coming from any internal source. The ACL illustrated in Box 5.12 is able to implement this rule and should be applied in both routers 101 and 102, specifically in the outgoing direction of their F2/1 interfaces.

Security Rules Set 3

Now, we want to restrict Datacenter access only to IP traffic coming from VLAN 11 terminals. The ACL illustrated in Box 5.13 is able to implement this rule and should be applied in both routers 11 and 2, specifically in the outgoing direction of their F1/15 interfaces.

Obviously, several other security rules sets could be designed and configured for this scenario. The sets that were presented are only illustrative examples.

6

Network Services

6.1 Introduction

This chapter covers some of the most common network services, describing their main functioning principles and presenting examples of their configuration and operation: Domain Name System (DNS), HyperText Transfer Protocol (HTTP), Trivial File Transfer Protocol (TFTP), File Transfer Protocol (FTP) and Email. Figure 6.1 shows how the different services are positioned in the protocol stack: HTTP, FTP and Email use the TCP transport protocol, while TFTP and DNS run over UDP. All these services follow a Client-Server architecture. A Server is a host that is running one or more server programs which share their resources with clients. A Client does not share any of its resources, but requests a Server's content or service function. Therefore, clients initiate communication sessions with servers, which await incoming requests.

For each service, we will present a brief description of its operation, together with an example of a typical configuration and experiment. All experiments will be conducted using the setup depicted in Figure 6.2, including a Linux PC Terminal and a Linux Server where all services will be deployed. Both PC and server are assumed to be running default installations of Ubuntu Linux 12.04LTS with kernel 3.2.0-29. Both PC and server have only one Ethernet card (eth0). All presented configuration/terminal commands assume that the user has previously acquired root (administration) rights with the terminal command:

```
# sudo su
```

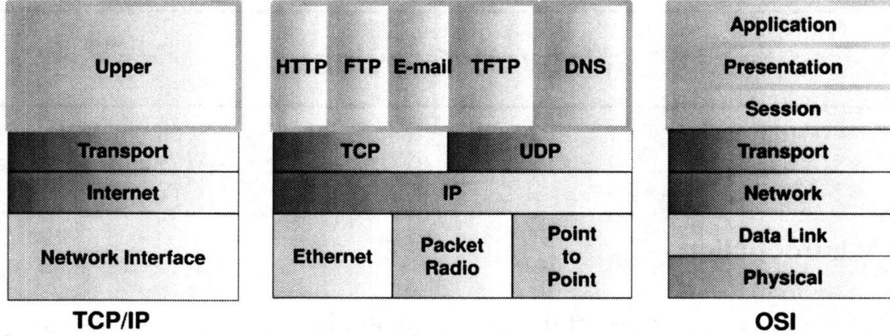

Figure 6.1 Positioning of the different network services in the protocol stack.

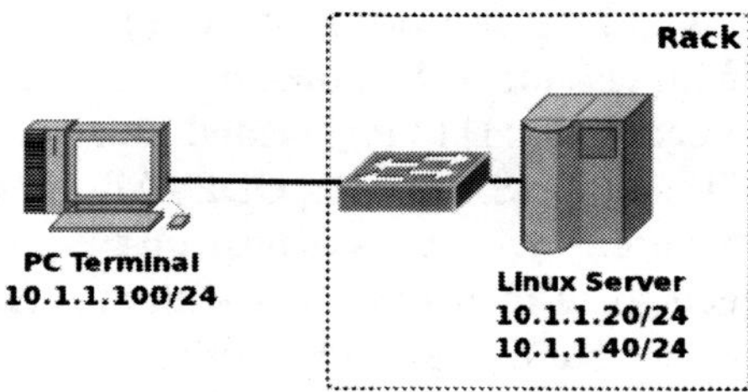

Figure 6.2 Setup used to test the various services.

6.2 DNS Service

The DNS service, dedicated to convert human readable names of hosts to IP addresses and vice versa, was already explained in Chapter 1. So, let us go immediately to the description of an experiment that illustrates its configuration and functioning principles.

6.2.1 DNS Experiment

To configure the Linux terminal with IPv4 address 10.1.1.100/24, the following command must be executed in the PC line terminal:

```
# ifconfig eth0 10.1.1.100 netmask 255.255.255.0
```

As shown in Figure 6.2, the Linux Server must have the IPv4 addresses 10.1.1.20/24 and 10.1.1.40/24. To configure these addresses the following commands must be executed in a Server line terminal:

```
# ifconfig eth0 10.1.1.20 netmask 255.255.255.0
# ifconfig eth0 add 10.1.1.40 netmask 255.255.255.0
```

In order to confirm the configuration of the interfaces, the following command can be used:

```
# ifconfig
```

The server should now have one interface (eth0, with IP address 10.1.1.20) and one sub-interface (eth0:0, with IP address 10.1.1.40).

The ping command can be used to test the connectivity between the terminal and both server addresses:

```
# ping 10.1.1.20
  PING 10.1.1.20 (10.1.1.20) 56(84) bytes of data.
  64 bytes from 10.1.1.20: icmp_req=1 ttl=64 time=0.449 ms
  64 bytes from 10.1.1.20: icmp_req=2 ttl=64 time=0.562 ms
  64 bytes from 10.1.1.20: icmp_req=3 ttl=64 time=0.598 ms
  ^C
  --- 10.1.1.20 ping statistics ---
  3 packets transmitted, 3 received, 0% packet loss
  rtt min/avg/max/mdev = 0.449/0.536/0.598/0.066 ms
# ping 10.1.1.40
  PING 10.1.1.40 (10.1.1.40) 56(84) bytes of data.
  64 bytes from 10.1.1.40: icmp_req=1 ttl=64 time=0.467 ms
  64 bytes from 10.1.1.40: icmp_req=2 ttl=64 time=0.504 ms
  64 bytes from 10.1.1.40: icmp_req=3 ttl=64 time=0.547 ms
  ^C
  --- 10.1.1.40 ping statistics ---
  3 packets transmitted, 3 received, 0% packet loss
  rtt min/avg/max/mdev = 0.467/0.506/0.547/0.032 ms
```

At the server, it is necessary to verify if the DNS (*bind9*) server is installed, by executing the command:

```
# aptitude show bind9 | grep State
```

If the *bind9* service is not installed, it can be installed executing the commands:

```
# aptitude update
# aptitude install bind9
```

Let us assume that a DNS server will serve as master server with authority over the domains *g1rs1.com* and *g1rs2.com*. The definitions of each domain name zone are shown in Box 6.1, which must be added to file */etc/bind/-named.conf.local*. Box 6.1 depict the zone file configurations for domain *g1rs1.com* [line 1-4] and domain *g1rs2.com* [line 5-8]. Configuration lines 1 and 5 define the domain name that the DNS zone will map, lines 2 and 6 define that this DNS server will be the master (all DNS data is obtained directly from configuration files) and lines 3 and 7 define the files that will contain the DNS records of the respective domain name.

The contents of the files */etc/bind/db.g1rs1.com* and */etc/bind/db.g1rs1.com* are depicted in Boxes 6.2 and 6.3, respectively. For both configuration files the following considerations hold: [line 1]

```
zone "g1rs1.com" {
   type master;
   file "/etc/bind/db.g1rs1.com";
};
zone "g1rs2.com" {
   type master;
   file "/etc/bind/db.g1rs2.com";
};
```

Box 6.1 Zone definitions for domains *g1rs1.com* and *g1rs2.com*.

```
1  $TTL   604800
2  $ORIGIN g1rs1.com.
3  @ IN   SOA ns1.g1rs1.com. adm.g1rs1.com. (
4      2   ; Serial
5      604800     ; Refresh
6      86400    ; Retry
7      2419200    ; Expire
8      604800) ; Negative Cache TTL
9    IN   NS   ns1.g1rs1.com.
10   IN   A 10.1.1.20
11 ns1 IN   A 10.1.1.20
```

Box 6.2 DNS server configuration for *g1rs1.com*

```
1  $TTL   604800
2  $ORIGIN g1rs2.com.
3  @ IN   SOA ns1.g1rs2.com. adm.g1rs2.com. (
4      2   ; Serial
5      604800     ; Refresh
6      86400    ; Retry
7      2419200    ; Expire
8      604800) ; Negative Cache TTL
9    IN   NS   ns1.g1rs2.com.
10   IN   A 10.1.1.40
11 ns1 IN   A 10.1.1.40
```

Box 6.3 DNS server configuration for *g1rs2.com*

defines the default TTL associated with the domain name resolution; [line 2] defines the domain name to which the DNS record will refer; [lines 3-8] define the SOA record with the default nameserver for the domain (*ns1.g1rs1.com* and *ns1.g1rs2.com*), the administrator email (*adm@g1rs1.com* and *adm@g1rs2.com*), the serial identifier (used for DNS server synchronization) and the remaining timing parameters; [line 9] defines the NS record (i.e. the name of a nameserver) for this domain, knowing that for each domain at least one NS record must be present; [lines 10-11] contain the A records that map IPv4 addresses with the domain name and associated nameserver name.

It is possible to verify the correctness of the zone definitions using the commands:

```
# named-checkzone g1rs1.com /etc/bind/db.g1rs1.com
zone g1rs1.com/IN: loaded serial 2
OK
# named-checkzone g1rs2.com /etc/bind/db.g1rs2.com
zone g1rs2.com/IN: loaded serial 2
OK
```

After all configurations have been completed, the DNS server must be restarted by executing the following command:

```
# service bind9 restart
```

If the bind9 service fails to restart, check the /var/log/syslog file for possible reasons.

The configurations of the DNS server can be tested by executing the following commands at the PC:

```
# dig @10.1.1.20 g1rs1.com
  (suppressed output)
  ;; QUESTION SECTION:
  ;g1rs1.com.        IN  A
  ;; ANSWER SECTION:
  g1rs1.com.     604800   IN   A 10.1.1.20
  ;; AUTHORITY SECTION:
  g1rs1.com.     604800   IN   NS   ns1.g1rs1.com.
  ;; ADDITIONAL SECTION:
  ns1.g1rs1.com.     604800   IN   A 10.1.1.20
  (suppressed output)
# dig @10.1.1.20 g1rs2.com
  (suppressed output)
  ;; QUESTION SECTION:
  ;g1rs2.com.        IN  A
  ;; ANSWER SECTION:
  g1rs2.com.     604800   IN   A 10.1.1.40
  ;; AUTHORITY SECTION:
  g1rs2.com.     604800   IN   NS   ns1.g1rs2.com.
  ;; ADDITIONAL SECTION:
  ns1.g1rs2.com.     604800   IN   A 10.1.1.40
  (suppressed output)
```

Figure 6.3 depicts the DNS packet exchanges resulting from issuing the *dig* commands. Packets 1 and 3 are the queries sent by the *dig* commands and packets 2 and 4 are the query responses sent by the DNS server. It is possible to observe that the query responses have three distinct answer sections: ANSWERS, AUTHORITATIVE NAMESERVERS and ADDITIONAL RECORDS. The ANSWERS section contains the direct response to the performed query, the AUTHORITATIVE NAMESERVERS section contains the NS records associated with the domain that is being queried (i.e. the domain name(s) of the DNS server(s) with authority over the domain)

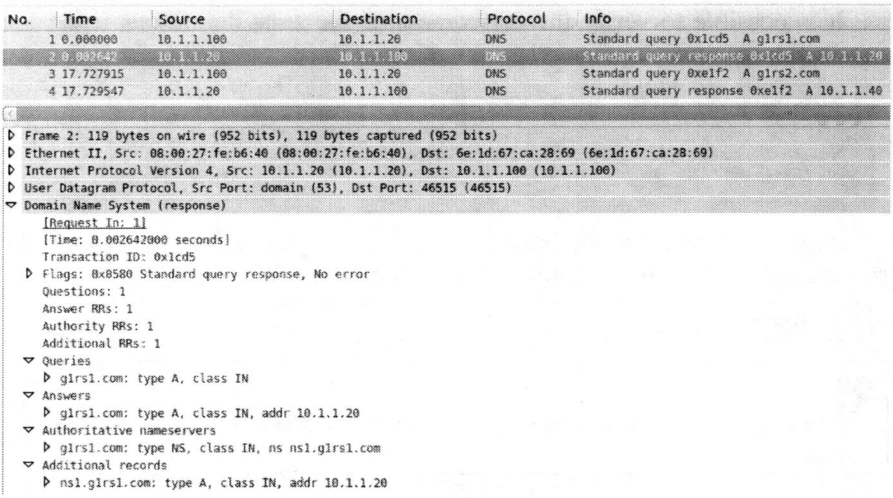

No.	Time	Source	Destination	Protocol	Info
1	0.000000	10.1.1.100	10.1.1.20	DNS	Standard query 0x1cd5 A glrs1.com
2	0.002642	10.1.1.20	10.1.1.100	DNS	Standard query response 0x1cd5 A 10.1.1.20
3	17.727915	10.1.1.100	10.1.1.20	DNS	Standard query 0xe1f2 A glrs2.com
4	17.729547	10.1.1.20	10.1.1.100	DNS	Standard query response 0xe1f2 A 10.1.1.40

```
▷ Frame 2: 119 bytes on wire (952 bits), 119 bytes captured (952 bits)
▷ Ethernet II, Src: 08:00:27:fe:b6:40 (08:00:27:fe:b6:40), Dst: 6e:1d:67:ca:28:69 (6e:1d:67:ca:28:69)
▷ Internet Protocol Version 4, Src: 10.1.1.20 (10.1.1.20), Dst: 10.1.1.100 (10.1.1.100)
▷ User Datagram Protocol, Src Port: domain (53), Dst Port: 46515 (46515)
▽ Domain Name System (response)
    [Request In: 1]
    [Time: 0.002642000 seconds]
    Transaction ID: 0x1cd5
  ▷ Flags: 0x8580 Standard query response, No error
    Questions: 1
    Answer RRs: 1
    Authority RRs: 1
    Additional RRs: 1
  ▽ Queries
    ▷ glrs1.com: type A, class IN
  ▽ Answers
    ▷ glrs1.com: type A, class IN, addr 10.1.1.20
  ▽ Authoritative nameservers
    ▷ glrs1.com: type NS, class IN, ns ns1.glrs1.com
  ▽ Additional records
    ▷ ns1.glrs1.com: type A, class IN, addr 10.1.1.20
```

Figure 6.3 DNS protocol exchanges.

and the ADDITIONAL RECORDS section contains the IP address(es) of the DNS server(s) with authority over the domain.

From now on, the Linux server will be the default DNS server for the PC. To configure the PC's default name server as the Linux server (10.1.1.20), define the contents of the file /etc/resolv.conf as:

```
# nameserver 10.1.1.20
```

6.3 HTTP Service

HTTP defines the interactions between "Web browsers" and "Web servers" and the formats of the corresponding messages. Each interaction consists of only two actions, as shown in Figure 6.4:

- The client sends a **Request** message identifying the file it wants to receive;
- The server sends a **Response** message with a negative or positive response (in this case, including the contents of the requested file).

A separate request is made for each file, even for Web pages with multiple files.

HTTP uses two types of TCP connections: non-persistent and persistent. In the first case, the client establishes a TCP connection to send the Request

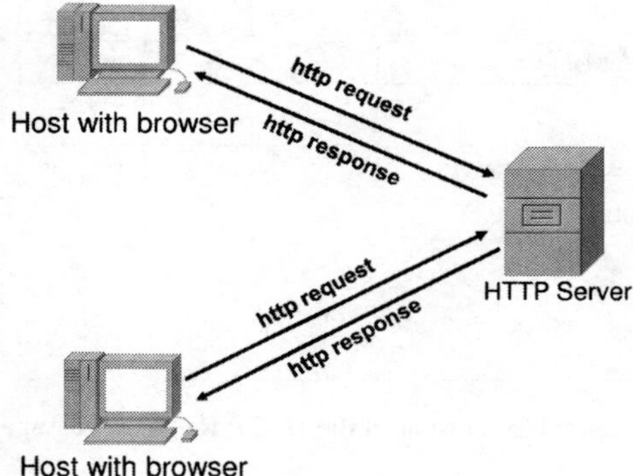

Figure 6.4 Interactions of the HTTP service.

and the server closes the TCP connection after sending the Response. The performance of this type of connection is penalized by the establishment time of each TCP connection and its slow start behavior. It is possible to use parallel TCP connections (whose number can be configured on the browsers) to decrease the server response time.

In the second case, the client establishes a TCP connection to send the Request including an explicit claim for the server to not close the TCP connection after sending the Response. The server waits for a timeout period (also configurable) in order to close unused connections. The client can use previously established connections with the pipelining mechanism (that is, it sends multiple requests without waiting for the arrival of the responses corresponding to previous requests) or without pipelining.

HTTP 1.0 version was defined in RFC 1945 [3] and only supports persistent TCP connections: it was developed for simple Web pages and servers with a limited processing capacity. Version 1.1 was defined in RFC 2616 [26] and works, by default, with persistent TCP connections and pipelining. It was developed for complex Web pages and servers with high processing capabilities. Both versions are compatible.

Each Web object is identified by a unique Uniform Resource Locator (URL), like for example:

```
http://www.someschool.edu:1024/somedir/page.html
```

The different parts of this URL have the following meaning: "http://" is the protocol used to communicate with the server (note that other methods can be supported, such as *file*, *ftp* or *mailto*); "www.someschool.edu" is the DNS

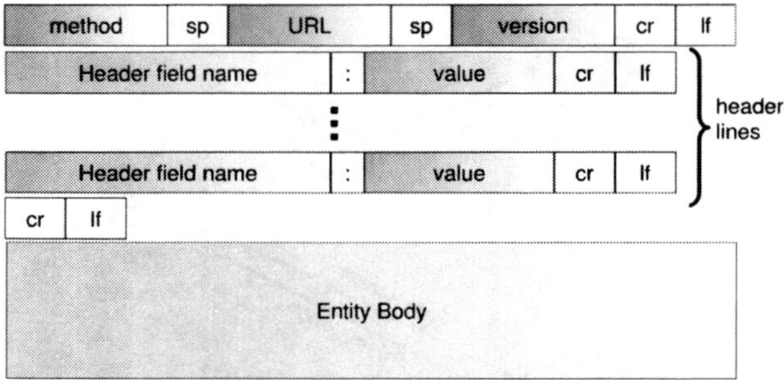

Figure 6.5 Format of the HTTP Request message.

```
GET /PageText.aspx?id=259 HTTP/1.1\r\n
    Host: www.ua.pt\r\n
    User-Agent: Mozilla/5.0 (X11; U; Linux i686; en-GB;
                rv:1.9.0.10) Gecko/2009042523 Ubuntu/9.04
                (jaunty) Firefox/3.0.10\r\n
    Accept: text/html,application/xhtml+xml,application/xml;
            \r\n
    Accept-Language: en-gb,en;q=0.5\r\n
    Accept-Encoding: gzip,deflate\r\n
    Accept-Charset: ISO-8859-1,utf-8;q=0.7,*;q=0.7\r\n
    Keep-Alive: 300\r\n
    Connection: keep-alive\r\n
    Referer: http://www.ua.pt/\r\n
```

Box 6.4 Example of a HTTP Request message

name corresponding to the IP address of the Web server station (it can obviously be replaced by the IP address); ":1024/" indicates the port number used on the Web server (this is optional, if not included the browser port used by default is port number 80); "somedir/" is the path to the intended document (optional, if it does not exist the document is assumed to be located in the main directory); "page.html" is the name of the intended file.

The format of the HTTP Request message is illustrated in Figure 6.5. Box 6.4 shows an example of a HTTP Request message. Messages are composed in ASCII format, start with a request line (GET, POST, HEAD, ...) and include a variable number of header lines. *Host* identifies the server address; *User-agent* specifies the browser type (in this case, Mozilla/5.0); *Connection* indicates if the server should close the connection (in this case, the *close* option should be used) or not (*keep-alive* option); *Referer* is the URL (or the more generic Uniform Resource Identifier (URI)) of the resource which links to the requested page. The different *Accept* lines specify which options are supported by the user agent that is making the Request.

```
HTTP/1.1 200 OK
Connection:close
Date: Thu, 06 Aug 1998 12:00:15 GMT
Server: Apache/1.3.0
Last-Modified: Mon, 22 Jun 1998 09:23:24 GMT
Content-Length: 6821
Content-Type: text/html
(carriage return, line feed)
(data, data, data, ...)
```

Box 6.5 Example of a HTTP Response message

```
GET /somedir/page.html HTTP/1.1
Host: www.someschool.edu
User-agent: Mozilla/4.0
If-modified-since: Mon, 22 Jun 1998 09:23:24 GMT
(carriage return, line feed)
```

Box 6.6 Conditional GET message

Box 6.5 shows an example of a HTTP Response message. Messages start with a response line, include a variable number of header lines and end with the contents of the requested file.

Several types of response lines are supported, like the following ones:

- 200 OK - The request was accepted and the contents of the file is included in the response;
- 301 Moved Permanently - The requested file was permanently moved and the answer includes a header line of the Location type with the new file location;
- 400 Bad Request - The request was not understood by the server;
- 404 Not Found - The file does not exist at the server;
- 505 HTTP Version Not Supported - The HTTP version of the request is not supported by the server.

If a browser supports Web caching, the conditional GET message allows to minimize the response times and the network traffic: if a specific file is stored in the terminal cache, the browser generates a request with a header line of the *If-modified-since* type. By doing this, the requested file will be downloaded only if the most recent version is newer than the version that is currently stored in the terminal Web cache. Box 6.6 illustrates a conditional GET message and Box 6.7 shows the corresponding answer when the requested file did not change (in this case, the response is of the type *304 Not Modified*).

```
HTTP/1.1 304 Not Modified
Date: Thu, 19 Aug 1998 12:00:15 GMT
Server: Apache/1.3.0
(carriage return, line feed)
```

Box 6.7 Conditional GET message

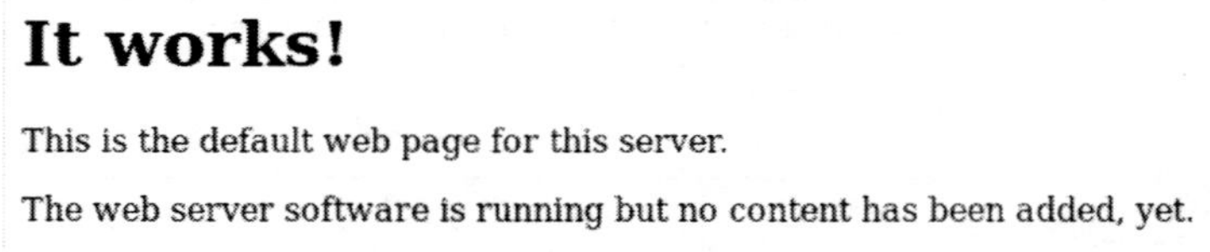

It works!

This is the default web page for this server.

The web server software is running but no content has been added, yet.

Figure 6.6 Apache2 default webpage.

6.3.1 HTTP Experiment

The Linux server must have a HTTP server installed. To verify if the HTTP (*apache2*) server is installed, the following commands can be executed in a Server line terminal:

```
# aptitude show apache2 | grep State
```

If the *apache2* service is not installed, it is necessary to install it by executing the following commands:

```
# aptitude update
# aptitude install apache2
```

To guarantee that *apache2* is running, the following command must be executed:

```
# service apache2 restart
```

Using a Web browser (at the PC) to access the following URL:

```
http://10.1.1.20
http://10.1.1.40
http://glar1.com
http://glar2.com
```

the default Apache2 web page, depicted in Figure 6.6, should appear.

Figures 6.7 and 6.8 depict the HTTP message exchange after requesting the URL *http://10.1.1.20*. Packets 1 to 3 are used to establish the TCP session between TCP port 48376 on the PC side and TCP port 80 on the server side, packet 4 (depicted in detail in Figure 6.7) is the HTTP GET message, packet 5 is the ACK sent by the server to confirm the reception of packet 4, packet 6 is the HTTP OK 200 response (depicted in detail in Figure 6.8) with the

```
No.  Time      Source      Destination  Protocol  Info
1 0.000000  10.1.1.100  10.1.1.20   TCP   48376 > http [SYN] Seq=0 Win=14600 Len=0 MSS=1460 SACK_PERM=1 TSval=3947425 TSecr=0 WS=64
2 0.000631  10.1.1.20   10.1.1.100  TCP   http > 48376 [SYN, ACK] Seq=0 Ack=1 Win=14480 Len=0 MSS=1460 SACK_PERM=1 TSval=5856783 TSecr=3947425 WS=8
3 0.000699  10.1.1.100  10.1.1.20   TCP   48376 > http [ACK] Seq=1 Ack=1 Win=14656 Len=0 TSval=3947425 TSecr=5856783
4 0.000876  10.1.1.100  10.1.1.20   HTTP  GET / HTTP/1.1
5 0.001295  10.1.1.20   10.1.1.100  TCP   http > 48376 [ACK] Seq=1 Ack=331 Win=15552 Len=0 TSval=5856783 TSecr=3947425
6 0.007458  10.1.1.20   10.1.1.100  HTTP  HTTP/1.1 200 OK  (text/html)
7 0.007511  10.1.1.100  10.1.1.20   TCP   48376 > http [ACK] Seq=331 Ack=484 Win=15680 Len=0 TSval=3947427 TSecr=5856784

▷ Frame 4: 396 bytes on wire (3168 bits), 396 bytes captured (3168 bits)
▷ Ethernet II, Src: 6e:1d:67:ca:28:69 (6e:1d:67:ca:28:69), Dst: CadmusCo_fe:b6:40 (08:00:27:fe:b6:40)
▷ Internet Protocol Version 4, Src: 10.1.1.100 (10.1.1.100), Dst: 10.1.1.20 (10.1.1.20)
▷ Transmission Control Protocol, Src Port: 48376 (48376), Dst Port: http (80), Seq: 1, Ack: 1, Len: 330
▽ Hypertext Transfer Protocol
  ▷ GET / HTTP/1.1\r\n
    Host: 10.1.1.20\r\n
    User-Agent: Mozilla/5.0 (X11; Ubuntu; Linux i686; rv:15.0) Gecko/20100101 Firefox/15.0.1\r\n
    Accept: text/html,application/xhtml+xml,application/xml;q=0.9,*/*;q=0.8\r\n
    Accept-Language: en-us,en;q=0.5\r\n
    Accept-Encoding: gzip, deflate\r\n
    Connection: keep-alive\r\n
    Pragma: no-cache\r\n
    Cache-Control: no-cache\r\n
    \r\n
    [Full request URI: http://10.1.1.20/]
```

Figure 6.7 HTTP message exchange and GET message.

```
No.  Time      Source      Destination  Protocol  Info
1 0.000000  10.1.1.100  10.1.1.20   TCP   48376 > http [SYN] Seq=0 Win=14600 Len=0 MSS=1460 SACK_PERM=1 TSval=3947425 TSecr=0 WS=64
2 0.000631  10.1.1.20   10.1.1.100  TCP   http > 48376 [SYN, ACK] Seq=0 Ack=1 Win=14480 Len=0 MSS=1460 SACK_PERM=1 TSval=5856783 TSecr=3947425 WS=8
3 0.000699  10.1.1.100  10.1.1.20   TCP   48376 > http [ACK] Seq=1 Ack=1 Win=14656 Len=0 TSval=3947425 TSecr=5856783
4 0.000876  10.1.1.100  10.1.1.20   HTTP  GET / HTTP/1.1
5 0.001295  10.1.1.20   10.1.1.100  TCP   http > 48376 [ACK] Seq=1 Ack=331 Win=15552 Len=0 TSval=5856783 TSecr=3947425
6 0.007458  10.1.1.20   10.1.1.100  HTTP  HTTP/1.1 200 OK  (text/html)
7 0.007511  10.1.1.100  10.1.1.20   TCP   48376 > http [ACK] Seq=331 Ack=484 Win=15680 Len=0 TSval=3947427 TSecr=5856784

▷ Frame 6: 549 bytes on wire (4392 bits), 549 bytes captured (4392 bits)
▷ Ethernet II, Src: CadmusCo_fe:b6:40 (08:00:27:fe:b6:40), Dst: 6e:1d:67:ca:28:69 (6e:1d:67:ca:28:69)
▷ Internet Protocol Version 4, Src: 10.1.1.20 (10.1.1.20), Dst: 10.1.1.100 (10.1.1.100)
▷ Transmission Control Protocol, Src Port: http (80), Dst Port: 48376 (48376), Seq: 1, Ack: 331, Len: 483
▽ Hypertext Transfer Protocol
  ▷ HTTP/1.1 200 OK\r\n
    Date: Wed, 07 Nov 2012 22:17:51 GMT\r\n
    Server: Apache/2.2.22 (Ubuntu)\r\n
    Last-Modified: Wed, 07 Nov 2012 15:57:10 GMT\r\n
    ETag: "419d4-b1-4cde9c571254a"\r\n
    Accept-Ranges: bytes\r\n
    Vary: Accept-Encoding\r\n
    Content-Encoding: gzip\r\n
  ▷ Content-Length: 146\r\n
    Keep-Alive: timeout=5, max=100\r\n
    Connection: Keep-Alive\r\n
    Content-Type: text/html\r\n
    \r\n
    Content-encoded entity body (gzip): 146 bytes -> 177 bytes
▽ Line-based text data: text/html
    <html><body><h1>It works!</h1>\n
    <p>This is the default web page for this server.</p>\n
    <p>The web server software is running but no content has been added, yet.</p>\n
    </body></html>\n
```

Figure 6.8 HTTP message exchange and response OK message.

HTML content of the Apache2 default page, and finally, packet 7 is the ACK sent by the client to confirm the reception of packet 6.

In order to have different webpages for different domain names, it is necessary to define VirtualHost containers, that maps a specific domain name request at a specific IP address and/or TCP port with the location of the webpage contents (local server directory). At the Server, in order to create a specific page for domain *g1rs1.com* three steps must be executed: (1) create a new directory to hold the unique contents; (2) configure the VirtualHost container and (3) activate the new VirtualHost container. The new directory may be created in the the root directory of the HTTP server (*/var/www/*). Therefore, the directory */var/www/g1rs1.com-80* will accommodate the contents of the web page associated to domain *g1rs1.com*. Inside this directory, it is necessary to create the default file *index.html* (the default page of each domain) with the following (simplistic) HTML content:

```
<html>
<body>
<h1>g1rs1.com</h1>
<h2>Port 80</h2>
</body>
</html>
```

In order to create a virtual host associated to domain *g1rs1.com*, file *g1rs1.com-80* must be created inside directory */etc/apache/sites-available/* with the following contents:

```
1 <VirtualHost *:80>
2   DocumentRoot /var/www/g1rs1.com-80
3   ServerName g1rs1.com
4 </VirtualHost>
```

A VirtualHost container includes: (line 1) the IP address and TCP port where requests will be processed (* represents all available IP addresses/interfaces); (line2) defines the directory that will hold the contents of the web page and (line 3) contains the ServerName directive that defines the domain name that will be associated with this VirtualHost (this name will be matched against the *Host* field of the incoming GET requests).

In order to activate the new VirtualHost domain and re-initiate the HTTP server the following commands must be executed:

```
# a2ensite g1rs1.com-80
# service apache2 restart
```

By accessing the following URL from a browser at the PC:

```
http://10.1.1.21
http://10.1.1.41
http://g1rs1.com
http://g1rs2.com
http://www.g1rs1.com
http://www.g1rs2.com
```

it is possible to observe that only the *http://g1rs1.com* request shows the new webpage; the requests to *www.g1rs1.com* and *www.g1rs2.com* are not completed because the domain names cannot be resolved (they are not configured at the DNS server), and the remaining requests show the default Apache webpage. In Figure 6.9, it is possible to observe the HTTP messages that were exchanged after requesting URL *http://g1rs1.com*. Moreover, packet 5 is a GET message with *Host: g1rs1.com*. The HTTP server, upon the reception

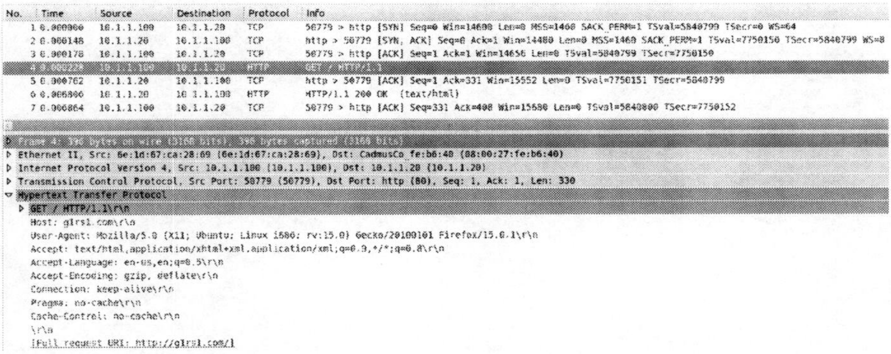

```
No.  Time      Source      Destination  Protocol  Info
1 0.000000  10.1.1.100  10.1.1.20    TCP   50779 > http [SYN] Seq=0 Win=14600 Len=0 MSS=1460 SACK_PERM=1 TSval=5840799 TSecr=0 WS=64
2 0.000148  10.1.1.20   10.1.1.100   TCP   http > 50779 [SYN, ACK] Seq=0 Ack=1 Win=14480 Len=0 MSS=1460 SACK_PERM=1 TSval=7750150 TSecr=5840799 WS=8
3 0.000178  10.1.1.100  10.1.1.20    TCP   50779 > http [ACK] Seq=1 Ack=1 Win=14656 Len=0 TSval=5840799 TSecr=7750150
4 0.000220  10.1.1.100  10.1.1.20    HTTP  GET / HTTP/1.1
5 0.000762  10.1.1.20   10.1.1.100   TCP   http > 50779 [ACK] Seq=1 Ack=331 Win=15552 Len=0 TSval=7750151 TSecr=5840799
6 0.006806  10.1.1.20   10.1.1.100   HTTP  HTTP/1.1 200 OK  (text/html)
7 0.006864  10.1.1.100  10.1.1.20    TCP   50779 > http [ACK] Seq=331 Ack=408 Win=15680 Len=0 TSval=5840800 TSecr=7750152

▷ Frame 4: 396 bytes on wire (3168 bits), 396 bytes captured (3168 bits)
▷ Ethernet II, Src: 6e:1d:67:ca:28:69 (6e:1d:67:ca:28:69), Dst: CadmusCo_fe:b6:40 (08:00:27:fe:b6:40)
▷ Internet Protocol Version 4, Src: 10.1.1.100 (10.1.1.100), Dst: 10.1.1.20 (10.1.1.20)
▷ Transmission Control Protocol, Src Port: 50779 (50779), Dst Port: http (80), Seq: 1, Ack: 1, Len: 330
▽ Hypertext Transfer Protocol
  ▷ GET / HTTP/1.1\r\n
    Host: g1rs1.com\r\n
    User-Agent: Mozilla/5.0 (X11; Ubuntu; Linux i686; rv:15.0) Gecko/20100101 Firefox/15.0.1\r\n
    Accept: text/html,application/xhtml+xml,application/xml;q=0.9,*/*;q=0.8\r\n
    Accept-Language: en-us,en;q=0.5\r\n
    Accept-Encoding: gzip, deflate\r\n
    Connection: keep-alive\r\n
    Pragma: no-cache\r\n
    Cache-Control: no-cache\r\n
    \r\n
    [Full request URI: http://g1rs1.com/]
```

Figure 6.9 HTTP message exchange and GET message with *Host: g1rs1.com*.

of the GET message, will respond with the contents of the VirtualHost that matches the Host field of the request. Note that if the Host field do not match any VirtualHost ServerName directive, the HTTP server name will respond with the contents of the default webpage.

At the DNS server, define for both domains a record that associates the name *www.g1rs1.com* with IP address 10.1.1.20 and the name *www.g1rs2.com* with IP address 10.1.1.40. The following A record should be added to file "/etc/bind/db.g1rs1.com":

```
www IN  A 10.1.1.20
```

and, the following A record should be added to file "/etc/bind/db.g1rs2.com":

```
www IN  A 10.1.1.20
```

After the changes, the DNS server must be restarted. The new configurations of the DNS server can be tested by executing the following commands:

```
# dig www.g1rs1.com
  (suppressed output)
  ;; ANSWER SECTION:
  www.g1rs2.com.     604800   IN   A 10.1.1.40
  (suppressed output)
# dig www.g1rs2.com
  (suppressed output)
  ;; ANSWER SECTION:
  www.g1rs2.com.     604800   IN   A 10.1.1.40
  (suppressed output)
```

In the definitions of the VirtualHost, located in file */etc/apache/sites-available/g1rs1.com-80*, add the directive:

```
ServerAlias www.g1rs1.com
```

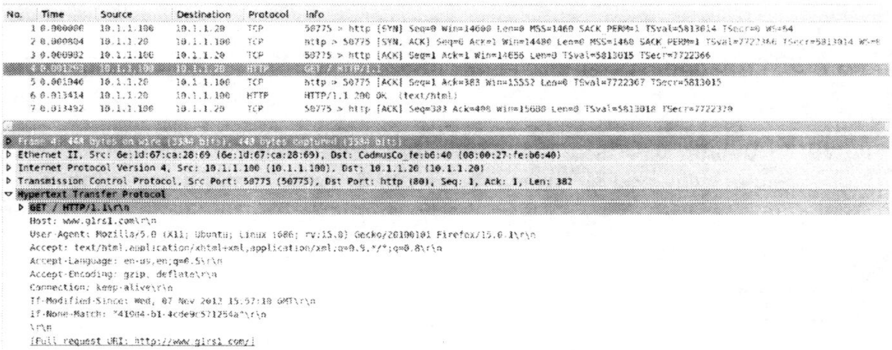

```
No.  Time      Source      Destination  Protocol  Info
  1 0.000000   10.1.1.100  10.1.1.20    TCP       50775 > http [SYN] Seq=0 Win=14600 Len=0 MSS=1460 SACK_PERM=1 TSval=5813014 TSecr=0 WS=64
  2 0.000804   10.1.1.20   10.1.1.100   TCP       http > 50775 [SYN, ACK] Seq=0 Ack=1 Win=14480 Len=0 MSS=1460 SACK_PERM=1 TSval=7722366 TSecr=5813014 WS=8
  3 0.000982   10.1.1.100  10.1.1.20    TCP       50775 > http [ACK] Seq=1 Ack=1 Win=14656 Len=0 TSval=5813015 TSecr=7722366
  4 0.001[illegible] 10.1.1.100  10.1.1.20    HTTP      GET / HTTP/1.1
  5 0.001946   10.1.1.20   10.1.1.100   TCP       http > 50775 [ACK] Seq=1 Ack=383 Win=15552 Len=0 TSval=7722367 TSecr=5813015
  6 0.013414   10.1.1.20   10.1.1.100   HTTP      HTTP/1.1 200 OK  (text/html)
  7 0.013492   10.1.1.100  10.1.1.20    TCP       50775 > http [ACK] Seq=383 Ack=408 Win=15680 Len=0 TSval=5813018 TSecr=7722370

▷ Frame 4: [illegible] bytes on wire ([illegible] bits), [illegible] bytes captured ([illegible] bits)
▷ Ethernet II, Src: 6e:1d:67:ca:28:69 (6e:1d:67:ca:28:69), Dst: CadmusCo_fe:b6:40 (08:00:27:fe:b6:40)
▷ Internet Protocol Version 4, Src: 10.1.1.100 (10.1.1.100), Dst: 10.1.1.20 (10.1.1.20)
▷ Transmission Control Protocol, Src Port: 50775 (50775), Dst Port: http (80), Seq: 1, Ack: 1, Len: 382
▽ Hypertext Transfer Protocol
  ▷ GET / HTTP/1.1\r\n
    Host: www.g1rs1.com\r\n
    User-Agent: Mozilla/5.0 (X11; Ubuntu; Linux i686; rv:15.0) Gecko/20100101 Firefox/15.0.1\r\n
    Accept: text/html,application/xhtml+xml,application/xml;q=0.9,*/*;q=0.8\r\n
    Accept-Language: en-us,en;q=0.5\r\n
    Accept-Encoding: gzip, deflate\r\n
    Connection: keep-alive\r\n
    If-Modified-Since: Wed, 07 Nov 2012 15:57:10 GMT\r\n
    If-None-Match: "41904-b1-4cde9c571254a"\r\n
    \r\n
    [Full request URI: http://www.g1rs1.com/]
```

Figure 6.10 HTTP message exchange and GET message with *Host: www.g1rs1.com*.

This directive defines additional Host values on the HTTP GET request that will redirect the server to this specific VirtualHost.

Re-initiate the HTTP server by typing:

```
# service apache2 restart
```

After accessing the following URLs:

```
http://g1rs1.com
http://www.g1rs1.com
```

it is possible to observe that both requests now show the specific page for *g1rs1.com* that was defined before. Figure 6.10 shows the HTTP messages that were exchanged after requesting URL "http://www.g1rs1.com". Note that packet 5 is a GET message with *Host: www.g1rs1.com* and, therefore, the returned webpage will be the one defined in a VirtualHost that contains a ServerName or ServerAlias directives with value *www.g1rs1.com*.

To define four distinct websites, *g1rs1.com* and *g1rs2.com* at port 80 and *g1rs1.com* and *g1rs2.com* at port 8080, it is necessary to create the directories */var/www/g1rs1.com-8080*, */var/www/g1rs2.com-80* and */var/www/g1rs2-8080* and define distinct default pages.

To force the HTTP server to also listen on port 8080, it is necessary to add the following directive to file */etc/apache2/ports.conf*

```
Listen 8080
```

and (re)define the virtual hosts for the different domains so that each site should be available only in the corresponding IP address and port number (see Box 6.8).

```
    <VirtualHost 10.1.1.21:80>
        DocumentRoot /var/www/g1rs1.com-80
        ServerName g1rs1.com
        ServerAlias www.g1rs1.com
    </VirtualHost>
###
    <VirtualHost 10.1.1.21:8080>
        DocumentRoot /var/www/g1rs1.com-8080
        ServerName g1rs1.com
        ServerAlias www.g1rs1.com
    </VirtualHost>
###
    <VirtualHost 10.1.1.41:80>
        DocumentRoot /var/www/g1rs2.com-80
        ServerName g1rs2.com
        ServerAlias www.g1rs2.com
    </VirtualHost>
###
    <VirtualHost 10.1.1.41:8080>
        DocumentRoot /var/www/g1rs2.com-8080
        ServerName g1rs2.com
        ServerAlias www.g1rs2.com
    </VirtualHost>
```

Box 6.8 Virtual hosts definition for the different domains

To activate all VirtualHost and restart the HTTP server, execute the following commands:

```
# a2ensite gXrs1.com-80
# a2ensite gXrs1.com-8080
# a2ensite gXrs2.com-80
# a2ensite gXrs2.com-8080
# service apache2 restart
```

6.4 TFTP Service

Trivial File Transfer Protocol (TFTP) is a simple file transfer service that was defined in RFC 1350 [92]. It runs over UDP and does not support any authentication mechanism. The initial packet is sent from the client to port 69 on the server side; the server then chooses a new port number for this service, freeing port 69 to attend a new request from another client.

Since TFTP runs over UDP, there is no flow control at the transport level: any flow control has to be implemented at the application level. TFTP implements a Stop and Wait flow control mechanism, based on five primitives: Read Request (RRQ), Write Request (WRQ), Data, Acknowledgment (ACK) and Error (ERR). Figure 6.11 shows the format of the different messages. The most important fields have the following meaning:

- FILENAME - ASCII character string that specifies the name of the file to read or write.

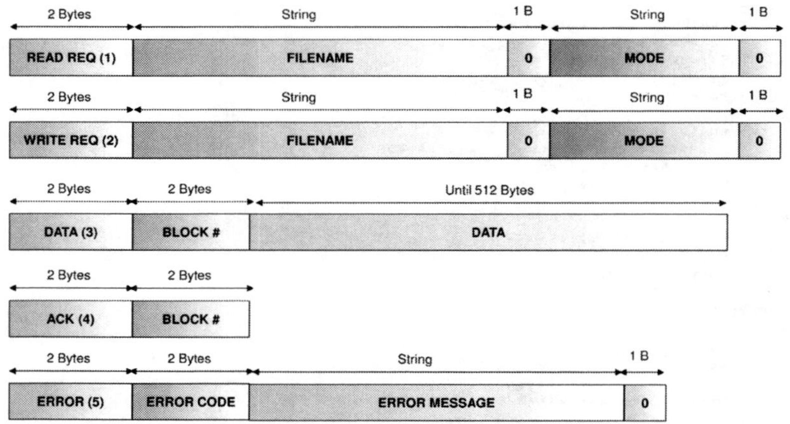

Figure 6.11 Format of the TFTP messages.

- MODE - ASCII character string that specifies the message mode.
- BLOCK # - In the Acknowledgment message this number is equal to the block number of the received message. The server uses the Acknowledgment to confirm the reception of the data blocks and the client uses the data blocks to confirm the reception of the Acknowledgments, except in the case of duplicate Acknowledgments and when an Acknowledgment closes a connection.
- ERROR - Acts as a Not-Acknowledgment (NACK); it can cause the retransmission of the message or the break of the connection.
- ERROR MESSAGE - ASCII string that specifies the type of error.
- ERROR CODE - Several codes are supported, like 00 - Not defined; 01 - File not found; 02 - Access violation; 03 - Disk full; 04 - Invalid operation code; 05 - Unknown port number; 06 - File already exists; 07 - No such user.

Figure 6.12 shows an example of a Write Request session. First, the client sends a WRQ message in order to ask for permission to send the specified file to the server; the server answers with an ACK confirming it is ready to receive the file; then, the client starts sending the file using DATA primitives, splitting data into blocks whose maximum size was previously configured (512 bytes, in this case); for each DATA message, the server answers with an ACK, confirming the reception of the corresponding block. Note that the client only sends a DATA message after receiving the ACK corresponding to the previous DATA message.

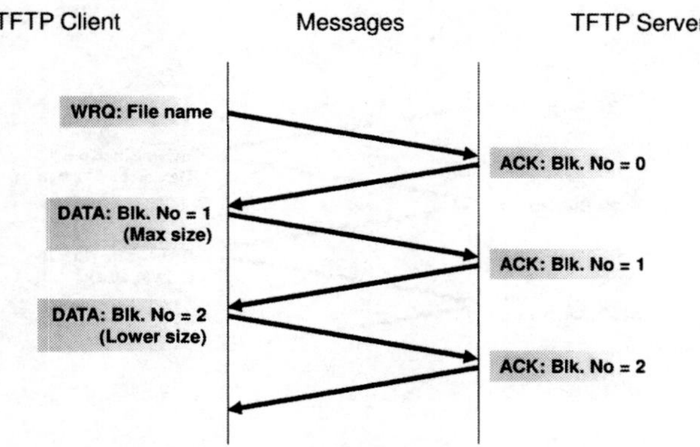

Figure 6.12 Write Request session.

If any DATA message does not arrive to its destination (or the corresponding ACK), the client has a TimeOut mechanism (also configurable in the application): if it does not receive the ACK until the maximum waiting time expires, it will resend the same data block.

The server detects the last DATA message when the data block size is smaller than the maximum size (there is not enough data to fulfill the entire DATA message).

Figure 6.13 shows an example of a Read Request session. The client requests for a specific file by sending a RRQ message, including the file name; then, the server starts transferring the file contents by sending a first data block of 512 bytes (the maximum block size that was previously configured) inside a DATA message; after receiving the Acknowledgment confirming the correct reception of the first block, the server sends another data block in a DATA message. Since the size of this data block is lower than the maximum block size, the client knows it is the last DATA primitive sent by the server.

But what happens when the data block size of the last DATA message is equal to maximum block size? In this case, the server has to send another DATA primitive with no data (data block size of 0) in order to notify the client that the transfer is complete.

6.4.1 TFTP Experiment

The Linux server must have a TFTP server installed. To verify if the TFTP (*atftpd*) server is installed, the following command can be executed in a Server line terminal:

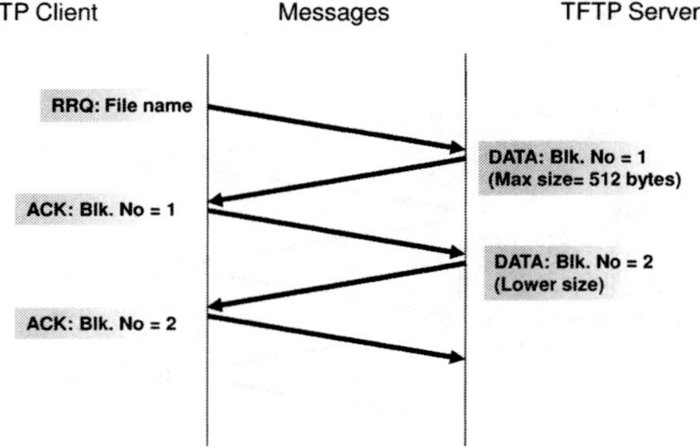

Figure 6.13 Read Request session.

```
# aptitude show atftd | grep State
```

If the *atftpd* server is not installed, it can be installed by executing the following commands:

```
# aptitude update
# aptitude install atftpd
```

In order to start the service (on UDP port 69), the following command should be executed:

```
# atftpd --deamon --port 69
```

The root directory of the TFTP server will be located at */srv/tftp*.

In the Linux PC terminal, in order to test the TFTP service we will require two files with random contents: one with 1500 bytes (named *file1500*) and the other with 1024 bytes (named *file1024*). To create them, the following *dd* commands can be used:

```
# dd if=/dev/urandom of=file1500 bs=1 count=1500
# dd if=/dev/urandom of=file1024 bs=1 count=1024
```

To test TFTP the following commands must be executed:

```
1 # tftp 10.1.1.20
2 tftp> binary
3 tftp> put file1500
4 tftp> put file1024
5 tftp> get file1500
6 tftp> get file1024
7 tftp> quit
```

where (command 1) activates the TFTP client defining the IPv4 address of the server; (command 2) defines the transference mode, which in this case is binary; (command 3) performs the TFTP upload of *file1500* to the server; (command 4) performs the TFTP upload of *file1024* to the server; (command 5) performs the TFTP download of *file1500* from the server; (command 6) performs the TFTP download of *file1024* from the server, and finally, (command 7) terminates the TFTP client.

The sequence of TFTP packets that were exchanged between client and server is depicted in Figures 6.14 to 6.17. The TFTP uploads in Figures 6.14 and 6.15 show that the first packet exchanged is the Write Request from the client to the server notifying the name of the file being uploaded and the transfer mode, followed by an Acknowledgment packet with block number 0 sent by the server to notify the reception of the Write Request; the following packets are Data packets followed by the respective Acknowledgment packets. All the Data packets have 512 bytes of data (maximum block data size), with the exception of the third packet (packet number 7 on both exchanges), which has 476 bytes in the upload of the 1500 bytes file and zero bytes in the upload of the 1024 bytes file. The last packet with zero bytes of data is justified by the protocol standard, which states that the last packet must have a size that is lower than the maximum block size to notify the end of the transmission. If no more data is available, a zero bytes packet is sent (see Figure 6.15).

The TFTP downloads in Figures 6.16 and 6.17 show that the first packet exchanged is a Read Request from the client to the server indicating the name of the file to download and the transfer mode, followed immediately by Data packets and the respective Acknowledgment packets. All the Data packets have 512 bytes of data (maximum block data size), with the exception of the third packet (packet number 6 on both exchanges), which has 476 bytes for the download of the 1500 bytes file and zero bytes for the download of the 1024 bytes file. Like in the upload case, the last packet with zero bytes of data is justified by the protocol standard, which states that the last packet must have a size lower than the maximum block size to notify the end of the transmission.

```
No.  Time      Source       Destination  Protocol Info
   1 0.000000  10.1.1.100   10.1.1.20    TFTP     Write Request, File: file1500, Transfer type: octet
   2 0.006835  10.1.1.20    10.1.1.100   TFTP     Acknowledgement, Block: 0
   3 0.006949  10.1.1.100   10.1.1.20    TFTP     Data Packet, Block: 1
   4 0.007690  10.1.1.20    10.1.1.100   TFTP     Acknowledgement, Block: 1
   5 0.007780  10.1.1.100   10.1.1.20    TFTP     Data Packet, Block: 2
   6 0.008529  10.1.1.20    10.1.1.100   TFTP     Acknowledgement, Block: 2
   7 0.008609  10.1.1.100   10.1.1.20    TFTP     Data Packet, Block: 3 (last)
   8 0.009204  10.1.1.20    10.1.1.100   TFTP     Acknowledgement, Block: 3

▷ Frame 7: 522 bytes on wire (4176 bits), 522 bytes captured (4176 bits)
▷ Ethernet II, Src: f2:68:7f:e7:d6:c3 (f2:68:7f:e7:d6:c3), Dst: CadmusCo_fe:b6:40 (08:00:27:fe:b6:40)
▷ Internet Protocol Version 4, Src: 10.1.1.100 (10.1.1.100), Dst: 10.1.1.20 (10.1.1.20)
▷ User Datagram Protocol, Src Port: 47664 (47664), Dst Port: 50186 (50186)
▽ Trivial File Transfer Protocol
    [DESTINATION File: file1500]
    Opcode: Data Packet (3)
    Block: 3
▷ Data (476 bytes)
```

Figure 6.14 TFTP upload of a file with 1500 bytes.

```
No.  Time      Source       Destination  Protocol Info
   1 0.000000  10.1.1.100   10.1.1.20    TFTP     Write Request, File: file1024, Transfer type: octet
   2 0.009428  10.1.1.20    10.1.1.100   TFTP     Acknowledgement, Block: 0
   3 0.009546  10.1.1.100   10.1.1.20    TFTP     Data Packet, Block: 1
   4 0.010167  10.1.1.20    10.1.1.100   TFTP     Acknowledgement, Block: 1
   5 0.010258  10.1.1.100   10.1.1.20    TFTP     Data Packet, Block: 2
   6 0.011226  10.1.1.20    10.1.1.100   TFTP     Acknowledgement, Block: 2
   7 0.011302  10.1.1.100   10.1.1.20    TFTP     Data Packet, Block: 3 (last)
   8 0.013259  10.1.1.20    10.1.1.100   TFTP     Acknowledgement, Block: 3

▷ Frame 7: 46 bytes on wire (368 bits), 46 bytes captured (368 bits)
▷ Ethernet II, Src: f2:68:7f:e7:d6:c3 (f2:68:7f:e7:d6:c3), Dst: CadmusCo_fe:b6:40 (08:00:27:fe:b6:40)
▷ Internet Protocol Version 4, Src: 10.1.1.100 (10.1.1.100), Dst: 10.1.1.20 (10.1.1.20)
▷ User Datagram Protocol, Src Port: 54587 (54587), Dst Port: 51354 (51354)
▽ Trivial File Transfer Protocol
    [DESTINATION File: file1024]
    Opcode: Data Packet (3)
    Block: 3
```

Figure 6.15 TFTP upload of a file with 1024 bytes.

6.5 FTP Service

The File Transfer Protocol (FTP), defined in RFC 959 [84], is a file transfer service that runs over TCP using two port numbers: a "data" port and a "command" port (also known as the control port). Traditionally, these ports are port 21 for the command port and port 20 for the data port. However, there are two FTP modes and the data port is not always 20.

In active mode FTP the client connects from a random port N (which should be higher than 1023) to the FTP server's command port 21. Then, the client starts listening to port N+1 and sends the FTP command *PORT IP_address.N+1* to the FTP server. The server will then connect back to the

No.	Time	Source	Destination	Protocol	Info
1	0.000000	10.1.1.100	10.1.1.20	TFTP	Read Request, File: file1500, Transfer type: octet
2	0.003544	10.1.1.20	10.1.1.100	TFTP	Data Packet, Block: 1
3	0.003628	10.1.1.100	10.1.1.20	TFTP	Acknowledgement, Block: 1
4	0.004179	10.1.1.20	10.1.1.100	TFTP	Data Packet, Block: 2
5	0.004260	10.1.1.100	10.1.1.20	TFTP	Acknowledgement, Block: 2
6	0.005340	10.1.1.20	10.1.1.100	TFTP	Data Packet, Block: 3 (last)
7	0.005418	10.1.1.100	10.1.1.20	TFTP	Acknowledgement, Block: 3

```
▷ Frame 6: 522 bytes on wire (4176 bits), 522 bytes captured (4176 bits)
▷ Ethernet II, Src: CadmusCo_fe:b6:40 (08:00:27:fe:b6:40), Dst: f2:68:7f:e7:d6:c3 (f2:68:7f:e7:d6:c3)
▷ Internet Protocol Version 4, Src: 10.1.1.20 (10.1.1.20), Dst: 10.1.1.100 (10.1.1.100)
▷ User Datagram Protocol, Src Port: 51036 (51036), Dst Port: 60156 (60156)
▽ Trivial File Transfer Protocol
    [Source File: file1500]
    Opcode: Data Packet (3)
    Block: 3
▷ Data (476 bytes)
```

Figure 6.16 TFTP download of a file with 1500 bytes.

No.	Time	Source	Destination	Protocol	Info
1	0.000000	10.1.1.100	10.1.1.20	TFTP	Read Request, File: file1024, Transfer type: octet
2	0.007965	10.1.1.20	10.1.1.100	TFTP	Data Packet, Block: 1
3	0.008088	10.1.1.100	10.1.1.20	TFTP	Acknowledgement, Block: 1
4	0.008788	10.1.1.20	10.1.1.100	TFTP	Data Packet, Block: 2
5	0.008833	10.1.1.100	10.1.1.20	TFTP	Acknowledgement, Block: 2
6	0.009340	10.1.1.20	10.1.1.100	TFTP	Data Packet, Block: 3 (last)
7	0.009378	10.1.1.100	10.1.1.20	TFTP	Acknowledgement, Block: 3

```
▷ Frame 6: 46 bytes on wire (368 bits), 46 bytes captured (368 bits)
▷ Ethernet II, Src: CadmusCo_fe:b6:40 (08:00:27:fe:b6:40), Dst: f2:68:7f:e7:d6:c3 (f2:68:7f:e7:d6:c3)
▷ Internet Protocol Version 4, Src: 10.1.1.20 (10.1.1.20), Dst: 10.1.1.100 (10.1.1.100)
▷ User Datagram Protocol, Src Port: 42968 (42968), Dst Port: 57130 (57130)
▽ Trivial File Transfer Protocol
    [Source File: file1024]
    Opcode: Data Packet (3)
    Block: 3
```

Figure 6.17 TFTP download of a file with 1024 bytes.

client's specified data port from its local data port, which is port 20. So, from the server side point of view, to support active FTP mode the following communication channels need to be opened: FTP server's port 21 from anywhere (the client initiates the connection); FTP server's port 21 to ports higher than 1023 (in this case, the server responds to client's control port); FTP server's port 20 to ports higher than 1023 (server initiates a data connection to the client's data port); FTP server's port 20 from ports higher 1023 (the client sends ACKs to the server's data port).

Once the data transfer is complete, the Server closes the data connection. The control connection is only closed when the FTP session ends. Figure 6.18 shows the general mechanism of the active FTP service.

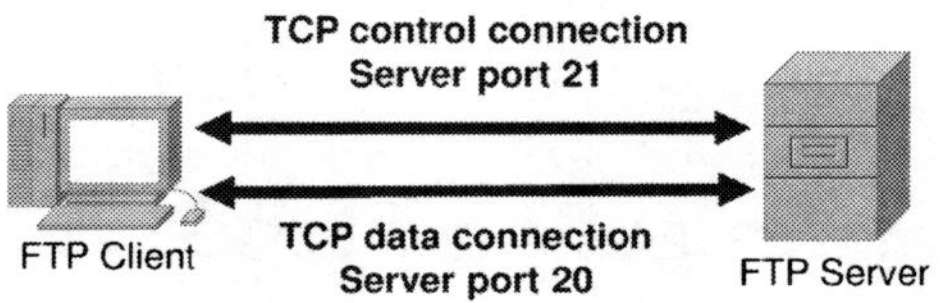

Figure 6.18 FTP service connections.

Active mode FTP poses a security problem to corporate networks. Note that the FTP client does not make the actual connection to the data port of the server; it simply tells the server what port it is listening on and the server connects back to the specified port on the client side. From the client side firewall this appears to be an outside system initiating a connection to an internal client, which is usually blocked. In order to resolve this issue, a different method for FTP connections was developed: the passive mode FTP.

In passive mode FTP the client initiates both connections to the server, solving the problem of firewalls filtering the incoming data port connection to the client from the server. When opening an FTP connection, the client opens two random unprivileged ports locally (N, higher than 1023, and N+1). The first port contacts the server on port 21, but instead of issuing a PORT command and allowing the server to connect back to its data port, the client issues a PASV command. As a result, the server opens a random unprivileged port (one that is higher than 1023) and sends the PORT command back to the client, specifying which port it choose. The client then initiates the connection from its port N+1 to the port number on the server side to transfer data.

So, from the server side point of view, to support passive FTP mode the following communication channels need to be opened: FTP server's port 21 from anywhere (the client initiates the connection); FTP server's port 21 to ports higher than 1023 (the server responds to the client's control port); FTP server's ports higher than 1023 from anywhere (the client initiates a data connection to the random port specified by server); FTP server's ports higher than 1023 to remote ports higher than 1023 (the server sends ACKs (and data) to the client's data port).

Since FTP runs over TCP, the flow control is automatically assured at the transport level. This service includes authentication mechanisms, supporting a "Username + Password" authentication or a simple "Username Anonymous" authentication mode. By default, the access credentials are transmitted in plain text.

FTP supports several commands, like the following examples:

- USER <username> - sends the username;
- PASS <password> - sends the password, completing the authentication process (these commands are sent in plain text);
- LIST - returns the list of files in current directory;
- RETR <filename> - retrieves (gets) the specified file;
- STOR <filename> - stores (puts) the specified file onto the remote host;
- PORT <IP_address.Port> - used by the Client to announce a port number that should be used by the Server in a subsequent data transfer operation.

The following are some of the possible return codes, each one consisting of a status code and an ASCII phrase:

- 331 Username OK, password required;
- 125 data connection already open; transfer starting;
- 425 Can't open data connection;
- 452 Error writing file.

6.5.1 FTP Experiment

The Linux server must have a FTP server installed. To verify if the FTP (*vsftpd*) server is installed, the following command can be executed in a Server line terminal:

```
# aptitude show vsftpd | grep State
```

If the *vsftpd* server is not installed, it can be installed executing the following commands:

```
# aptitude update
# aptitude install vsftpd
```

To guarantee that *vsftpd* is running, the following command must be executed:

```
# service vsftpd restart
```

In the Linux PC terminal, in order to test the FTP service we will require one file with random contents and 15 Kbytes of size (named *file15K*). To create it the following *dd* command can be used:

```
# dd if=/dev/urandom of=file15K bs=1k count=15
```

No.	Time	Source	Destination	Protocol	Info
1	0.000000	10.1.1.100	10.1.1.20	TCP	55105 > ftp [SYN] Seq=545103718
2	0.001091	10.1.1.20	10.1.1.100	TCP	ftp > 55105 [SYN, ACK] Seq=1546274767 Ack=545103719
3	0.001179	10.1.1.100	10.1.1.20	TCP	55105 > ftp [ACK] Seq=545103719 Ack=1546274768
6	0.034431	10.1.1.20	10.1.1.100	FTP	Response: 220 (vsFTPd 2.3.5)
8	2.595897	10.1.1.100	10.1.1.20	FTP	Request: USER salvador
10	2.597815	10.1.1.20	10.1.1.100	FTP	Response: 331 Please specify the password.
12	5.139913	10.1.1.100	10.1.1.20	FTP	Request: PASS 12345678
14	5.213459	10.1.1.20	10.1.1.100	FTP	Response: 230 Login successful.
16	5.213633	10.1.1.100	10.1.1.20	FTP	Request: SYST
18	5.214215	10.1.1.20	10.1.1.100	FTP	Response: 215 UNIX Type: L8
20	11.083979	10.1.1.100	10.1.1.20	FTP	Request: TYPE I
21	11.085236	10.1.1.20	10.1.1.100	FTP	Response: 200 Switching to Binary mode.

Figure 6.19 FTP message exchanges during login (some TCP control packets were suppressed).

In order to test FTP, the following commands must be executed:

```
1 # ftp 10.1.1.20
2 Name (10.1.1.20:salvador): salvador
3 Password: *****
4 ftp> put file15K
5 ftp> get file15K
6 ftp> passive
7 ftp> put file15K
8 ftp> quit
```

where (command 1) activates the FTP client defining the IPv4 address of the server; (command 2) is used to enter the username (in this case *salvador*); (command 3) is used to enter the password (in this case *12345678*); (command 4) performs the FTP non passive upload of file *file15K* to the server; (command 5) performs the FTP download of file *file15K* from the server; (command 6) activates the passive mode of file transfer; (command 7) performs the FTP passive upload of file *file15K* to the server, and finally, (command 8) terminates the FTP client. The sequence of TCP/FTP packets that were exchanged between client and server is depicted in Figures 6.19 to 6.23. Note that some TCP control packets (TCP window size adjustment [packets 4-5] and acknowledgments) and data packets were suppressed from the list.

The messages exchanged during the FTP login process, depicted Figure 6.19, are the following: [packets 1-3] show that initially a FTP control TCP session (port 20/ftp) should be openned. Upon the opening of the TCP session, the server sends a response with code 220 [packet 6], notifying that the server is ready for a new user. The client will prompt the user to enter the username and will send the user input to the server using a Request with the USER command [packet 8]. The server, upon the acceptance of the user name, will request for a password using a Response 331 [packet 10], to which the client responds with a Request PASS message [packet 12] containing the

No.	Time	Source	Destination	Protocol	Info
23	11.085433	10.1.1.100	10.1.1.20	FTP	Request: PORT 10,1,1,100,216,176
24	11.087075	10.1.1.20	10.1.1.100	FTP	Response: 200 PORT command successful. Consider using PASV.
25	11.087175	10.1.1.100	10.1.1.20	FTP	Request: STOR file15K
26	11.095320	10.1.1.20	10.1.1.100	TCP	ftp-data > 55472 [SYN] Seq=3745950757
27	11.095381	10.1.1.100	10.1.1.20	TCP	55472 > ftp-data [SYN, ACK] Seq=2078295126 Ack=3745950758
28	11.095611	10.1.1.20	10.1.1.100	TCP	ftp-data > 55472 [ACK] Seq=3745950758 Ack=2078295127 Win=1
29	11.097812	10.1.1.20	10.1.1.100	FTP	Response: 150 Ok to send data.
30	11.099598	10.1.1.100	10.1.1.20	FTP-DATA	FTP Data: 1448 bytes
31	11.099673	10.1.1.100	10.1.1.20	FTP-DATA	FTP Data: 1448 bytes
46	11.102111	10.1.1.100	10.1.1.20	TCP	55472 > ftp-data [FIN, ACK] Seq=2078310487 Ack=3745950758
53	11.103654	10.1.1.20	10.1.1.100	TCP	ftp-data > 55472 [FIN, ACK] Seq=3745950758 Ack=2078310488
55	11.106060	10.1.1.20	10.1.1.100	FTP	Response: 226 Transfer complete.

```
▷ Frame 23: 91 bytes on wire (728 bits), 91 bytes captured (728 bits)
▷ Ethernet II, Src: f2:68:7f:e7:d6:c3 (f2:68:7f:e7:d6:c3), Dst: CadmusCo_fe:b6:40 (08:00:27:fe:b6:40)
▷ Internet Protocol Version 4, Src: 10.1.1.100 (10.1.1.100), Dst: 10.1.1.20 (10.1.1.20)
▷ Transmission Control Protocol
▽ File Transfer Protocol (FTP)
  ▽ PORT 10,1,1,100,216,176\r\n
      Request command: PORT
      Request arg: 10,1,1,100,216,176
      Active IP address: 10.1.1.100 (10.1.1.100)
      Active port: 55472
```

Figure 6.20 FTP (non passive) upload of a file (some TCP control and data packets were suppressed).

password (in clear-text). Packets 16 to 21 are used by the client to query the server about its system (Request SYST) and request a binary mode of transmission (Request TYPE I); the server will respond with messages 215 and 200, respectively.

Figure 6.20 shows the messages exchanged during a FTP (non passive) session to upload a file. After issuing the *put* command, the client will send a Request PORT message [packet 23] to the server indicating the IP address and TCP port to which the server must open a new TCP session to exchange data. The "PORT 10,1,1,100,216,176" is composed by 6 numbers sent in clear-text, where the first four numbers represent the four bytes of the IPv4 address (i.e. 10.1.1.100) and the last two numbers represent the two bytes of the proposed TCP port, where the first number represents the most significant byte. The port number can be calculated by multiplying the penultimate number of the PORT command by 256 and adding the last number of the PORT command (i.e. 216*256+176=55472). Packet 24 is a response with code 200, used to notify that the previous request was successfully completed. The client will issue a Request STOR command [packet 25] to request the server the storage of a specific file. Upon the reception of the PORT and STOR requests, the server is able to open a new TCP session (FTP-DATA) from is local port 21/ftp-data to the TCP port announced by the client (in this case 55472) [packets 26-28]. After opening the new FTP-DATA session, the server, using

No.	Time	Source	Destination	Protocol	Info
57	15.987994	10.1.1.100	10.1.1.20	FTP	Request: PORT 10,1,1,100,194,214
58	15.989917	10.1.1.20	10.1.1.100	FTP	Response: 200 PORT command successful. Consider using PASV.
59	15.990021	10.1.1.100	10.1.1.20	FTP	Request: RETR file15K
60	15.996401	10.1.1.20	10.1.1.100	TCP	ftp-data > 49878 [SYN] Seq=3753655288
61	15.996436	10.1.1.100	10.1.1.20	TCP	49878 > ftp-data [SYN, ACK] Seq=1828975282 Ack=3753655289
62	15.996760	10.1.1.20	10.1.1.100	TCP	ftp-data > 49878 [ACK] Seq=3753655289 Ack=1828975283
63	15.997439	10.1.1.20	10.1.1.100	FTP	Response: 150 Opening BINARY mode data connection for file15K
64	15.998141	10.1.1.20	10.1.1.100	FTP-DATA	FTP Data: 1448 bytes
65	15.998168	10.1.1.20	10.1.1.100	FTP-DATA	FTP Data: 1448 bytes
86	15.998985	10.1.1.20	10.1.1.100	TCP	ftp-data > 49878 [FIN, ACK] Seq=3753670649 Ack=1828975283
87	15.999131	10.1.1.100	10.1.1.20	TCP	49878 > ftp-data [FIN, ACK] Seq=1828975283 Ack=3753670650
89	16.000494	10.1.1.20	10.1.1.100	FTP	Response: 226 Transfer complete.

```
▷ Frame 57: 91 bytes on wire (728 bits), 91 bytes captured (728 bits)
▷ Ethernet II, Src: f2:68:7f:e7:d6:c3 (f2:68:7f:e7:d6:c3), Dst: CadmusCo_fe:b6:40 (08:00:27:fe:b6:40)
▷ Internet Protocol Version 4, Src: 10.1.1.100 (10.1.1.100), Dst: 10.1.1.20 (10.1.1.20)
▷ Transmission Control Protocol
▽ File Transfer Protocol (FTP)
  ▽ PORT 10,1,1,100,194,214\r\n
      Request command: PORT
      Request arg: 10,1,1,100,194,214
      Active IP address: 10.1.1.100 (10.1.1.100)
      Active port: 49878
```

Figure 6.21 FTP (non passive) download of a file (some TCP control and data packets were suppressed).

the control TCP session, will send a response message with code 150 [packet 29] to notify the client that the data transmission may start. Packets 30 to 54 are data packets, TCP acknowledgments and TCP FIN flagged packets to close the FTP-DATA session, which happens after all data has being transmitted. The last packet [55] is a response packet sent by the server (using the control TCP session) with code 226 to notify the client that data transfer were completed and successful.

Figure 6.21 shows the messages exchanged during a non passive FTP download of a file. The flow of messages is very similar to the one observed in the FTP upload, with the obvious change on the direction of the data packets and the request issued by the client. In this case the client sends a Request RETR [packet 59] to signal the server about the file that it wants to retrieve.

The passive FTP upload of a file is depicted in Figure 6.22: the main difference from a non passive upload is that the new opened TCP session is not started from the server side (upon the reception of the PORT command), but started from the client side. To activate the passive mode, the client send a Request message with code PASV [packet 91], to which the server responds with message code 227 [packet 92], associated with 6 numbers in clear-text that announce to the client the IP address and port that are waiting for the establishment of a new TCP session. The format of the IP and port definition is the same used in the PORT command: in this case, it is "10,1,1,20,209,70",

No.	Time	Source	Destination	Protocol	Info
91	23.227966	10.1.1.100	10.1.1.20	FTP	Request: PASV
92	23.231460	10.1.1.20	10.1.1.100	FTP	Response: 227 Entering Passive Mode (10,1,1,20,209,70)
93	23.231621	10.1.1.100	10.1.1.20	TCP	45081 > 53574 [SYN] Seq=4226148531
94	23.232586	10.1.1.20	10.1.1.100	TCP	53574 > 45081 [SYN, ACK] Seq=1880799843 Ack=4226148532
95	23.232571	10.1.1.100	10.1.1.20	TCP	45081 > 53574 [ACK] Seq=4226148532 Ack=1880799844
96	23.232615	10.1.1.100	10.1.1.20	FTP	Request: STOR file15K
97	23.238599	10.1.1.20	10.1.1.100	FTP	Response: 150 Ok to send data.
98	23.238739	10.1.1.100	10.1.1.20	FTP-DATA	FTP Data: 1448 bytes
99	23.238787	10.1.1.100	10.1.1.20	FTP-DATA	FTP Data: 1448 bytes
115	23.239987	10.1.1.100	10.1.1.20	TCP	45081 > 53574 [FIN, ACK] Seq=4226163892 Ack=1880799844
121	23.241418	10.1.1.20	10.1.1.100	TCP	53574 > 45081 [FIN, ACK] Seq=1880799844 Ack=4226163893
123	23.242308	10.1.1.20	10.1.1.100	FTP	Response: 226 Transfer complete.

```
▷ Frame 92: 113 bytes on wire (904 bits), 113 bytes captured (904 bits)
▷ Ethernet II, Src: CadmusCo_fe:b6:40 (08:00:27:fe:b6:40), Dst: f2:68:7f:e7:d6:c3 (f2:68:7f:e7:d6:c3)
▷ Internet Protocol Version 4, Src: 10.1.1.20 (10.1.1.20), Dst: 10.1.1.100 (10.1.1.100)
▷ Transmission Control Protocol
▽ File Transfer Protocol (FTP)
  ▽ 227 Entering Passive Mode (10,1,1,20,209,70).\r\n
      Response code: Entering Passive Mode (227)
      Response arg: Entering Passive Mode (10,1,1,20,209,70).
      Passive IP address: 10.1.1.20 (10.1.1.20)
      Passive port: 53574
```

Figure 6.22 FTP (passive) upload of a file (some TCP control and data packets were suppressed).

No.	Time	Source	Destination	Protocol	Info
125	24.579908	10.1.1.100	10.1.1.20	FTP	Request: QUIT
126	24.581184	10.1.1.20	10.1.1.100	FTP	Response: 221 Goodbye.
127	24.581538	10.1.1.20	10.1.1.100	TCP	ftp > 55105 [FIN, ACK] Seq=1546275242 Ack=545103867
128	24.581585	10.1.1.100	10.1.1.20	TCP	55105 > ftp [FIN, ACK] Seq=545103867 Ack=1546275242

Figure 6.23 FTP message exchanges during quit (some TCP control packets were suppressed).

which means that the server will be waiting at IP address 10.1.1.20 for a connection to port 53574 (53574=209*256+70). Packets 93 to 95 are used to open the new FTP-DATA session to transfer data. From that point on the behavior is exactly the same as in the non passive upload.

Finally, Figure 6.23 shows the end of the FTP client-server interaction. After a Quit command issued by the user, the client will send a Request QUIT message [packet 125] to the server, to which the server responds with a message with code 221 [packet 126] indicating that it will close the control connection. Packets 127 and 128 are used to close the FTP control TCP session.

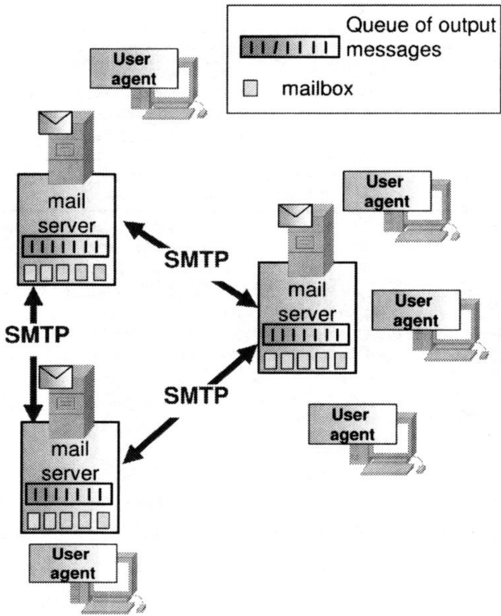

Figure 6.24 Mail service components.

6.6 Email Service

The Email (a short for electronic mail) service consists on the transmission of messages over communications networks. There are two basic elements: user agents, which are user applications used to write, send, receive and read electronic mail messages; mail servers, which are server applications that send and receive messages to/from clients.

User agents exchange messages with their servers. A mail server includes a mailbox for each client to store its messages and a queue of output messages where still undelivered messages are temporarily stored. Figure 6.24 illustrates these basic Mail service concepts.

The following protocols are involved in the different transfers of electronic mail messages:

- Exchange of messages between mail servers - SMTP (Simple Mail Transfer Protocol);
- Send messages from the user agent to the mail server of the sender - SMTP or HTTP (Hyper-Text Transport Protocol);

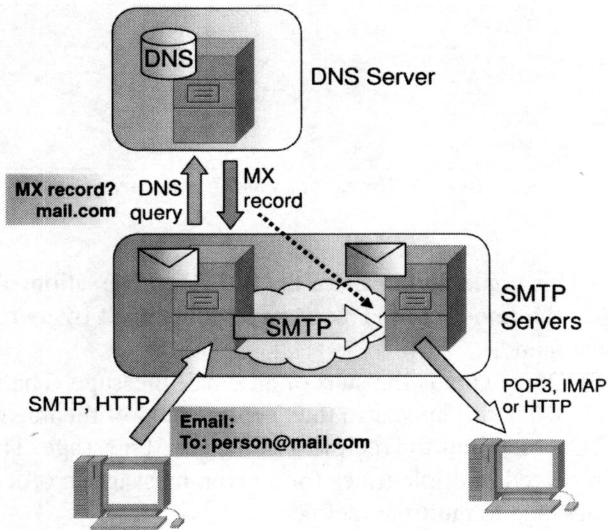

Figure 6.25 Email protocols and their roles.

- Access to the mailbox (sending messages from the mail server to the user agent) - POP3 (Post Office Protocol, version 3), IMAP (Internet Mail Access Protocol) or HTTP.

The different protocols and their roles are illustrated in Figure 6.25.

6.6.1 SMTP

This protocol was defined in RFC 2821 [62] and runs over TCP, using port number 25 on the server side. This is a push-type protocol, that is, communication is established by the entity that wants to send information. By default, the sender mail server sends messages directly to the receivers mail servers. SMTP follows a client-server architecture, where the client sends commands that are answered by the server. Similarly to HTTP, each response is composed by a 3 number code followed by an optional phrase. All email messages are sent in ASCII format, ending with "CRLF.CRLF", where CR is the carriage return character and LF is the line feed character.

SMTP interaction between mail servers rely on several commands exchanged between the mail servers:

```
From: salvador@av.it.pt
To: nogueira@ua.pt
Subject: Are you hungry?

Hi Nogueira!
Let us lunch?
Salvador!
.
```

Box 6.9 *Example of a SMTP message.*

- HELO - This command initiates the SMTP conversation. The host connecting to the remote SMTP server identifies itself by its fully qualified DNS host name.
- MAIL FROM - This is the start of an email message. The source email address is what will appear in the "From:" field of the message.
- RCPT TO - Identifies the recipient of the email message. This command can be repeated multiple times for a given message in order to deliver a single message to multiple recipients.
- DATA - This command notifies that the email message body will follow. The stream of data is terminated by a "." on a line by itself.
- QUIT - Terminates an SMTP connection. Multiple email messages can be transfered during a single TCP/IP connection, which allows for a more efficient transfer of email.

The final format of the message is shown in Box 6.9. As can be seen, messages are sent in ASCII format, starting with a set of header lines that is followed by an empty line and by the body of the message: some header lines are mandatory (From and To), while others (Subject, etc...) are optional. Other header lines (Received, etc...) are inserted by the mail servers.

In order to support email messages messages with different types of information, RFCs 2045 and 2046 [29] [30] defined MIME (Multipurpose Internet Mail Extensions), an extension that is appropriate to transmit non-ASCII data. This extension introduces the following header lines:

- Content-Transfer-Encoding - Specifies the algorithm that is used to code the contents in ASCII format (*base64* is a possible algorithm);
- Content-type: type/subtype; parameters - Specifies the information type (for example, text/plain; charset="ISO-8859-1", text/html, image/gif, image/jpeg, video/mpeg, video/quicktime, application/msword).

Box 6.10 show an example of a SMTP message with MIME. In this case, a JPEG image is included in the Email message (as stated in the Content-Type field). At the receiver user agent, the message content is decoded by

```
From: nogueira@ua.pt
To: salvador@av.it.pt
Subject: Cool picture!
MIME-Version: 1.0
Content-Transfer-Encoding: base64
Content-Type: image/jpeg

(base64 encoded data ...........
...............................
.......... base 64 encoded data)
```

Box 6.10 Example of a SMTP message with a MIME extension.

```
From: nogueira@ua.pt
To: salvador@av.it.pt
Subject: Another cool picture!
MIME-Version: 1.0
Content-Type: multipart/mixed; Boundary=98766789

--98766789
Content-Type: text/plain
Hi Junior,
Please find enclosed the agreed photo.
--98766789
Content-Transfer-Encoding: base64
Content-Type: image/jpeg
(base64 encoded data ...........
...............................
.......... base 64 encoded data)
--98766789
Content-Type: text/plain
Salvador
.
```

Box 6.11 Example of a MIME extension of the multipart type.

the *base64* algorithm in order to obtain the original content; then, a JPEG decoder is applied to visualize the image.

Box 6.11 shows a MIME extension of the multipart type. The multipart/mixed header line allows to compose a message with multiple information types. The Boundary parameter is simply a separator between the different information types that are included in the message body.

Now, let us see the SMTP message that is received at the destination (Box 6.12): each mail server inserts a header line of the Received type that identifies the server that sent the message, the server that received it and the time instant of the message reception. In this case, the *salvador* user has configured his server, "mail.av.it.pt", to forward messages to the "smtp.googlemail.com" server (account "pjsalvador@gmail.com").

```
Received: from mail.av.it.pt
          by smtp.googlemail.com; Thu, 5 Jul 2012 16:26:31
Received: from mail.ua.pt
          by mail.av.it.pt; Thu, 5 Jul 2012 16:25:38
From: nogueira@ua.pt
To: pjsalvador@gmail.com
Subject: Another cool picture!
MIME-Version: 1.0
Content-Transfer-Encoding: base64
Content-Type: image/jpeg

(base64 encoded data ...........
........... base 64 encoded data)
```

Box 6.12 Received SMTP message.

6.6.2 SMTP Experiment

The Linux server must have a SMTP server installed. To verify if the SMTP (*postfix*) server and local e-mail delivery tools (*maildrop*) are installed, the following command should be executed in a Server line terminal:

```
# aptitude show postfix maildrop | grep State
```

If the *postfix* server or *mailutils* are not installed, they can be installed executing the following commands:

```
# aptitude update
# aptitude install postfix maildrop
```

If *postfix* has already been installed, run the following command to reset the configuration:

```
# dpkg-reconfigure postfix
```

During the reconfiguration of the service, choose the type of configuration as "Internet Site", which will force the server to send and receive e-mail using only SNMP. The remaining options can be left with the default configuration (answering OK to all prompts).

To configure and restart *postfix*, the following commands must be executed:

```
1 # maildirmake /etc/skel/Maildir
2 # postconf -e "home_mailbox = Maildir/"
3 # useradd -m -s /bin/bash mailtest
4 # passwd mailtest
5    12345678
6 # service postfix restart
```

where (line 1 command) defines a template directory to save the email of each user (Maildir/), (line 2 command) updates *postfix* configuration to use each user's *Maildir* directory to store the e-mails, (line 3 command) creates a new directory for testing with username equal to *mailtest*, (line 4 and 5 commands) define the new user password as *12345678* and, finally, (line 6 command) restarts *postfix*. Note that directory */etc/skel/* is used as the template for creating the base directory of each new user.

In order to test the basic SMTP server setup, it is possible to send an email to interact with the server by using an interactive command line. To access the server interactive command line, it is necessary to open a *netcat* session to the server's TCP port 25/smtp. Upon connection, it is possible to construct an e-mail message and send it using the following individual commands:

```
1  # netcat 10.1.1.20 25
2    220 ubuntu ESMTP Postfix (Ubuntu)
3  MAIL FROM: root@localhost
4    250 2.1.0 Ok
5  RCPT TO: mailtest@localhost
6    250 2.1.5 Ok
7  DATA
8    354 End data with <CR><LF>.<CR><LF>
9  Subject: First Test
10 Hi everybody!
11 .
12   250 2.0.0 Ok: queued as 3525741ACE
13 QUIT
14   221 2.0.0 Bye
```

These client commands and server answers have the following meaning: (line 2) 220 answer from the server notifying the client that the server is ready; (line 3) command that defines the source address of the e-mail with the "MAIL FROM:" directive; (lines 4, 6 and 12) are 220 answers from the server notifying the client that the request actions were completed; (line 5) command defines the destination address of the e-mail with the "RCPT TO:" directive; (line 7) is a DATA command to signal the start of the e-mail body; (line 8) is a 354 server answer notifying the client to start the e-mail body input and how to stop (with a carriage return, a dot and a carriage return); (lines 9 and 10) contain the e-mail body, noting that the e-mail subject is defined here with the directive "Subject:"; (line 11) is a dot signaling the end of the e-mail body input; finally, in lines 13 and 14 the client exits this interaction with the "QUIT" command, and the server answers with a 221 message to signal the end of the server-client interaction. SMTP exchanged packets are depicted in Figure 6.26 and reflect directly the text commands that were introduced.

To check if the message was delivered, it is necessary to inspect the contents of the directory *Maildir/new/* inside the *mailtest* user home directory (*/home/mailtest/*). Each file inside the *Maildir/new/* contains one (new) e-mail received by the user.

No.	Time	Source	Destination	Protocol	Info
4	28.045353	10.1.1.20	10.1.1.100	SMTP	S: 220 ubuntu ESMTP Postfix (Ubuntu)
6	37.244892	10.1.1.100	10.1.1.20	SMTP	C: mail from: root@localhost
8	37.254027	10.1.1.20	10.1.1.100	SMTP	S: 250 2.1.0 Ok
10	48.252858	10.1.1.100	10.1.1.20	SMTP	C: rcpt to: mailtest@localhost
12	48.338937	10.1.1.20	10.1.1.100	SMTP	S: 250 2.1.5 Ok
14	53.060848	10.1.1.100	10.1.1.20	SMTP	C: data
16	53.064284	10.1.1.20	10.1.1.100	SMTP	S: 354 End data with <CR><LF>.<CR><LF>
18	60.172841	10.1.1.100	10.1.1.20	SMTP	C: DATA fragment, 20 bytes
20	65.748875	10.1.1.100	10.1.1.20	SMTP	C: DATA fragment, 14 bytes
22	66.676877	10.1.1.100	10.1.1.20	SMTP	C: DATA fragment, 2 bytes
24	66.696697	10.1.1.20	10.1.1.100	SMTP	S: 250 2.0.0 Ok: queued as 315C541ACE
26	67.932866	10.1.1.100	10.1.1.20	SMTP	C: DATA fragment, 5 bytes
27	67.939289	10.1.1.20	10.1.1.100	SMTP	S: 221 2.0.0 Bye

Figure 6.26 SMTP message exchanges during the e-mail sent (TCP control packets were suppress).

To inspect these files, perform the following commands:

```
# su - mailtest
# cd /home/mailtest/Maildir/new/
# ls
  1352355970.V801I22fa8M936316.ubuntu
# cat 1352355970.V801I22fa8M936316.ubuntu
  Return-Path: <root@localhost>
  X-Original-To: mailtest@localhost
  Delivered-To: mailtest@localhost
  Received: from unknown (unknown [10.1.1.100])
    by ubuntu (Postfix) with SMTP id A589941AD3
    for <mailtest@localhost>; Thu, 12 Jul 2012 16:25:38 (WET)
  SUBJECT: First Test
  Hi everybody!
```

The meaning of these commands is: (command 1) is used to start a shell interaction as user *mailtest*; (command 2) changes directory to the new messages inside the base mail directory; (command 3) lists all files inside the new messages directory and, finally, (command 4) displays the contents of the e-mail that was previously sent.

6.6.3 POP3

Post Office Protocol - version 3 (POP3) was defined in RFC 1939 [72] and runs over TCP, using port number 110 at the server side. Communication is established by the entity (user agent) that wants to receive information, so this is a pull type protocol. Message transfer can occur using one of the following two modes:

- Send and remove: Messages are removed from the server mailbox after being sent;
- Send and store: Messages are kept in the mailbox after being sent.

The protocol is executed in three phases:

- Authentication: The user agent sends the user name and a password;
- Transaction: The mail server sends all messages that are in the user mailbox; for each message, the user agent indicates if it should be removed from the mailbox or not;
- Update: The mail server removes from the mailbox all messages that were identified for removal by the user agent.

The following commands are supported by the POP3 service:

- USER <userid> - Specifies the user identifier;
- <PASS password> - Specifies the password;
- STAT - The answer contains the number of messages and their total size in bytes, like for example "+OK 3 345910";
- LIST - The answer is a list where each line identifies the number and size, in bytes, of each message, ending with a line containing only a dot;
- RETR msg# - Recovers message number # (for example, RETR 2);
- DELE msg# - Deletes message number # (for example, DELE 3);
- RSET - Clears any mark of deletable messages;
- QUIT - Quits the session.

6.6.4 IMAP

The Internet Mail Access Protocol (IMAP) was defined in RFC 2060 [13] and runs over TCP, using port number 143 on the Server side. Compared to POP3, IMAP provides the user relevant additional functionalities:

- The ability to create and manage a system of message directories at the server;
- The ability to perform search operations in the system of directories, which can be useful to users running the service on multiple terminals;
- The possibility of requesting only parts of the mail messages, which can be useful when the terminal is connected through low bit rate connections.

On an IMAP session, the server can be into one of four different states:

- Non-authenticated: The initial state, just before the user agent sends the username and the corresponding password;
- Authenticated: In this state the user agent should identify a directory before sending any command that affects mail messages;

- Selected: The user agent can send message management commands (to visualize, remove, transfer, etc);
- Logout: When the session ends.

The IMAP protocol supports a lot of commands. Let us see some of the commands supported by the IMAP client:

- CAPABILITY - Checks the capabilities of a server;
- LOGIN - Enables login, specifying a username and a password;
- SELECT - Selects a specific directory of the mailbox;
- CREATE/DELETE/RENAME - Creates/deletes/renames directories of the mailbox;
- LIST - Lists the directories of the mailbox;
- SUBSCRIBE/UNSUBSCRIBE - Places/removes a directory from the active state;
- STATUS - Verifies the state of a specific directory of the mailbox;
- FETCH - Allows recovering the whole mailbox or parts of mail messages or directories;
- LOGOUT - Enables to logout from the server.

Since this is a sophisticated service, there are many useful commands besides the ones listed above: APPEND (appends an argument as a new message in the specified destination mailbox), EXPUNGE (permanently removes from the currently selected mailbox all messages that have the *Deleted* flag set), SEARCH (searches the mailbox for messages that match a given searching criteria), STORE (alters data associated with a message in the mailbox), COPY (copies the specified message(s) to the specified destination mailbox), etc.

6.6.5 POP3 and IMAP Experiments

The Linux server must have POP3 and IMAP servers installed. To verify if the POP3 (*courier-pop*) server and the IMAP (*courier-imap*) server are installed, the following command can be executed in a Server line terminal:

```
# aptitude show courier-pop courier-imap | grep State
```

If the *courier-pop* or *courier-imap* servers are not installed, they can be installed by executing the following commands:

```
# aptitude update
# aptitude install courier-pop courier-imap
```

In order to guarantee that *courier-pop* or *courier-imap* servers are running, the following commands must be executed:

```
# service courier-pop restart
# service courier-imap restart
```

In order to fully test the POP3 server, a new e-mail must be sent to user *mailtest*:

```
netcat localhost 25
MAIL FROM: root@localhost
RCPT TO: mailtest@localhost
DATA
Subject: Second Test
Hello and good morning!
.
quit
```

To access the e-mailbox of the *mailtest* user at the server, using the POP3 protocol, it is possible to interact with the server using an interactive command line. In order to access the server interactive command line, it is necessary to open a *netcat* session to the server's TCP port 110/pop3. Upon connection, it is possible to access the e-mailbox using the following individual commands:

```
1  # netcat 10.1.1.20 110
2    +OK Hello there.
3  USER mailtest
4    +OK Password required.
5  PASS 12345678
6    +OK logged in.
7  STAT
8    +OK 2 604
9  LIST
10   +OK POP3 clients that break here, they violate STD53.
11   1 297
12   2 307
13   .
14 RETR 1
15   +OK 297 octets follow.
16     Return-Path: <root@localhost>
17     X-Original-To: mailtest@localhost
18     Delivered-To: mailtest@localhost
19     Received: from unknown (unknown [10.1.1.100])
20       by ubuntu (Postfix) with SMTP id A589941AD3
21       for <mailtest@localhost>; Thu, 12 Jul 2012 16:25:38 (WET)
22     SUBJECT: First Test
23
24     Hi everybody!
25     .
26 QUIT
27   +OK Bye-bye.
```

Upon connection, and after each user's command, the server responds with "OK" messages (lines 2, 4, 5, 8, 10, 15 and 27); (line 3) is the USER command sent by the client to identify the e-mailbox that is being accessed; (line 5) is the PASS command is used to send the password of the e-mailbox; (line 7) is the STAT command used to retrieve the total number of messages and their overall size in bytes; (line 8) associated with the OK answer from the server is a string with two numbers, the first indicates the number of messages on the e-mailbox (2 messages) and the second indicates the overall size in bytes (604 bytes); (line 9) is the LIST command used to retrieve the list of messages and their individual sizes in bytes; (lines 10 to 13) are the OK answers from the server associated with the list of messages located on the server itself (including an identification number and individual size per line), knowing that in this example there are two messages, one with 297 bytes and another with 307 bytes - the dot is used to signal the end of the list; (line 14) is the RETR command used to request the contents of a particular message identified by the number that is inserted after the RETR command; (lines 15 to 25) are the OK answers from the server associated with the contents of the requested message and, finally, (lines 26 and 27) contain the QUIT command sent by the client, and the respective OK answer from the server, used to terminate the POP3 interaction.

In order to access the e-mailbox of the *mailtest* user at the server, using the IMAP protocol, it also possible to interact with the server using an interactive command line. To access the server interactive command line it is necessary to open a *netcat* session to the server's TCP port 143/imap. Upon connection, it is possible to access the e-mail box using the following individual commands (note that all commands are preceded by ". " dot-space) :

```
# netcat 10.1.1.20 143
  * OK [CAPABILITY ...]
. LOGIN mailtest 12345678
  . OK LOGIN Ok.
. LIST "." "*"
  * LIST (\Marked \HasNoChildren) "." "INBOX"
  . OK LIST completed
. STATUS INBOX (messages)
  * STATUS "INBOX" (MESSAGES 2)
  . OK STATUS Completed.
. SELECT INBOX
  * FLAGS (\Draft \Answered ...)
  * OK [PERMANENTFLAGS (\* \Draft \Answered ...)] Limited
  * 2 EXISTS
  * 2 RECENT
. FETCH 1 BODY [TEXT]
  * 1 FETCH (BODY[TEXT] {15}
  Hi everybody! )
  . OK FETCH completed.
. FETCH 1 BODY[HEADER]
  * 1 FETCH (BODY[HEADER] {282}
  Return-Path: <root@localhost>
  X-Original-To: mailtest@localhost
  Delivered-To: mailtest@localhost
  Received: from unknown (unknown [10.1.1.100])
    by ubuntu (Postfix) with SMTP id A589941AD3
    for <mailtest@localhost>; Thu, 12 Jul 2012 06:25:38 (WET)
  SUBJECT: First Test )
  . OK FETCH completed.
. LOGOUT
  * BYE Courier-IMAP server shutting down
  . OK LOGOUT completed
```

Upon connection, the server responds with a "OK [CAPABILITY ...]" message (lines 2) including the capabilities of the server; (line 3) is the LOGIN command sent by the client to identify the e-mailbox that is being accessed and the respective password; (line 4) is a LOGIN OK message from the server confirming the successful login; (line 5) is a LIST command sent by the client to request all folders on the root of the e-mailbox; (line 6 and 7) are the responses from the server listing all folders - in this case there is only the INBOX folder; (line 8) is a STATUS command sent by the client to retrieve the number of messages in the INBOX folder; (lines 9 and 10) are the response from the server announcing the number of messages in the INBOX folder; (lines 11) is a SELECT command to enter the INBOX folder; (lines 12 to 15) are the response from the server, after selecting the INBOX folder, announcing the flags associated with the INBOX folder and the total and recent messages of the INBOX folder; (line 16) is a FETCH command sent by the client to request the e-mail body of a particular message (for message 1: 1 BODY [TEXT]); (lines 17 to 19) are the response of the server with the e-mail content; (line 20) is a FETCH command sent by the client to request the e-mail header of a particular message (for message 1: 1 BODY [HEADER]); (lines 21 to 29) are the response from the server with the e-mail header and, finally, (lines 30 to 32) contain the LOGOUT command sent by the client, and the respective BYE and OK answers from the server, that are used to terminate the IMAP interaction.

Bibliography

[1] A. Adams, J. Nicholas, and W. Siadak. RFC 3973 - Protocol Independent Multicast - Dense Mode: Protocol Specification (Revised). http://tools.ietf.org/html/rfc3973, January 2005.

[2] L. Berger, I. Bryskin, A. Zinin, and R. Coltun. RFC 5250 - The OSPF Opaque LSA Option. http://tools.ietf.org/html/rfc5250, July 2008.

[3] T. Berners-Lee, R. Fielding, and H. Frystyk. RFC 1945 - Hypertext Transfer Protocol - HTTP/1.0. http://www.ietf.org/rfc/rfc1945.txt, 1996.

[4] Y. Bernet, R. Yavatkar, P. Ford, F. Baker, and L. Zhang. Internet Draft - A Framework for End-to-End QoS Combining RSVP/Intserv and Differentiated Services. http://www.ietf.org/proceedings/43/I-D/draft-bernet-intdiff-00.txt, March 1998.

[5] S. Blake, D. Black, M. Carlson, E. Davies, Z. Wang, and W. Weiss. RFC 2475 - An Architecture for Differentiated Services. http://tools.ietf.org/html/rfc2475, December 1998.

[6] R. Braden, D. Clark, and S. Shenker. RFC 1633 - Integrated Services in the Internet Architecture: an Overview. http://tools.ietf.org/html/rfc1633, June 1994.

[7] R. Braden and L. Zhang. RFC 2209 - Resource ReSerVation Protocol (RSVP) - Version 1 Message Processing Rules. http://tools.ietf.org/html/rfc2209, September 1997.

[8] R. Braden, L. Zhang, S. Berson, S. Herzog, and S. Jamin. RFC 2205 - Resource ReSerVation Protocol (RSVP) - Version 1 Functional Specification. http://tools.ietf.org/html/rfc2205, September 1997.

[9] B. Cain, S. Deering, I. Kouvelas, B. Fenner, and A. Thyagarajan. RFC 3376 - Internet Group Management Protocol, Version 3. http://tools.ietf.org/html/rfc3376, October 2002.

[10] Vinton G. Cerf and Robert E. Kahn. A Protocol for Packet Network Intercommunication. *IEEE Transactions on Communications*, 22(5):637–648, 1974.

[11] David Clark. RFC 815: IP Datagram Reassembly Algorithms. http://tools.ietf.org/html/rfc815, July 1982.

[12] R. Coltun. RFC 2370 - The OSPF Opaque LSA Option. http://tools.ietf.org/rfc/rfc2370.txt, July 1998.

[13] M. Crispin. RFC 2060 - Internet Message Access Protocol - Version 4. http://tools.ietf.org/html/rfc2060, 1996.

[14] W. Daley and R. Kammer. FIPS 46-3: The official document describing the DES standard. http://csrc.nist.gov/publications/fips/fips46-3/fips46-3.pdf, 1999.

[15] S. Deering. RFC 1112 - Host Extensions for IP Multicasting. http://tools.ietf.org/html/rfc1112, August 1999.

[16] S. Deering and R. Hinden. RFC 2460: Internet Protocol, Version 6 (IPv6) Specification. http://tools.ietf.org/html/rfc2460, December 1998.

[17] H. Demirci, E. Türe, and A. Selçuk. A new meet in the middle attack on the IDEA block cipher. In *10th Annual Workshop on Selected Areas in Cryptography*, 2004.

[18] E. W. Dijkstra. A note on two problems in connexion with graphs. *Numerische Mathematik*, 1:269–271, 1959. 10.1007/BF01386390.

[19] J. Doyle. *Routing TCP/IP, Volume I (CCIE Professional Development)*. Cisco Press, 1998.

[20] J. Doyle and J. Carroll. *Routing TCP/IP, Volume II (CCIE Professional Development)*. Cisco Press, 1st Edition, 2001.

[21] R. Droms. RFC 2131: Dynamic Host Configuration Protocol. http://www.ietf.org/rfc/rfc2131.txt, March 1997.

[22] A. Durand, P. Fasano, I. Guardini, and D. Lento. RFC 3053: IPv6 Tunnel Broker. http://www.ietf.org/rfc/rfc3053.txt, January 2001.

[23] D. Eastlake and P. Jones. US Secure Hash Algorithm 1 (SHA1). http://tools.ietf.org/html/rfc3174, 2001.

[24] D. Estrin, D. Farinacci, A. Helmy, D. Thaler, S. Deering, M. Handley, V. Jacobson, C. Liu, P. Sharma, and L. Wei. RFC 2362 - Protocol Independent Multicast-Sparse Mode: Protocol Specification. http://www.ietf.org/rfc/rfc2362.txt, June 1998.

[25] B. Fenner, M. Handley, H. Holbrook, and I. Kouvelas. RFC 4601 - Protocol Independent Multicast - Sparse Mode: Protocol Specification (Revised). http://tools.ietf.org/html/rfc4601, August 2006.

[26] R. Fielding, J. Gettys, J. Mogul, H. Frystyk, L. Masinter, P. Leach, and T. Berners-Lee. RFC 2616 - Hypertext Transfer Protocol - HTTP/1.1. http://www.ietf.org/rfc/rfc2616.txt, 1999.

[27] R. Finlayson, T. Mann, J. Mogul, and M. Theimer. RFC 903: A Reverse Address Resolution Protocol. http://www.ietf.org/rfc/rfc903.txt, June 1984.

[28] B. Fraser. RFC 2196 - Site Security Handbook. http://www.ietf.org/rfc/rfc2196.txt, September 1997.

[29] N. Freed and N. Borenstein. RFC 2045 - Multipurpose Internet Mail Extensions (MIME) Part One: Format of Internet Message Bodies. http://www.ietf.org/rfc/rfc2045.txt, 1996.

[30] N. Freed and N. Borenstein. RFC 2046 - Multipurpose Internet Mail Extensions (MIME) Part Two: Media Types. http://www.ietf.org/rfc/rfc2046.txt, 1996.

[31] M. Garey, R. Graham, and D. Johnson. The complexity of computing Steiner minimal trees. *SIAM Journal on Applied Mathematics*, 34:477–495, 1978.

[32] R. Gilligan and E. Nordmark. RFC 2893: Transition Mechanisms for IPv6 Hosts and Routers. http://www.ietf.org/rfc/rfc2893.txt, August 2000.

[33] IEEE Network Working Group. 802.1Q - Virtual LANs. http://standards.ieee.org/getieee802/.

[34] IEEE Network Working Group. 802.1s - Multiple Spanning Trees. http://standards.ieee.org/getieee802/.

[35] IEEE Network Working Group. 802.1w - Rapid Reconfiguration of Spanning Tree. http://standards.ieee.org/getieee802/.

[36] IETF Network Working Group. RFC 1058: Routing Information Protocol. http://tools.ietf.org/rfc/rfc1058.txt, June 1988.

[37] IETF Network Working Group. RFC 2453: RIP Version 2, Carrying Additional Information. http://tools.ietf.org/html/rfc1723, November 1994.

[38] IETF Network Working Group. RFC 213: IP Encapsulation within IP. http://tools.ietf.org/html/rfc2003, October 1996.

[39] IETF Network Working Group. RFC 2080: RIPng for IPv6. http://tools.ietf.org/html/rfc2080, January 1997.

[40] IETF Network Working Group. RFC 2236: Internet Group Management Protocol, Version 2. http://www.rfc-editor.org/rfc/rfc2236.txt, November 1997.

[41] IETF Network Working Group. RFC 2328: OSPF Version 2. http://www.ietf.org/rfc/rfc2328.txt, April 1998.

[42] IETF Network Working Group. RFC 2710: Multicast Listener Discovery (MLD) for IPv6. http://www.ietf.org/rfc/rfc2710.txt, October 1999.

[43] IETF Network Working Group. RFC 2740: OSPF for IPv6. http://www.ietf.org/rfc/rfc2740.txt, December 1999.

[44] IETF Network Working Group. RFC 3315: Dynamic Host Configuration Protocol for IPv6 (DHCPv6). http://www.ietf.org/rfc/rfc3315.txt, July 2003.

[45] IETF Network Working Group. RFC 3736: Stateless Dynamic Host Configuration Protocol Service for IPv6. http://tools.ietf.org/html/rfc3736, April 2004.

[46] J. Hagino and K. Yamamoto. RFC 3142: An IPv6-to-IPv4 Transport Relay Translator. http://www.ietf.org/rfc/rfc3142.txt, June 2001.

[47] S. Hakimi. Steiner's problem in graphs and its implications. *Networks*, 1(2):113–133, 1971.

[48] D. Harkins and D. Carrel. RFC 2409 - The Internet Key Exchange (IKE). http://www.ietf.org/rfc/rfc2409.txt, 1998.

[49] J. Heinanen, F. Baker, W. Weiss, and J. Wroclawski. RFC 2597 - Assured Forwarding PHB Group. http://tools.ietf.org/html/rfc2597, June 1999.

[50] H. Holbrook, B. Cain, and B. Haberman. RFC 4604 - Using Internet Group Management Protocol Version 3 and Multicast Listener Discovery Protocol Version 2 (MLDv2) for Source-Specific Multicast. http://tools.ietf.org/html/rfc4604, August 2006.

[51] C. Huitema. RFC 3068: An Anycast Prefix for 6to4 Relay Routers. http://www.ietf.org/rfc/rfc3068.txt, June 2001.

[52] IEEE. IEEE Std. 802.11, Part 11: Wireless LAN Medium Access Control (MAC) and Physical Layer (PHY) Specifications. http://standards.ieee.org/about/get/802/802.11.html, 1997.

[53] IEEE. IEEE Std. 802.11a, Supplement to Part 11: Wireless LAN Medium Access Control (MAC) and Physical Layer (PHY) Specifications: Higher-Speed Physical Layer Extension in the 5 GHz Band. http://standards.ieee.org/about/get/802/802.11.html, 1999.

[54] IEEE. IEEE Std. 802.11b, Supplement to Part 11: Wireless LAN Medium Access Control (MAC) and Physical Layer (PHY) Specifications: Higher-Speed Physical Layer Extension in the 2.4 GHz Band. http://standards.ieee.org/about/get/802/802.11.html, 1999.

[55] IEEE. IEEE Std. 802.11g, Supplement to Part 11: Wireless LAN Medium Access Control (MAC) and Physical Layer (PHY) Specifications: Further Higher-Speed Physical Layer Extension in the 2.4 GHz Band. http://standards.ieee.org/about/get/802/802.11.html, 2003.

[56] K. Ingham and S. Forrest. A history and survey of network firewalls. http://www.cs.unm.edu/ treport/tr/02-12/firewall.pdf, 2002.

[57] ISO/IEC. ISO/IEC standard 7498-1:1994. http://standards.iso.org/, 1994.

[58] V. Jacobson, K. Nichols, and K. Poduri. RFC 2598 - An Expedited Forwarding PHB. http://tools.ietf.org/html/rfc2598, June 1999.

[59] S. Kent and R. Atkinson. RFC 2402 - IP Authentication Header. http://www.ietf.org/rfc/rfc2402.txt, 1998.

[60] S. Kent and R. Atkinson. RFC 2406 - IP Encapsulating Security Payload (ESP). http://www.ietf.org/rfc/rfc2406.txt, 1998.

[61] H. Kitamura. RFC 3089: A SOCKS-based IPv6/IPv4 Gateway Mechanism. http://www.ietf.org/rfc/rfc3089.txt, April 2001.

[62] J. Klensin. RFC 2821 - Simple Mail Transfer Protocol. http://tools.ietf.org/html/rfc2821, 2001.

[63] Charles Kozierok. TCP/IP Guide - Version 3.0. http://www.tcpipguide.com/free/t_IPDatagramEncapsulation.htm, 2012.

[64] J. Kurose and K. Ross. *Computer Networking: A Top-down Approach*. Addison Wesley, 6th Edition, 2012.

[65] Naval Research Laboratory. Multi-Generator (MGEN). http://cs.itd.nrl.navy.mil/work/mgen/index.php, 2012.

[66] G. Lee, M. Shin, and H. Kim. Implementing NAT-PT/SIIT, ALGs and consideration to the mobility support in NAT-PT environment. In *Proceedings of the 6th International Conference on Advanced Communication Technology*, pages 433–439, 2004.

[67] S. Lee, S. Shin, Y. Kim, E. Nordmark, and A. Durand. RFC 3338: Dual Stack Hosts Using Bump-in-the-API (BIA). http://www.ietf.org/rfc/rfc3338.txt, October 2002.

[68] T. Li and Y. Rekhter. RFC 2430 - A Provider Architecture for Differentiated Services and Traffic Engineering. http://tools.ietf.org/html/rfc2430, October 1998.

[69] D. Maughan, M. Schertler, M. Schneider, and J. Turner. RFC 2408 - Internet Security Association and Key Management Protocol (ISAKMP). http://www.ietf.org/rfc/rfc2408.txt, 1998.

[70] P. Mockapetris. RFC 1034: An IPv6-to-IPv4 Transport Relay Translator. http://tools.ietf.org/html/rfc1034, November 1987.

[71] J. Moy. RFC 1584 - Multicast Extensions to OSPF. https://tools.ietf.org/rfc/rfc1584.txt, March 1994.

[72] J. Myers and M. Rose. RFC 1939 - Post Office Protocol - Version 3. http://tools.ietf.org/html/rfc1939, 1996.

[73] K. Nichols, S. Blake, F. Baker, and D. Black. RFC 2474 - Definition of the Differentiated Services Field (DS Field) in the IPv4 and IPv6 Headers. http://tools.ietf.org/html/rfc2474, December 1998.

[74] E. Nordmark. RFC 2765: Stateless IP/ICMP Translation Algorithm (SIIT). http://tools.ietf.org/html/rfc2765, February 2000.

[75] University of Southern California. RFC 791: Internet Protocol - DARPA Internet Program Protocol Specification. http://tools.ietf.org/pdf/rfc791.pdf, September 1981.

[76] United States National Institute of Standards and Technology. Announcing the Advanced Encryption Standard (AES). http://csrc.nist.gov/publications/fips/fips197/fips-197.pdf, 2001.

[77] Priscilla Oppenheimer. *Top-Down Network Design*. Cisco Press, 3rd Edition, 2010.

[78] H. Orman. RFC 2412 - The OAKLEY Key Determination Protocol. http://www.ietf.org/rfc/rfc2412.txt, 1998.

[79] C. Perkins. IP Mobility Support for IPv4. http://www.ietf.org/rfc/rfc3344.txt, 2002.

[80] Radia Perlman. An algorithm for distributed computation of a spanning tree in an extended lan. In *Proceedings of the ninth symposium on Data communications*, SIGCOMM '85, pages 44–53, New York, NY, USA, 1985. ACM.

[81] Radia Perlman. *Interconnections, Second Edition*. Addison-Wesley, 2000.

[82] David C. Plummer. RFC 826: An Ethernet Address Resolution Protocol. http://tools.ietf.org/html/rfc826, November 1982.

[83] J. Postel. RFC 768: User Datagram Protocol. http://tools.ietf.org/html/rfc768, August 1980.

[84] J. Postel and J. Reynolds. RFC 959 - File Transfer Protocol (FTP). http://tools.ietf.org/html/rfc959, 1985.

[85] E. Rescorla. Diffie-Hellman Key Agreement Method. http://www.ietf.org/rfc/rfc2631.txt, 1999.

[86] A. Retana, D. Slice, and R. White. *Advanced IP Network Design (CCIE Professional Development)*. Cisco Press, 1999.

[87] R. Rivest. The MD5 Message-Digest Algorithm. http://tools.ietf.org/html/rfc1321, 1992.

[88] R. Rivest, A. Shamir, and L. Adleman. A Method for Obtaining Digital Signatures and Public-Key Cryptosystems. *Communications of the ACM*, 21(2):120–126, 1978.

[89] Bruce Schneier. Official Blowfish website. http://www.schneier.com/blowfish.html, 2012.

[90] S. Shenker, C. Partridge, and R. Guerin. RFC 2212 - Specification of Guaranteed Quality of Service. http://tools.ietf.org/html/rfc2212, September 1997.

[91] IEEE Computer Society. 802.1D: Media Access Control (MAC) Bridges. www.dcs.gla.ac.uk/ lewis/teaching/802.1D-2004.pdf, June 2004.

[92] K. Sollins. RFC 1350 - The TFTP Protocol (Revision 2). http://tools.ietf.org/html/rfc1350, 1992.

[93] P. Srisuresh and M. Holdrege. RFC 2663: IP Network Address Translator (NAT) Terminology and Considerations. http://tools.ietf.org/html/rfc2663, August 1999.

[94] W. Stallings. *Data and Computer Communications*. Pearson Education, 2007.

[95] W. Stevens. RFC 2001: TCP Slow Start, Congestion Avoidance, Fast Retransmit, and Fast Recovery Algorithms. http://tools.ietf.org/html/rfc2001, January 1997.

[96] Cisco Systems. Inter-Switch Link and IEEE 802.1Q Frame Format - Document ID 17056. http://www.cisco.com/.

[97] Cisco Systems. Understanding Multiple Spanning Tree Protocol (802.1s). http://www.cisco.com/.

[98] Cisco Systems. Understanding Rapid Spanning Tree Protocol (802.1w). http://www.cisco.com/.

[99] A. Tanenbaum and D. Wetherall. *Computer Networks*. Prentice Hall, 5th Edition, 2010.

[100] D. Thaler and C. Ravishankar. Distributed center-location algorithms. *IEEE Journal on Selected Areas in Communications*, 15(3):291–303, 1997.

[101] G. Tsirtsis and P. Srisuresh. RFC 2767: Network Address Translation - Protocol Translation. http://www.ietf.org/rfc/rfc2766.txt, February 2000.

[102] K. Tsuchiya, H. Higuchi, and Y. Atarashi. RFC 2767: Dual Stack Hosts using the Bump-In-the-Stack Technique (BIS). http://tools.ietf.org/html/rfc2767, February 2000.

[103] D. Waitzman, C. Partridge, and S. Deering. RFC 1075 - Distance Vector Multicast Routing Protocol. http://www.ietf.org/rfc/rfc1075.txt, November 1998.

[104] B. Waxman. Routing of multipoint connections. *IEEE Journal on Selected Areas in Communications*, 6(9):1617–1622, 1988.

[105] Cryptography World. The Cryptography Guide: Triple DES. http://www.cryptographyworld.com/des.htm, 2010.

[106] J. Wray. RFC 2744 - Generic Security Service API Version 2: C-bindings. http://tools.ietf.org/html/rfc2744, January 2000.

[107] J. Wroclawski. RFC 2210 - The Use of RSVP with IETF Integrated Services. http://tools.ietf.org/html/rfc2210, September 1997.

[108] J. Wroclawski. RFC 2211 - Specification of the Controlled-Load Network Element Service. http://tools.ietf.org/html/rfc2211, September 1997.

[109] L. Zhou, V. Renesse, and M. Marsh. Implementing ipv6 as a peer-to-peer overlay network. In *Proceedings of the 21st IEEE Symposium on Reliable Distributed Systems*, pages 347–351, 2002.

Index

About the Authors

António Nogueira joined the Department of Electronics, Telecommunications and Informatics of University of Aveiro, Portugal, in 1999, being an Assistant Professor since 2005. During these years, he has been actively involved in the genesis of several networking courses, all of them following a practical paradigm. His research activities have been mainly devoted to the following areas: traffic modeling, traffic engineering, network modeling and planning, network security and service design for the Future Internet. He had the opportunity to supervise several Master and PhD students, whose works were always tightly linked to the goals of different research projects. António Nogueira is co-author of more than eighty scientific papers published in peer-reviewed international conferences and magazines, being also involved in several national and international research projects. He holds a PhD in Electrical Engineering since 2005.

Paulo Salvador was born in Coimbra, Portugal, on February 17, 1975. He graduated in Electronics and Telecommunications Engineering in 1998 and received the PhD degree in Electrical Engineering in 2005, both from University of Aveiro, Portugal. In 1998, he joined Institute of Telecommunications, Aveiro, Portugal, as a researcher. In 2003 he joined the Department of Electronics, Telecommunications and Informatics of University of Aveiro, Portugal, being an Assistant Professor since 2006. During his teaching career has created, restructured and taught multiple networking courses at graduate and post-graduate level, all of them based on the practical approach paradigm. His research activities have been focused on the following areas: (1) traffic, network and user behavior modeling, (2) network design, management and security and (3) service design for the Future Internet. He supervised multiple Master and PhD students, whose works were always an integrated part of several research projects. Paulo Salvador is co-author of more than ninety scientific papers published in peer-reviewed international conferences and journals, being also involved in several national and international research projects. He has also served as consultant to different network operators in areas related with network management and traffic monitoring.

Lightning Source UK Ltd.
Milton Keynes UK
UKOW02n0841151214

243137UK00001B/15/P